T0205957

Introduction to Artificial Intelligence

Mariusz Flasiński

Introduction to Artificial Intelligence

Mariusz Flasiński
Information Technology Systems
Department
Jagiellonian University
Kraków
Poland

ISBN 978-3-319-82016-3 ISBN 978-3-319-40022-8 (eBook)
DOI 10.1007/978-3-319-40022-8

Translation from the Polish language edition: *Wstęp do sztucznej inteligencji* by Mariusz Flasiński,

Softcover reprint of the hardcover 1st edition 2016

Printed on acid-free paper

This Springer imprint is published by Springer Nature
The registered company is Springer International Publishing AG Switzerland

Preface

There are a variety of excellent monographs and textbooks on Artificial Intelligence. Let us mention only such classic books as: *Artificial Intelligence. A Modern Approach* by S.J. Russell and P. Norvig, *Artificial Intelligence* by P.H. Winston, *The Handbook of Artificial Intelligence* by A. Barr, P.R. Cohen, and E. Feigenbaum, *Artificial Intelligence: Structures and Strategies for Complex Problem Solving* by G. Luger and W. Stubblefield, *Artificial Intelligence: A New Synthesis* by N. Nilsson, *Artificial Intelligence* by E. Rich, K. Knight, and S.B. Nair, and *Artificial Intelligence: An Engineering Approach* by R.J. Schalkoff. Writing a (very) concise introduction to Artificial Intelligence that can be used as a textbook for a one-semester introductory lecture course for computer science students as well as for students of other courses (cognitive science, biology, linguistics, etc.) has been the main goal of the author.

As the book can also be used by students who are not familiar with advanced models of mathematics, AI methods are presented in an intuitive way in the second part of the monograph. Mathematical formalisms (definitions, models, theorems) are included in appendices, so they can be introduced during a lecture for students of computer science, physics, etc. Appendices A–J relate to Chaps. 4–13, respectively.

Short biographical notes on researchers that influenced Artificial Intelligence are included in the form of footnotes. They are presented to show that AI is an interdisciplinary field, which has been developed for more than half a century due to the collaboration of researchers representing such scientific disciplines as philosophy, logics, mathematics, computer science, physics, electrical engineering, biology, cybernetics, biocybernetics, automatic control, psychology, linguistics, neuroscience, and medicine.

Acknowledgments

The author would like to thank the reviewer Prof. Andrzej Skowron Ph.D., D.Sc., Faculty of Mathematics, Informatics, and Mechanics, University of Warsaw for very useful comments and suggestions.

Kraków
December 2015

Mariusz Flasiński

Contents

Part I
Fundamental Ideas of Artificial Intelligence

Chapter 1
History of Artificial Intelligence

Many fundamental methodological issues of Artificial Intelligence have been of great importance in philosophy since ancient times. Such philosophers as Aristotle, St. Thomas Aquinas, William of Ockham, René Descartes, Thomas Hobbes, and Gottfried W. Leibniz have asked the questions: "What are basic cognitive operations?", "What necessary conditions should a (formal) language fulfill in order to be an adequate tool for describing the world in a precise and unambiguous way?", "Can reasoning be automatized?". However, the first experiments that would help us to answer the fundamental question: "Is it possible to construct an *artificial intelligence* system?" could not be performed until the twentieth century, when the first computers were constructed.

Certainly, the question: "When can we say that a system constructed by a human designer is intelligent?" is a key problem in the AI field. In 1950 Alan M. Turing[1] proposed a solution of this problem with the help of the so-called *imitation game* [307]. The imitation game is, in fact, an *operational* test of artificial intelligence and it can be described in the following way. Let us assume that a human interrogator has a conversation with both another human being and a computer at the same time. This conversation is performed with the help of a device which makes the simple identification of an interlocutor impossible. (For example, both interlocutors send their statements to a computer monitor.) The human interrogator, after some time, should guess which statements are sent by the human being and which ones are sent by the computer. According to Turing, if the interrogator cannot make such a distinction, then the (artificial) intelligence of the computer is the same as the intelligence of the human being. Let us note that intelligence is, somehow, considered equivalent to linguistic competence in the *Turing test*. As we will see further on, such an equivalence between intelligence and linguistic competence occurs in some AI models.

---

[1] Alan Mathison Turing—outstanding mathematician, logician, and computer scientist, a researcher at the University of Cambridge, University of Manchester, and the National Physical Laboratory in London. He is considered as one of the "fathers" of computer science. Turing defined the universal (Turing) machine, which is a mathematical model of computation.

M. Flasiński, *Introduction to Artificial Intelligence*,
DOI 10.1007/978-3-319-40022-8_1

Although 1956 is usually taken to be the year of the birth of Artificial Intelligence, because it is the year of the famous conference at Dartmouth College, the author would consider the preceding year to be the beginning of AI. In 1955 the first AI system, called *Logic Theorist*, was designed by Allen Newell,[2] Herbert A. Simon,[3] and implemented by J. Clifford Shaw[4] at Carnegie Mellon University [197]. The system proved nearly 40 theorems included in Alfred N. Whitehead and Bertrand Russell's fundamental monograph *Principia Mathematica*. The designers of the system tried to publish their results in the prestigious *Journal of Symbolic Logic*. The editors rejected the paper, claiming it contained just new proofs of elementary theorems and overlooking the fact that a computer system was a co-author. Pamela McCorduck writes in her monograph [197] that for Herbert Simon designing Logic Theorist meant a solution of the Cartesian mind-body problem.[5]

The further research of Simon, Newell, and Shaw into constructing systems possessing mental abilities resulted in the implementation of *General Problem Solver, GPS* in 1959. The system solved a variety of formal problems, for example: symbolic integration, finding paths in Euler's problem of the Königsberg bridges, playing the Towers of Hanoi puzzle, etc. Defining the paradigm of *cognitive simulation*,[6] which says that in AI systems general schemes of human ways of problem solving should be simulated, was a methodological result of their research. This paradigm is a basic assumption of advanced research projects into *cognitive architectures*, which are presented in Chap. 14.

While Newell and Simon constructed their systems on the basis of cognitive simulation, John McCarthy[7] led research at MIT into another crucial area of Artificial Intelligence. In 1958 he presented the important paper [193] at the Symposium on the Mechanization of Mental Processes, which was organized at Teddington. McCarthy proposed the paradigm of solving *common sense* problems with the use of formal logic-based models of reasoning. At that time such an idea seemed peculiar. During a discussion Yehoshua Bar-Hillel[8] called the paper "half-baked", claiming that logic-based (deductive) inference is not an adequate model for human common-sense reasoning. This criticism did not put McCarthy off the idea to use mathematical

---

[2] Allen Newell—a professor of computer science and cognitive psychology at Carnegie Mellon University. He received the Turing Award in 1975. Newell is a co-designer of the cognitive architecture Soar and the programming language IPL.

[3] Herbert Alexander Simon—a professor at the Carnegie Mellon University. His excellent work concerns economics, sociology, psychology, political science, and computer science. He received the Turing Award in 1975 and the Nobel Prize in Economics in 1978.

[4] John Clifford Shaw—a researcher at RAND Corporation, a co-designer of a programming langauge IPL.

[5] Philosophical views which are important for Artificial Intelligence are presented in Sect. 15.1.

[6] Methodological assumptions of *cognitive simulation* are discussed in Sect. 2.1.

[7] John McCarthy—a professor of the Massachusetts Institute of Technology, Stanford University, and Princeton University. He introduced the term *Artificial Intelligence, AI*. McCarthy received the Turing Award in 1971.

[8] Yehoshua Bar-Hillel—an eminent philosopher (a disciple of Rudolf Carnap), mathematician, and computer scientist (the Bar-Hillel pumping lemma for context-free languages).

logic in Artificial Intelligence. He continued research into a logic-based programming language for an implementation of intelligent systems. McCarthy was inspired by one of the modern logic calculi, namely *lambda calculus* (*λ-calculus*), which was introduced by Alonzo Church[9] and Stephen C. Kleene[10] in the 1930s. His research was successful and he constructed the *Lisp* language[11] in 1958–1960 [194]. Lisp and its dialects, such as Scheme, and Common Lisp, is still used for constructing AI systems. At the beginning of the 1970s a new research approach based on *First-Order Logic, FOL,* appeared within the logic-based paradigm. It resulted in the construction of the second classic AI language, namely *Prolog*, by Alain Colmerauer and Philippe Roussel (Université d'Aix-Marseille II) [57]. This approach originated the popular model of *Constraint Logic Programming, CLP*. More detailed discussion of the *logic-based approach in AI* is included in Sect. 2.2 and in Chap. 6.

In the middle of the 1960s the third approach in symbolic AI appeared, namely the *knowledge-based approach*. It was originated by a project to implement the Dendral expert system conducted by Edward Feigenbaum[12] and Joshua Lederberg[13] at Stanford University. The identification of unknown organic molecules on the basis of an analysis of their mass spectra and knowledge of chemistry was the goal of the system. The system was very helpful for organic chemists. The experience gained by Feigenbaum led him to introduce a new AI paradigm that differs from both cognitive simulation and the logic-based approach. This paradigm can be described in the following way. Firstly, instead of solving general problems, intelligent systems should focus on well-defined application areas (as did the Dendral system). Secondly, an intelligent system should be equipped with all the knowledge that human experts possess in a particular field. Therefore, such systems are often called *expert systems*. Thirdly, this knowledge should be treated as a kind of data and it should be stored in the *knowledge base* of the system. A general inference mechanism[14] should be implemented in the system in order to reason over knowledge.[15] In this last aspect, the knowledge-based approach is somehow similar to the logic-based approach, especially as analogous inference schemes are used in both approaches. The last assumption concerns a way of verifying correctness of the system functionality.

[9] Alonzo Church—a professor of logic, mathematics, and philosophy at Princeton University and University of California, Los Angeles. He was a doctoral advisor for such famous scientists as Alan Turing, Stephen C. Kleene, Michael O. Rabin, J. Barkley Rosser, and Dana Scott.

[10] Stephen C. Kleene—a professor of mathematics at Princeton University. His work concerns recursion theory, theory of computable functions, and regular expressions.

[11] Lisp is one of the two oldest languages, along with Fortran, which are still used nowadays.

[12] Edward Albert Feigenbaum—a professor of computer science at Stanford University. He is considered one of the AI pioneers. Herbert Simon was his doctoral advisor. In 1994 he received the Turing Award.

[13] Joshua Lederberg—a professor at Stanford University and Rockefeller University, molecular geneticist, microbiologist, and one of the AI pioneers. In 1958 he received the Nobel Prize in medicine.

[14] *General* means here independent from specific knowledge.

[15] In expert systems knowledge is often formalized as a set of rules. We call such systems *rule-based systems*. Rule-based and expert systems are introduced in Chap. 9.

The implemented system should be embedded in an environment in which human experts solve problems. Its testing consists of checking whether it *simulates* experts well. If so, then it means that the system works correctly. (This is analogous to cognitive simulation, in which the system should simulate the intelligent behavior of a human being.) These assumptions are fulfilled to a high degree in the knowledge representation model in the form of *rule-based systems*, which are introduced in Sect. 2.3 and Chap. 9.

Successes in constructing symbolic AI systems encouraged Newell and Simon to formulate in 1976 a fundamental view of *Strong Artificial Intelligence*, namely the *physical symbol system hypothesis* [209]. A *physical symbol system* consists of a set of elements called symbols that are used by the system to construct symbolic structures called expressions and a set of processes for their modification, reproduction, and destruction. In other words, the system transforms a certain set of expressions. The hypothesis is formulated as follows [209]:

> A physical symbol system has the necessary and sufficient means for general intelligent action.

This radical view of Newell and Simon that a system transforming symbolic structures can be considered intelligent was criticized strongly by John R. Searle[16] in his paper "*Minds, Brains, and Programs*" [269] published in 1980. Searle divided views concerning AI into two basic groups. Adherents of *Weak Artificial Intelligence* treat a computer as a convenient device for testing hypotheses concerning brain and mental processes and for simulating brain performance. On the other hand, adherents of *Strong Artificial Intelligence* consider a properly programmed computer equivalent to a human brain and its mental activity. Of course, the physical symbol system hypothesis belongs to the latter approach. In order to reject this hypothesis Searle proposed a thought experiment, which challenges the Turing test (the imitation game) as well. One of the various versions of this experiment, called the *Chinese room*, can be described in the following way. Let us assume that a native English speaker, who does not speak Chinese, is closed in a room. In the first phase of the experiment he receives pieces of paper with Chinese characters on them and he is asked to respond to them. However, he does not understand these notes. Fortunately, in the room there is a book which contains instructions on how to respond to Chinese characters by writing other Chinese characters on a piece of paper. So, each time he receives a piece of paper with Chinese characters on it, he "produces" a response to it according to these instructions. In the second phase of the experiment he receives pieces of paper with English sentences, e.g., some questions, and he is asked to respond to them. So, he responds... Let us assume that the quality of responses in both cases is the same. Now, Searle asks the question, Is there any difference between the two phases of the experiment? Then, he answers, Yes. In the first phase the man in the room does not understand what he is doing, however in the second phase he does.

[16] John Rogers Searle—a professor of philosophy at the University of California, Berkeley. His work concerns mainly the philosophy of language, philosophy of mind, and social philosophy. In 2000 he was awarded the *Jean Nicod Prize*.

One can easily notice that, beginning from the Turing test, in Artificial Intelligence an immense significance is attached to natural language and an intelligent systems ability to use it. Three linguistic theories, namely Noam Chomsky's theory of generative grammars [47], Roger Schank's conceptual dependency theory [263, 264], and George Lakoff's cognitive linguistics [175] are especially important in Artificial Intelligence. Therefore, we now present them.

Noam Chomsky[17] introduced the first version of *generative grammars* in 1957.[18] According to his theory, the ability to learn a grammar of natural language, called the *universal grammar*, is a property of the human brain. The universal grammar is "parameterized" when a child is exposed to a specific language, which creates a grammar of this language in his/her brain. Such a grammar has the form of a formal system, which consists of a *finite set of rules* that are used to generate an infinite set of sentences belonging to the language. Thus, the grammar is a generator of the language. Chomsky distinguishes not only a syntax level but also a semantic level in the structure of a generative grammar. Moreover, syntax constitutes a basic, primary level, whereas semantics is somehow derived from syntax. Thus, there are some analogies here to the physical symbol system model. Although the initial great (let us say *excessive*) expectations of natural language processing (understanding) with the help of this theory have not been completely fulfilled, it has become a model of fundamental importance in computer science. The birth of this theory can be considered as the origin of the dynamic development of *mathematical linguistics*.[19] It focuses on defining mathematical formalisms which can be used for representing structures of both formal and natural languages. The basic assumptions of mathematical linguistics are presented in Sect. 2.5 and methods based on the theory of generative grammars are introduced in Chap. 8.

Roger Schank,[20] in his *conceptual dependency theory* developed at the end of the 1960s [263], claimed that it is not syntax which should be a starting point for defining language semantics but *concepts*, strictly speaking *dependencies (relations) among concepts*. In order to formalize various types of dependencies among concepts, he defined a model of a representation of relations in the form of *conceptual dependency graphs*. These graphs are defined in such a way that two language structures having the same meaning are represented by the same graph. In Artificial Intelligence conceptual dependency theory constitutes a paradigm for *structural models of*

[17] Avram Noam Chomsky—a professor of linguistics at the Massachusetts Institute of Technology. His research was very important for the foundations of computer science (mainly the theory of formal languages, programming languages, and computer linguistics), linguistics, and theory of mind. He received a lot of honorary doctorates from prestigious universities, including University of Bologna, Harvard University, Cambridge University, Uppsala University, McGill University, and Loyola University Chicago.

[18] Our considerations concern Chomsky's *standard model*.

[19] Although the *idea* of a grammar generating a language, i.e., a grammar as a rewriting system, can be found in the work of logicians Axel Thue and Emil Post.

[20] Roger Schank—a professor of psychology and computer science at Yale University and Northwestern University. His work concerns Artificial Intelligence (Natural Language Processing, case-based reasoning) and cognitive psychology.

knowledge representation. In these models knowledge is represented by graph-like or hierarchical structures. *Semantic networks* introduced by Allan M. Collins[21] and Ross Quillian [56], *frames* defined by Marvin Minsky[22] in [203], and *scripts* proposed by Schank and Robert P. Abelson[23] [264] are the most popular models in this approach. The research in this area concerned *Natural Language Processing, NLP*, initially. It resulted in the construction of well-known NLP systems such as ELIZA[24] simulating a psychotherapist, which was constructed by Joseph Weizenbaum[25] in 1966, SHRDLU, the first system which understood a natural language in the context of a simplified world-like environment (Terry Winograd,[26] 1970), MARGIE (Roger Schank 1973), and SAM (Richard E. Cullingford, 1978), which were based on conceptual dependency theory, and PAM, constructed by Robert Wilensky[27] in 1978, which was able to interpret simple stories. The approach based on structural models of knowledge representation has been applied for constructing AI systems beyond Natural Language Processing as well. For the implementation of such systems a variety of programming languages and environments have been developed including KRL defined by Terry Winograd and Daniel G. Bobrow[28] and KL-ONE introduced by Ronald J. Brachman[29] and James G. Schmolze.[30] The approach based on structural models of knowledge representation is discussed in Sect. 2.4, whereas models of semantic networks, frames, and scripts are presented in Chap. 7.

[21] Allan M. Collins—a psychologist, a professor of Northwestern University. His work concerns Artificial Intelligence and cognitive psychology.

[22] Marvin Minsky—a mathematician, a professor of computer science at the Massachusetts Institute of Technology. For his contribution to AI he received the Turing Award in 1969.

[23] Robert P. Abelson—a professor of psychology at Yale University. His work concerns applications of statistical analysis and logic in psychology and political science.

[24] In fact, ELIZA was the first well-known *chatbot*, that is, a program which is able to simulate a conversation with a human being.

[25] Joseph Weizenbaum—a professor of computer science at the Massachusetts Institute of Technology. He is considered one of the AI pioneers.

[26] Terry Winograd—a professor of computer science at Stanford University. His work influenced Artificial Intelligence, the theory of mind, and Natural Language Processing.

[27] Robert Wilensky—a professor of computer science at the University of California, Berkeley. His work concerns systems of natural language understanding, knowledge representation, and AI planning systems.

[28] Daniel Gureasko Bobrow—one of the AI pioneers, a research fellow at the Palo Alto Research Center (PARC). He was the President of the American Association for Artificial Intelligence and the editor-in-chief of the prestigious journal *Artificial Intelligence*.

[29] Ronald Jay Brachman—a head of AT&T Bell Laboratories Artificial Intelligence Principles Research Department and the DARPA Information Processing Techniques Office. His work concerns structural models of knowledge representation and description logic.

[30] James G. Schmolze—a professor of computer science at Tufts University. His work concerns Artificial Intelligence, especially knowledge representation and reasoning with incomplete knowledge.

In the area of linguistics, *cognitive linguistics* represented by George Lakoff[31] seems to oppose the physical symbol system hypothesis to the largest degree. In his famous book "*Women, Fire, and Dangerous Things: What Categories Reveal About the Mind*" published in 1987 [175], he launched the most effective attack on Strong AI. Instead of considering whether a machine which manipulates symbolic expressions can be called intelligent, G. Lakoff asked the following question: Is it possible to define a mapping from our real-world environment into a set of such symbolic expressions? He pointed out that an assumption about the possibility of the construction of such a mapping was somehow hidden in the physical symbol system hypothesis. This assumption was, however, based on the classic Aristotelian approach to creating concepts. According to this approach concepts have crisp boundaries, i.e., an object belongs to a category, which relates to a concept or it does not. However, Lakoff has claimed that such an approach should not have been assumed in the light of modern psychology and philosophy. Ludwig Wittgenstein [316] and Eleanor Rosch [245] have shown that there are "better" and "worse" representative members of a category and both human (bodily) experience and imagination play a basic role in the process of categorization (conceptualization). Therefore, categories should not be treated as "boxes" including some objects and not the others, but they should be defined taking into account their fuzziness. This fuzziness results from the fact that our cognition is determined by the form of our (human) body, especially by the form of our brain (*embodied mind thesis*). Therefore, cognitive linguists often claim that *connectionist models*, which include mainly models of artificial neural networks, are more adequate for constructing AI systems than methods which are based on logic and symbolic processing.

Connectionist models,[32] in which mental phenomena are modeled as emergent processes[33] occurring in networks that consist of elementary components, have a long history in AI. The first model of an *artificial neuron* which could be used as an elementary component of such a network was defined by Warren S. McCulloch[34] and

[31]George Lakoff—a professor of linguistics at the University of California, Berkeley. At the end of the 1960s he contributed to the fundamentals of *cognitive linguistics*, which was in opposition to the *Chomsky theory*. His work influenced the development of cognitive science, theory of mind, Artificial Intelligence, and political science.

[32]The characteristics presented in this chapter concern models of *distributed connectionist networks*. A more general approach is presented in Sect. 3.1.

[33]For our considerations we assume that a process is *emergent* if it cannot be described on the basis of its elementary components. In other words, a (simplified) description of a process at a lower level is not sufficient for its description at a higher level.

[34]Warren Sturgis McCulloch—a neurophysiologist and cybernetician working at the Massachusetts Institute of Technology, Yale University, and University of Illinois at Chicago. His and Walter Pitts' research into treating a nervous system as a universal computing device (inspired by the views of Gottfried Leibniz) resulted in the definition of an artificial neuron. He was the President of the American Society for Cybernetics.

Walter Pitts[35] in 1943 [198]. *Neural networks, NN*, are used for simulating processes occurring in biological neural networks in a brain. They consist of interconnected simple components, which process information in parallel. From a functional point of view a neural network is a kind of "black box" which can be trained in order to learn adequate responses to various stimuli. Such learning consists of modifying parameters of a neural network structure. As a result this knowledge is somehow coded in such a parameterized structure. Thus, a connectionist approach differs completely from a symbolic approach. Therefore, one of the "hottest" discussions in AI took place between adherents of these two approaches. In 1957 Frank Rosenblatt[36] defined the *perceptron*, which was at that time[37] a simple one-layer neural network [247]. Then, Marvin Minsky and Seymour Papert[38] published their famous book "Perceptrons" [204] in 1969, in which they showed the strong limitations of perceptrons, e.g., the inability to compute some logical functions like XOR. As a result, many AI researchers concluded that the study of neural networks is not promising.

Although such negative opinions of eminent AI authorities caused limited financing of research into neural networks, it continued in the 1970s and the 1980s and resulted in many successful results. In 1972 Teuvo Kohonen[39] constructed the *associative network* [164]. Three years later Kunihiko Fukushima, a senior research scientist at the Science and Technical Research Laboratories, Japan Broadcasting Corporation, constructed the *Cognitron*, which was a multi-layer neural network [107]. In 1982 two important models were developed: the *Hopfield*[40] *recurrent network* [140] and *Self-Organizing Maps, SOM*, by Kohonen [165]. David E. Rumelhart[41] and Geoffrey E. Hinton[42] published, with their collaborators, a paper on learning

---

[35]Walter Harry Pitts, Jr.—a logician and mathematician working at the Massachusetts Institute of Technology and the University of Chicago (together with W.S. McCulloch, J. Lettvin and N. Wiener).

[36]Frank Rosenblatt—a psychologist and computer scientist, a professor at Cornell University. His work concerns neural networks, neurodynamics, and cognitive systems. For his contribution to computational intelligence IEEE established the *Frank Rosenblatt Award* in 2004.

[37]Later, more complex multi-layer perceptrons were constructed.

[38]Seymour Papert—a mathematician and computer scientist. He is known for designing the LOGO programming language.

[39]Teuvo Kohonen—a professor at Helsinki University of Technology and a prominent researcher of neural networks. In 1991 he was elected the first President of the European Neural Network Society.

[40]John J. Hopfield—an eminent researcher of neural networks, a professor of physics and molecular biology at California Institute of Technology, Princeton University, and the University of California, Berkeley. He was the President of the American Physical Society.

[41]David Everett Rumelhart—a professor of psychology at Stanford University and the University of California, San Diego. His work concerns applications of mathematics in psychology and artificial intelligence, and connectionist models. For his contribution to Artificial Intelligence the Rumelhart Prize was established in 2000.

[42]Geoffrey E. Hinton—a professor of computer science at the University of Toronto, Cambridge University, and Carnegie Mellon University, and a psychologist. His work concerns applied mathematics and neural networks. In 2005 he received the IJCAI Award for Research Excellence and in 2001 the Rumelhart Prize.

multi-layer networks with the *backpropagation* method in 1986 [252]. The next year Stephen Grossberg[43] and Gail Carpenter[44] defined neural networks based on *Adaptive Resonance Theory, ART* [42]. After such a series of successful research results, the comeback of the connectionist model and neural networks occurred in 1987, in which the *First International Conference on Neural Networks* was organized. During the conference its chairman Bart Kosko[45] announced the victory of the neural networks paradigm. The connectionist approach is discussed in Sect. 3.1, whereas models of neural networks are presented in Chap. 11.

Adherents of artificial neural networks have assumed that simulating nature at its biological layer is a proper methodological principle for constructing AI systems. They have tried to simulate the brain in both its anatomical/structural and physiological/functional aspects. A simulation of nature in its evolutionary aspect is a basic paradigm for defining *biology-inspired models*, which include an important group of AI methods called *evolutionary computing*. These methods are used in the crucial area of searching for an optimum solution of a problem. The fundamental principles of this approach include generating and analyzing many potential solutions in parallel, treating a set of such potential solutions as a *population* of *individuals*, and using operations on these individuals-solutions which are analogous to biological evolution operations such as crossover, mutation, and natural selection. Alex Fraser[46] published the first paper presenting the ideas of evolutionary computing in 1957 [102]. He can be considered the pioneer of this paradigm. Then, four basic groups of methods were developed within this approach: *genetic algorithms*, which became popular, when John Holland[47] published his excellent monograph in 1975 [139], *evolution strategies*, which were introduced in the 1960s by Ingo Rechenberg[48] [236] and Hans-Paul Schwefel[49] [267], *evolutionary programming*, originated by Lawrence J. Fogel[50] at the University of California, Los Angeles in the 1960s [99] and *genetic programming*,

[43]Stephen Grossberg—a professor of mathematics, psychology, and biomedical engineering at Boston University. He was the first President of the International Neural Network Society and a founder of the prestigious journal *Neural Networks*.

[44]Gail Carpenter—a professor of mathematics at Boston University. In the 1970s she published excellent papers on the use of dynamical systems for a generalization of Hodgkin-Huxley models (see Sect. 11.1).

[45]Bart Kosko—a professor at the University of Southern California. His research contribution concerns neural networks and fuzzy logic.

[46]Alex S. Fraser—an eminent researcher working in New Zealand, Australia, and the United States. His work concerns computer modeling in biology.

[47]John Henry Holland—a professor of psychology and a professor of electrical engineering and computer science at the University of Michigan. He developed the *(Holland's) schema model*, which is fundamental in the theory of genetic algorithms.

[48]Ingo Rechenberg—a professor of computer science at the Technical University of Berlin, one of the pioneers of bionics.

[49]Hans-Paul Schwefel—a professor of computer science at Dortmund University of Technology. His work concerns optimization theory (fluid dynamics) and system analysis.

[50]Lawrence J. Fogel—a researcher and an author of patents on control theory, telecommunications, cybernetics, and biomedical engineering.

popularized by John Koza[51] in the 1990s [172]. The biology-inspired approach is discussed in Sect. 3.3 and its models are presented in Chap. 5.

Methods constructed on the basis of various mathematical theories have played a fundamental role in the development of Artificial Intelligence models since the very beginning. The most important ones include pattern recognition, cluster analysis, Bayesian inference and Bayes networks, models based on fuzzy set theory, and models based on rough set theory. *Pattern recognition* is one of the oldest fields of Artificial Intelligence. In fact, it was originated as a research area before the term *Artificial Intelligence* was coined, because the first method within this approach was developed by Ronald A. Fisher[52] in 1936 [90]. Pattern recognition methods are used for the classification of unknown objects/phenomena, which are represented by sets of features. The complementary issue of *cluster analysis* consists of grouping a set of objects/phenomena into classes (categories). These models are presented in Chap. 10.

Bayesian theory was used in Artificial Intelligence very early.[53] Its role in AI increased even more when Judea Pearl[54] introduced an inference model based on *Bayes networks* [222]. The use of this model for reasoning with imperfect knowledge is presented in Chap. 12.

The imperfection of a system of notions that is used to describe the real world is a crucial problem, if we apply mathematical models of reasoning. Since the world is multi-aspect and complex, the concepts used in its representation are usually unambiguous and imprecise, which is unacceptable. In order to solve this dilemma, in 1965 Lotfi A. Zadeh[55] introduced *fuzzy set theory* [321] and in the 1980s Zdzisław Pawlak[56] developed *rough set theory* [216]. Both theories are introduced in Chap. 13.

Artificial Intelligence is a very interesting interdisciplinary research area. It is also a place of heated disputes between adherents of different schools. In fact, we can find scientists in every school who claim that their approach is the only one that is

---

[51] John Koza—a professor of computer science at Stanford University. His work concerns cellular automata, multi-agent systems, applications of computer science in molecular biology, and AI applications in electrical engineering, automatic control, and telecommunication.

[52] Ronald Aylmer Fisher—a professor of genetics at University College London and the University of Cambridge. He was the principal founder of mathematical statistics. For his scientific contribution he was elected to the Royal Society in 1929.

[53] For example, statistical pattern recognition, which is presented in Sect. 10.5.

[54] Judea Pearl—a professor at the University of California, Los Angeles, computer scientist and philosopher. In 2000 he published an excellent monograph on causality and its role in statistics, psychology, medicine, and social sciences entitled *Causality: Models, Reasoning, and Inference* (Cambridge University Press). For this book he received the prestigious Lakatos Award in 2001.

[55] Lotfi Asker Zadeh—a professor at the University of California, Berkeley, electrical engineer, computer scientist, and mathematician. His work concerns Artificial Intelligence, control theory, logics, and linguistics.

[56] Zdzisław Pawlak—a professor at Warsaw University of Technology and the Mathematics Institute of the Polish Academy of Sciences. His work concerns mathematical foundations of computer science, mathematical linguistics, automated theorem proving, and formal models in genetics.

right.[57] Fortunately, in spite of the disputes, which in principle are of a philosophical nature, the practice of constructing AI systems has consisted of *integrating* various approaches since the 1980s. For such an integration *cognitive architectures* are especially convenient. The best known cognitive architectures include Soar (originally called SOAR, for *Symbols, Operators, And Rules*) [210] developed by Allen Newell in 1983–1987, ACT* (*Adaptive Control of Thought*) [5] and ACT-R (*Adaptive Control of Thought—Rational*) [6] designed by John Robert Anderson[58] in 1983 and 1993, respectively. *Multi-agent systems (MASs)* are good examples of such architectures. Cognitive architectures are presented in Chap. 14.

Bibliographical Note

Fundamental monographs and papers on Artificial Intelligence include [18, 19, 55, 147, 168, 189, 211, 241, 256, 257, 261, 262, 273, 299, 315]. The history of Artificial Intelligence is presented in [62, 197].

[57]Because of the introductory nature of the monograph, we have only mentioned the fundamental dispute between *Strong AI* and *Weak AI*. Other important disputes include *computationalists versus connectionists*, *neats versus scruffies*, etc. For the reader who is interested in the history of AI, the books included in the *Bibliographic note* are recommended.

[58]John Robert Anderson—a professor of psychology and computer science at Carnegie Mellon University, one of the pioneers of cognitive science, and a designer of intelligent tutoring systems. He was the President of the Cognitive Science Society.

[illegible]

Chapter 2
Symbolic Artificial Intelligence

Basic methodological assumptions of methods which belong to *symbolic Artificial Intelligence* are presented in this chapter. The three following fundamental beliefs are common to these methods:

- a model representing an intelligent system can be defined in an *explicit* way,
- knowledge in such a model is represented in a symbolic way,[1]
- mental/cognitive operations can be described as formal operations[2] over symbolic expressions and structures which belong to a knowledge model.

In symbolic AI two subgroups of methods can be distinguished. Within the first subgroup researchers try to define *general (generic)* models of knowledge representation and intelligent operations. This subgroup includes cognitive simulation and logic-based reasoning.

By contrast, defining models which concern *specific* application areas is the goal of research within the second group. Thus, such models are based on representations of domain knowledge. This subgroup includes rule-based knowledge representation, structural knowledge representation, and an approach based on mathematical linguistics.

These groups of methods are presented briefly in the following sections.

---

[1]For example, knowledge is defined in the form of graphs, logic formulas, symbolic rules, etc. Methods of symbolic AI are developed on the basis of logic, theory of formal languages, various areas of discrete mathematics, etc.

[2]For example, operations in the form of inference rules in logic, productions in the theory of formal languages, etc.

M. Flasiński, *Introduction to Artificial Intelligence*,
DOI 10.1007/978-3-319-40022-8_2

2.1 Cognitive Simulation

Cognitive simulation is one of the earliest approaches to Artificial Intelligence. The approach was introduced by Newell and Simon, who used it to design their famous systems *Logic Theorist* and *General Problem Solver, GPS*. The main idea of cognitive simulation consists of defining heuristic algorithms[3] in order to simulate human cognitive abilities, e.g., reasoning, problem solving, object recognition, and learning. During such a simulation a sequence of elementary steps, which are analogous to those made by a human being, is performed by a computer. Therefore, in order to design such algorithms we try to discover elementary concepts and rules which are used by a human being for solving generic problems. Now, we introduce four basic concepts of cognitive simulation, namely state space, problem solving as searching state space, Means-Ends Analysis, and problem reduction.

Let us begin with the concept of *state space*. Its initial state represents the situation in which we begin problem solving. Let us consider the example of chess. The initial state represents the initial position of the pieces at the start of a game. Goal states (or the goal state, if the problem has only one solution) represent a problem at the moment of finding a solution. Thus, in chess goal states represent all the situations in which we checkmate the opponent. The remaining (intermediate) states represent all situations that are possible on the way to solving the problem. Thus, in chess they represent all situations that are allowed taking into account the rules of the game. A state space is a graph in which nodes correspond to states (initial, goal, and intermediate) and edges represent all allowable transitions from one state to another. Thus, for chess a state space can be defined in the following way. From the node of the initial state we define transitions (edges) to nodes that correspond to situations after the first "white" move. Then, for each such intermediate state we define transitions (edges) to nodes that correspond to situations after the first "black" move in response to a situation caused by the first "white" move, etc.

Problem solving as searching a state space is the second concept of cognitive simulation. The idea is straightforward: if we do not know how to formulate algorithmic rules of problem solving, we can try to solve this problem with the method of *"trial and error" ("generate and test", "guess and check")*. Of course, the use of this method in its "pure form" is not a good idea for, e.g., playing chess. However, sometimes this is the only method we can use in our everyday life. For example, we have forgotten the combination lock code of our suitcase or we have promised our fiancée to make a delicious omelet in the evening and we have lost the recipe. (However, we have a lot of eggs in the refrigerator, so we can make some experiments.) Let us notice that such generation of potential solutions can be "blind" (we do not

[3] A heuristic algorithm is an algorithm which can generate an accepted solution of a problem although we cannot formally prove the adequacy of the algorithm w.r.t. the problem. The term *heuristics* was introduced by the outstanding mathematician George Pólya. Newell attended lectures delivered by Pólya at Stanford University.

use codes which represent important dates) or can be limited to a certain subarea of the state space (we use important dates as codes). What is more, we can have a certain measure, called here a *heuristic function*, which tells us how close we are to a satisfactory solution (e.g., the "tasting measure", which tells us how close we are to a delicious omelet). Then, if our experimental omelet is almost delicious, we should modify the recent result only a little bit. A correct procedure to generate possible solutions should have three properties. Firstly, it should be *complete*, i.e., it should be able to generate *every* potential solution. Secondly, it should not generate a potential solution more than once. Thirdly, it should make use of information that restricts the state space to its subarea as well as measures/functions of the quality of the potential solutions generated.

If we have a function which assesses the quality of potential solutions, then we can use a technique called *Means-Ends Analysis, MEA*, for controlling the process of generating potential solutions. If we are at a certain state, then we use such a function for determining the difference/distance between this state and a goal state and we use a *means* (i.e., the means is usually in the form of an operator or a procedure) which reduces this difference. For example, if our experimental omelet is nearly as tasty as the ideal one and the only difference is that it is not sweet enough, then we should use a "sweetening operator" to reduce this difference. We use the MEA technique iteratively, starting from the initial state until we reach a goal state.

Problem reduction is the last concept. The idea consists of replacing a complex problem, which is difficult to solve at once, by a sequence of simpler subproblems. For example, if obtaining a driving license is our problem, then we should divide it into two subproblems: passing a theory test and passing a road (practical) test. Let us notice that for many problems their reduction to a sequence of simpler subproblems is necessary, because different state spaces must be defined for these subproblems and in consequence different quality functions and different operators must be determined.

2.2 Logic-Based Approach

As we have mentioned in a previous chapter, John McCarthy, who introduced a logic-based approach to AI, claimed that intelligent systems should be designed on the basis of formalized models of logical reasoning rather than as *simulators* of heuristic rules of human mental processes. This idea was revolutionary in computer science at the end of the 1950s, because of the following facts. At that time the standard methodology of designing and implementing information systems was based on the *imperative paradigm*.[4] It consists of defining a program as a sequence of commands[5] that should be performed by a computer. Thus, a computer programmer

[4]The Object-Oriented paradigm is, nowadays, the second standard approach.

[5]In Latin *imperativus* means *commanded*.

has to determine *how* computations should be performed in order to solve a problem. Meanwhile, according to McCarthy, a programmer who implements an intelligent program should determine only the required properties of a solution. In other words, he/she should specify *what* the solution of the problem is and not *how* the solution is to be obtained. The solution should be found by a universal problem solver (a generic program for solving problems). We say that such a methodology of designing and implementing information systems is based on the *declarative paradigm*.[6]

For example, if we want to write a declarative-paradigm-based program which solves the problem of determining whether two people are siblings, we can do it as follows:

siblings(X, Y) :– parent(Z, X) <u>and</u> parent(Z, Y),

which is interpreted in the following way:

X and Y are siblings if there exists Z such that Z is a parent of X <u>and</u> Z is a parent of Y.

As we can see, in our program we declare only the required properties of a solution, i.e., *what it means* that two people are siblings, and we leave reasoning to a computer. Of course, we should define a database containing characteristics of people who can be objects of such reasoning.

There are two main approaches when we use the declarative paradigm in Artificial Intelligence,[7] namely logic programming and functional programming. In *logic programming* a specification of the required properties of a solution is expressed as a set of *theorems* in a logic language, usually in the language of First-Order Logic, FOL. A universal problem solver uses these theorems to perform reasoning according to FOL rules of inference. So, if we want to solve a specific problem with the help of such a solver, e.g., we want to check whether John Smith and Mary Smith are siblings, then we formulate a hypothesis of the form siblings(John Smith, Mary Smith) and the system tries to verify this hypothesis making use of facts stored in its knowledge base as well as theorems (like the one introduced above and saying what being siblings means).

Functional programming is based on *lambda calculus*, which was introduced by Alonzo Church and Stephen C. Kleene. Lambda calculus is a formal system of mathematical logic which is used for specifying computation functions defined with the help of highly formalized expressions. In functional programming a specification of the required properties of a solution is defined by just such a complex function. In this case, a universal problem solver should be able to interpret these expressions in order to perform an *evaluation* (i.e., symbolic computation) of the corresponding functions according to the principles of lambda calculus.[8]

---

[6]We *declare* required properties of a solution of a problem.

[7]This paradigm is also used nowadays beyond AI. For example, such programming languages as SQL and HTML are also based on the declarative paradigm.

[8]A more detailed description of the functional approach is included in Sect. 6.5.

2.3 Rule-Based Knowledge Representation

Newell and Simon continued research into models of cognitive processes after introducing their theory of cognitive simulation. In 1972 they proposed a *production system* model [208]. It consists of two basic components: a *production memory* and a *working memory*.

A production memory corresponds to *long-term memory* in psychology, and working memory to *short-term memory*. In long-term memory knowledge is represented with the help of *productions/rules*. A rule is of the following form: *If a certain* condition *holds, then perform a certain* action, where an action can be either a conclusion drawn from the condition or a certain action performed on the system environment (of course, only if the condition is fulfilled). Since rules are stored in long-term memory, we assume that they are available constantly.

A working memory contains information which changes in time. This is mainly information concerning the environment of the system. In one mode of the system reasoning information is checked continuously with respect to conditional parts of rules. If the condition of some rule is fulfilled,[9] then this rule is applied. If a rule application consists of drawing a certain conclusion, then this conclusion is stored in the working memory. If a rule application consists of performing a certain action on the system environment, then the system initializes this action, e.g., it switches off some device by sending a command to a processor which controls this device, sending a command to a robot actuator, which causes a movement of the robot arm, etc.

In spite of the fact that the production system model was introduced by Newell and Simon as a general theory to simulate generic cognitive processes, in Artificial Intelligence it is known mainly in a more specific version, namely as a model of an expert rule-based system. This model is presented in Chap. 9.

2.4 Structural Knowledge Representation

According to the theory of Gilbert Ryle[10] our taxonomy of knowledge includes *declarative knowledge*, which is a static knowledge concerning facts ("knowing that") and *procedural knowledge*, which is knowledge about performing tasks ("knowing how"). For example, a genealogical tree is a representation of declarative knowledge, and a heuristic algorithm, which simulates problem solving by a human being, corresponds to procedural knowledge.

Structural models of knowledge representation are used for defining declarative knowledge. They are usually in the form of graph-like hierarchical structures.

---

[9] In such a situation we say that the system has *matched* a certain fact (facts) stored in the working memory to the rule.

[10] Gilbert Ryle—a professor of philosophy at Oxford University, one of the most eminent representatives of analytic philosophy, the editor of the prestigious journal *Mind*.

Although originally they were used for *Natural Language Processing, NLP*, it turned out that they could be used in other AI application areas as well. *Conceptual dependency theory*, developed by Schank [263] at the end of the 1960s was one of the first such models. Schank claimed, contrary to the Chomsky generative grammar theory,[11] that a language syntax was rather a set of pointers to semantic information that could be used as a starting point for a direct semantic analysis. Conceptual dependency theory was formulated just for delivering convenient (structural) formalisms for performing a semantic analysis automatically. Since sentences of a (natural) language could not be used for performing an (automatic) semantic analysis in a direct way, Schank introduced a canonical, normalized representation[12] of semantic dependencies between language constructs (phrases, sentences).

Such a canonical representation is defined with the help of *conceptual dependency graphs*. Labeled nodes of such graphs correspond to *conceptual primitives*, which can be used for defining semantic representations. For example, a *primitive act PTRANS* means a transfer of the physical location of an object and *SPEAK* means an action generating a sound by a living object. Labeled edges of such graphs represent various relations (dependencies) between conceptual primitives. Conceptual dependency graphs are defined in an unambiguous way according to precise principles. These graphs can be analyzed in an automatic way, which allows the system to perform a semantic analysis of corresponding constructs of a natural language. Summing up, Schank has verified a hypothesis saying that we can try to perform semantic analysis automatically if we define concept-based language representations in an explicit and precise way.

Semantic networks, introduced by Collins and Quillian [56] is one of the first graph models defined on the basis of the assumptions presented above. Their nodes represent objects or concepts (classes of abstraction, categories) and their edges represent relations between them. Two specific relations are distinguished: *is subclass of* (e.g., *Triangle is subclass of Polygon*) and *is*—for denoting that some object belongs to a certain class (e.g., *John Smith is Human Being*).

Frames (frame-based systems), introduced by Minsky [203], can be treated as a substantial extension of semantic networks. A node of such a network is called a frame, and has a complex internal structure. It allows one to characterize objects and classes in a detailed way. Frame theory has the following psychological observation as a basic assumption. If somebody encounters a new unknown situation, then he/she tries to get a structure called a frame out of his/her memory. This structure, which represents a stereotyped situation similar to the current situation, can be then used for generating an adequate behavior. In AI systems a graph-like structure of frames is defined according to precise principles which allow the system to process and analyze it in an automatic way.

---

[11] The Chomsky theory is presented in the next section.

[12] A normalization of a semantic representation is necessary if it is to be performed automatically, because all sentences which have the same meaning, e.g., *John has lent a book to Mary.*, *Mary has borrowed a book from John.* should have the same representation.

Scripts have been proposed by Schank and Abelson [264] as a method for Natural Language Processing (NLP). Scripts can be defined with the help of conceptual dependency graphs introduced above. The model is also based on a certain observation in psychology. If one wants to understand a message concerning a certain event, then one can refer to a generalized pattern related to the type of this event. The pattern is constructed on the basis of similar events that one has met previously. Then it is stored in the human memory. One can easily notice that the concept of scripts is similar conceptually to the frame model.

In the past, structural models of knowledge representation were sometimes criticized for being not formal enough. This situation changed in the 1980s, when a dedicated family of formal systems based on mathematical logic called *description logics* were defined for this purpose. The foundations of description logics are presented in Appendix D, whereas a detailed presentation of semantic networks, frames, and scripts is included in Chap. 7.

2.5 Mathematical Linguistics Approach

As we have seen in previous sections, constructing a description of the physical world that can be used for intelligent system reasoning is one of the fundamental issues of Artificial Intelligence. We can try to solve this problem in a similar way as in cognitive simulation, that is to simulate a human being, which generates such a description with the help of natural language. However, the following crucial problem arises in such a case. Any natural language is an *infinite* set of sentences, which are constructed according to the principles of the corresponding grammar. Thus, automatically checking whether the syntax of a sentence is correct is a non-trivial problem.

A solution to this problem has been proposed by Noam Chomsky, who has defined *generative grammars* for the purpose of syntactic analysis of natural languages. Generative grammars are just *generators* of infinite languages. On the other hand, in Artificial Intelligence sometimes we are interested more in constructing systems that can be used for analysis of sentences than in their generation. Fortunately, *formal automata* can be used for this purpose. They are defined on the basis of generative grammars in mathematical linguistics. The task of a formal automaton can be described (in the minimalist case) as determining whether an expression is a sentence of a given language. In other words: Is the expression defined according to the principles of the grammar corresponding to the given language? In fact, a pair *generative grammar—formal automaton* can be treated as a (simple) physical symbol system.

Such a minimalist formulation of the task of an automaton (as a syntax analyzer only) can be generalized in two respects. Firstly, we can use a formal automaton for (semantic) interpretation of language sentences. Although the initial great expectations of *natural language* interpretation have not been fulfilled, in the case of *formalized languages* the Chomsky model has turned out to be very useful for an interpretation task. For the purpose of interpretation of formalized languages the descriptive power of generative grammars has been increased considerably in Artificial Intelligence. Their standard modifications (extensions) have consisted of adding attributes to language components (*attributed grammars*) and defining "multi-dimensional" generative grammars. Standard Chomsky grammars generate sequential (string) structures, since they were defined originally in the area of linguistics. As we have discussed in the previous section, graph-like structures are widely used in AI for representing knowledge. Therefore, in the 1960s and 1970s grammars generating graph structures, called *graph grammars*, were defined as an extension of Chomsky grammars.

The second direction of research into generalizations of the formal language model concerns the task of formal language translation. A translation means here a *generalizing translation*, i.e., performing a kind of abstraction from expressions of a lower-level language to expressions of a higher-level language. Formal automata used for this purpose should be able to read expressions which belong to the basic level of a description and produce as their output expressions which are generalized interpretations of the basic-level expressions. Such automata are often called *transducers*.

The problem of automatic synthesis of formal automata is very important in Artificial Intelligence. To solve this problem *automata synthesis* algorithms, which generate the rules of an automaton on the basis of a generative grammar, have been defined. The successes in this research area have been achieved due to the development of the theory of programming language translation. At the same time, an even more fundamental problem, namely the problem of automatic *induction (inference) of a grammar* on the basis of a sample of language sentences has appeared. This problem is still an open problem in the area of Artificial Intelligence. All the issues mentioned in this section are discussed in detail in Chap. 8.

Chapter 3
Computational Intelligence

Computational Intelligence, CI, is the second group of methods in Artificial Intelligence. It is a complementary approach with respect to symbolic AI. Although a precise definition of this paradigm is difficult to formulate [76], we can try to list the common features of CI methods as follows:

- numeric information is *basic* in a knowledge representation,
- knowledge processing is based *mainly* on numeric computation,
- *usually* knowledge is not represented in an *explicit* way.

Of course, there are models of Computational Intelligence which do not fulfill such characteristics entirely, especially where the third item is concerned. The Bayes networks model, which is introduced in Chap. 12, is a good example of such an exception. Therefore, in order to avoid a misunderstanding about which methods can be included in Computational Intelligence in this monograph, we list them in an *explicit* way. So, we assume that the following methods can be treated as being defined on the basis of the CI paradigm: (artificial) neural networks, pattern recognition, cluster analysis, Bayesian inference, models based on fuzzy sets, models based on rough sets, evolutionary computing (genetic algorithms, evolution strategies, evolutionary programming, and genetic programming), swarm intelligence, and artificial immune systems.

Now we present these methods in a general way, dividing them into three groups, namely: connectionist models, mathematics-based models, and biology-based models.

3.1 Connectionist Models

In the nineteenth century the *associationist approach* appeared in psychology. Its representatives claimed that the association of mental states is a basic mechanism of mental processes. In this approach the nature of complex mental phenomena is

M. Flasiński, *Introduction to Artificial Intelligence*,
DOI 10.1007/978-3-319-40022-8_3

explained by the interaction of simpler ones. This general idea of associationism was developed by Edward L. Thorndike[1] as the *connectionist approach* [301]. According to this approach, learning is a result of associations between stimuli and responses to stimuli. Associations become strengthened if an organism is trained with *stimulus-response* exercises, and they become weakened if such training is discontinued. Responses which are rewarded become strengthened and after some time become habitual responses.

These ideas of connectionism have been assimilated in Artificial Intelligence for the purpose of describing mental processes, which has led to *connectionist models* in AI. In these models associations are represented with the help of *connectionist networks*. There are two main types of these networks [21].

In *localist connectionist networks* each component of knowledge (concept, object, attribute, etc.) is stored in a single element of a network. Therefore, we can include, e.g., semantic networks [21] in this model, although we have ascribed them to symbolic AI in a previous chapter. Although they are not treated in AI as typical connectionist networks, in fact they fulfill the conditions of their definition. For example, in the ACT-R model [6], mentioned in Chap. 1, each node of a semantic network has an activity parameter (a weight), which is used to stimulate strengthening/weakening mechanisms described above.[2] Bayes networks are another example of localist connectionist networks.[3]

Distributed connectionist networks are the second type of such networks. In this case knowledge is stored in a distributed way, i.e., it is distributed among many elements of a network. *Artificial neural networks* are the best example of such networks. According to custom only neural networks are associated with the connectionist approach in Artificial Intelligence. Later, our considerations of connectionist models will be limited to distributed connectionist networks only.

In a distributed connectionist approach, mental states are modeled as *emergent processes*, which take place in networks consisting of elementary processing units. As we have mentioned in Chap. 1, a process is *emergent* if it cannot be described on the basis of its elementary sub-processes. This results from the fact that the nature and the functionality of an emergent process is something more than just the simple sum of functionalities of its sub-processes.[4]

[1]Edward Lee Thorndike—a professor of psychology at Columbia University. His work concerns zoopsychology and educational psychology. He was the President of the American Psychological Association.

[2]In fact, ACT-R is a hybrid AI system, which is based on both the symbolic approach and the sub-symbolic (CI) approach.

[3]Bayes networks are presented in Chap. 12.

[4]Any human mental process is a good example here. Although a single biological neuron does not think, a brain treated as a network consisting of neurons thinks.

The fundamentals of distributed connectionism were established by David E. Rumelhart and James L. McClelland[5] in [253,254]. Apart from the characteristics of this approach mentioned above, we assume that mental states in a network are modeled in such a way that network units process them in a parallel way. The units perform numeric operations. As a consequence of such operations any processing unit can be *activated*. Then, the result of such an activation is propagated to all units which are connected to this unit. The network acquires and stores knowledge in an *implicit* way by modifying the parameters (weights) of connections among the processing units. The process of modifying these parameters is treated as *network learning*.

A model of artificial neural networks as a representative of the (distributed) connectionist approach is presented in detail in Chap. 11.

3.2 Mathematics-Based Models

As we have mentioned in Chap. 1, models defined on the basis of various mathematical theories play a fundamental role in Artificial Intelligence.

The first methods used for solving one of the key problems of AI, namely the automatic recognition of objects, appeared at the beginning of computer science. This is the field of *pattern recognition*. Objects (phenomena, processes, etc.) are represented by sets of features (attributes). Recognition of an unknown object/phenomenon is performed by a *classifier*, which ascribes the object to one of a number of predefined categories.[6] In order to construct a classifier a set of example objects with their correct classifications should be available.[7] Such a set, called a *learning (training) set*, is a kind of knowledge base of a classifier. The main idea of a classification process can be defined as the task of finding the object X in the learning set which is "similar" most to the unknown object. If the classifier finds such an object X, it ascribes the unknown object to the class that the object X belongs to. In fact, this (simplified here) general idea of the classification is implemented with the help of advanced mathematical models such as the Bayesian probability model, discriminating functions, minimum-distance models, etc. These models are presented in detail in Chap. 10. The complementary issue of *cluster analysis*, which consists of grouping a set of objects/phenomena into classes (categories), is discussed in Chap. 10 as well.

The second important group of mathematics-based methods relates to the crucial issue of the *possibility of formally specifying*:

---

[5]James Lloyd "Jay" McClelland—a professor of psychology at Carnegie Mellon University and Stanford University. His work concerns psycholinguistics and applications of connectionist models in pattern recognition, speech understanding, machine learning, etc.

[6]The *categories*, also called *classes*, should be defined earlier, i.e., when we formulate the problem. For example, if a classifier is constructed to support medical diagnosis, then disease entities are the categories.

[7]Such a set corresponds to human experience in solving a given classification problem. For example, in medical diagnosis it corresponds to the diagnostic experience of a physician.

- vague notions that are used for a description of the world, and
- an inference process, when only imperfect knowledge is available.

Imperfect knowledge can result from various causes. For example, input information can be uncertain (uncertainty of knowledge), measurements of signals received by an AI system can be imprecise (imprecision of knowledge) or the system may not know all required facts (incompleteness of knowledge).

The model of *Bayes networks* is used for inference that is based on propositions to which the probability of an event occurring is added. The probability measure expresses our uncertainty related to the knowledge rather than the degree of truthfulness of a specific proposition. There are usually a lot of possible factors which influence the result of such probabilistic reasoning. An assessment of probabilities of these factors as well as their combinations is often impossible in real-world applications. Therefore, in this model we construct a graph which represents connections between only those factors which are essential for reasoning.

If our knowledge is incomplete, we can use *Dempster-Shafer theory* for reasoning. In this theory we use specific measures to express the degree of our ignorance. If we acquire new knowledge, these measures are modified to express the fact that our ignorance is reduced.

Knowledge is continuously acquired by intelligent systems. In a classical logic we assume that after adding new propositions to a model the set of its consequences does not decrease. However, this assumption is not true in the case of AI systems which reason over the real-world environment. To put it simply, new facts can cause old assumptions not to be true any more. To solve this problem we use *non-monotonic logics* such as default logic, autoepistemic logic, or circumscription, or specific models like the Closed-World Assumption model.

Bayes networks, Dempster-Shafer Theory, and non-monotonic logics are presented in Chap. 12.

The problem of defining formal specifications of concepts which are used for a description of the world seems to be even more difficult. On one hand, we have vague notions, which are used in everyday life. On the other hand, mathematics-based models require notions which should be precise and unambiguous.

The vagueness of notions can be considered in two ways. First of all, a notion can be ambiguous, which usually results from its subjective nature. Notions relating to the height of a human being (e.g., tall, short) are good examples of such notions. In this case we should take into account the subjective nature of a notion by introducing a measure, which grades "being tall (short)". This is the main idea of *fuzzy set theory*.

The vagueness of a notion can relate to the *degree of precision (detail, accuracy)* which is required during a reasoning process. This degree should be adequate with respect to the problem to be solved, i.e., it should be determined in such a way that our AI system should distinguish between objects which are considered to belong to different categories and it should not distinguish between objects which are treated as belonging to the same category. This is the main idea of *rough set theory*.

Both theories which are used to solve the problem of vagueness of notions are introduced in Chap. 13.

3.3 Biology-Based Models

As we have mentioned in Sect. 2.1 cognitive simulation consists of discovering an optimum solution to a problem by searching a state space which contains potential solutions. In *biology-based models* such a search is performed by simulating evolutionary aspects of nature.

Evolutionary computing is a basic group of biology-based models. Generating many potential solutions, called a *population*, is the main idea of these methods. These potential solutions in a state space are treated as individuals, by analogy to biological evolution. The operations which are analogous to genetic operations such as *crossover* and *mutation* are applied to individuals-solutions. The best-fitted individuals are selected for "breeding offsprings". The fitness of individuals is evaluated with the help of a *fitness function*, which plays an analogous role to a heuristic function in cognitive simulation. The probabilistic nature of evolutionary computing is its essential feature. The succeeding populations, called *generations*, are generated as long as some individuals represent an accepted solution.[8] As we have mentioned in Chap. 1, there are four basic groups of methods within this approach: *genetic algorithms*, *evolution strategies*, *evolutionary programming*, and *genetic programming*. Generally, these methods differ in the way individuals are represented (e.g., binary coding in genetic algorithms, real number vectors in evolution strategies, and tree structures in genetic programming), the importance of various genetic operations (e.g., the fundamental role of crossover in genetic algorithms and genetic programming, the fundamental role of mutation in evolution strategies, and the occurrence of only mutation in evolutionary programming) and the way of generating a new population. Evolutionary computing is presented in detail in Chap. 5.

Within the biology-inspired approach we also distinguish methods which are constructed on the basis of models other than evolutionary theory. They include mainly *swarm intelligence* and *Artificial Immune Systems, AISs*. These methods are presented in Sect. 5.5.

[8]There can be other conditions for algorithm termination, e.g., a fixed number of generations, computation time, etc.

Part II
Artificial Intelligence Methods

Chapter 4
Search Methods

We begin our presentation of AI models with search methods not only for chronological reasons, but also because of their methodological versatility. As we have mentioned in the first chapter and in Sect. 2.1, these methods are based on an approach called *cognitive simulation*, which was introduced by Newell and Simon. The main idea of this approach consists of constructing heuristic algorithms simulating elementary rules of human mental/cognitive processes. In the following sections we discuss basic ideas of search in a state space, blind search, heuristic search, adversarial search, search in a constraint satisfaction problem, and special methods of a heuristic search.

4.1 State Space and Search Tree

In Sect. 2.1 the basic ideas of state space search have been introduced. In this section we discuss two of them: state space and solving problems by heuristic search. Let us present them using the following example.

Let us imagine that we are in the labyrinth shown in Fig. 4.1a and we want to find an exit. (Of course, we do not have a plan of the labyrinth.) If we want to solve this problem with the help of search methods, we should, firstly, define an *abstract model* of the problem. The word *abstract* means here taking into account only such aspects of the problem that are essential for finding a solution.[1] Let us notice that in the case of a labyrinth the following two elements, shown in Fig. 4.1b, are essential: characteristic points (crossroads, ends of paths) and paths. After constructing such a model, we can abstract from the real world and define our method on the basis of

[1]Deciding which aspects are essential and which should be neglected is very difficult in general. This phase of the construction of a solution method is crucial and influences both its effectiveness and efficiency. On the other hand, an abstract model is given for some problems, e.g., in the case of games.

M. Flasiński, *Introduction to Artificial Intelligence*,
DOI 10.1007/978-3-319-40022-8_4

the model alone (cf. Fig. 4.1c).[2] We assume that the starting situation is denoted by a small black triangle and the final situation with a double border (cf. Fig. 4.1c).

After constructing an abstract model of a problem, we can define a *state space*. As we have discussed in Sect. 2.1, it takes the form of a graph. Nodes of the graph represent possible phases (steps) of problem solving and are called *states*. Graph edges represent transitions from one phase of problem solving to another. Some nodes have a special meaning, namely: the *starting node* represents a situation in which we begin problem solving, i.e., it represents the *initial state* of a problem, and *final nodes* represent situations corresponding to problem solutions i.e., they are *goal states*.[3] Thus, solving a problem consists of finding a path in a graph that begins at the starting node and finishes at some final node. This path represents the way we should go from one situation to another in the state space in order to solve the problem.

A state space usually contains a large number of states. Therefore, instead of constructing it in an explicit way, we generate and analyze[4] step by step only those states which should be taken into account in the problem-solving process. In this way, we generate a *search tree* that represents only the interesting part of the state space. For our labyrinth a fragment of such a tree is shown in Fig. 4.2a.

Let us notice that nodes of a search tree represent the possible phases of problem solving (problem states) defined by an abstract model of a problem. Thus, each node of a search tree, being also a node of a state space, corresponds to a situation during movement through a labyrinth. This is denoted[5] in Fig. 4.2a with a bold characteristic point corresponding to the place we are at present and a bold path corresponding to our route from our initial position to our present position. Thus, the first (uppermost) node[6] of the search tree corresponds to the initial situation, when we are at a point A of the labyrinth. We can go along two paths from this point: either right[7] to a point B (to the left node of the tree) or left to a point C (to the right node of the tree), etc.

Till now we have considered the abstract model of the problem based on the labyrinth plan (from the "perspective of Providence"). Let us notice that firstly, we do not have such a plan in a real situation. Secondly, we like to simplify visualizing a state space, and in consequence visualizing a search tree. In fact, wandering around in the labyrinth we know only the path we have gone down. (We assume that we make signs A, B, C, etc. at characteristic points and we mark the path with chalk.) Then, a fragment of a search tree corresponding to the one shown in Fig. 4.2a can be

---

[2] Of course, in explaining the idea of an abstract model of problem, we will assume a "perspective of Providence" to draw a plan of the labyrinth. In fact, we do not know this plan - we know only the types of elements that can be used for constructing this plan.

[3] The remaining nodes correspond to intermediate states of problem solving.

[4] According to *Means-Ends Analysis, MEA*, discussed in Sect. 2.1.

[5] Again, from the "perspective of Providence", not from our perspective.

[6] Let us recall that such a node is called the *root* of the tree.

[7] Let us remember that a "black triangle" is behind us, cf. Fig. 4.1a.

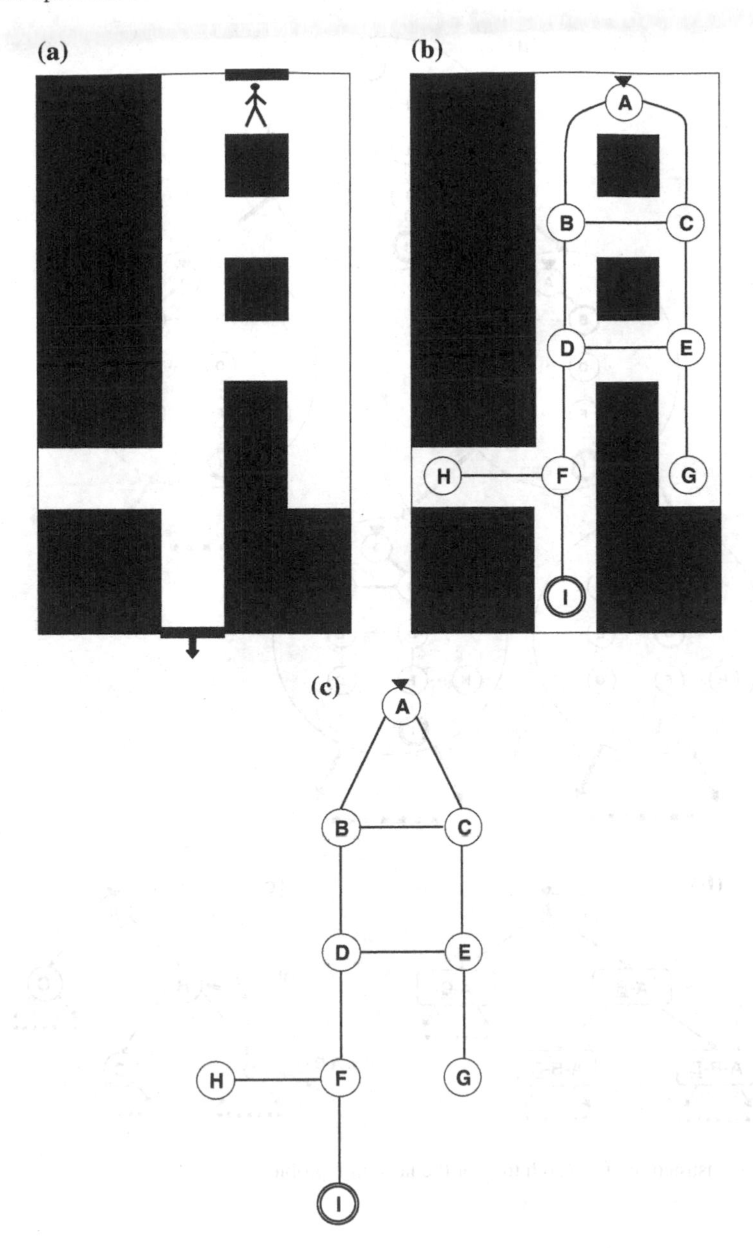

Fig. 4.1 An abstract model of a problem—a representation of a labyrinth

depicted as in Fig. 4.2b. Now, the path we have gone through till the present moment, i.e., the sequence of characteristic points we have visited, is written into each tree node, and the place we are in at present is underlined.

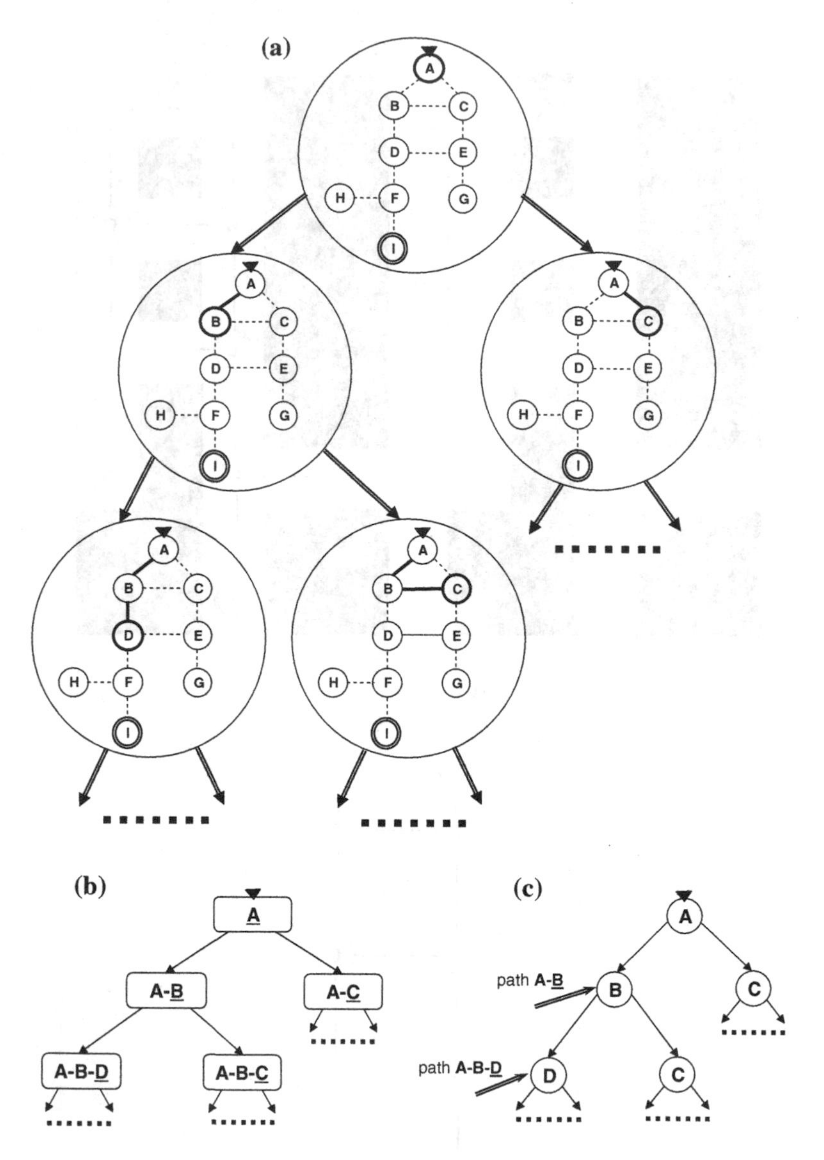

Fig. 4.2 Construction of a search tree for the labyrinth problem

In fact, we can simplify the labels of the nodes of the search tree even more. Let us notice that a label A-B-D means that we have reached the node D visiting (successively) nodes A and B. On the other hand, such information can be obtained by going from the root of the tree to our current position. Thus, we can use the tree shown in Fig. 4.2c instead of the one depicted in Fig. 4.2b. We will use such a representation in our further considerations.

In the remaining part of this chapter, we discuss basic methods of generating a search tree. The specificity of these methods consists in the order in which nodes of such a tree (representing states of a corresponding state space) are generated and looked through. This order determines the so-called *search strategy*, and it is a criterion of the taxonomy of techniques of state space searching.

In general, we can divide these techniques into two basic groups: *blind search* methods and *heuristic search* methods. In the first group we mainly use information concerning the structure of the state space, i.e., information about possible transitions between states. Knowledge concerning the specifics of the problem to be solved is used in a minimum degree. In the second group such knowledge is used to define a heuristic function assessing the *quality* of a state. The heuristic function says how far a given state is from a goal state. These two groups of search methods are discussed in the following two sections.

4.2 Blind Search

In the first section we have shown how a search tree is generated for a state space (cf. Fig. 4.2c). Of course, in real-world problems such a tree is very big. Therefore, we generate a part of it only, i.e., we generate it till we reach a goal state representing a solution. In the case of our labyrinth problem, however, we can define a search tree representing all the possible tours from the initial state to the goal state as is shown in Fig. 4.3a.[8] In Fig. 4.3a we marked additionally the *root node* (we start constructing our tree at this node), (some) *leaf nodes* (nodes which do not have successors) and *tree levels* determined recursively from the predecessor of each node. The number of child nodes of a given node v is called the *rank* of v. For example, the rank of the root node A equals 2, because this node has two child nodes (labeled B and C). The rank of any leaf node equals 0.

One can easily notice in Fig. 4.3a that there are four possible routes to the exit (the state I marked with a double border),[9] namely A-B-C-E-D-F-I, A-B-D-F-I, A-C-B-D-F-I, and A-C-E-D-F-I. (The reader can compare it with the labyrinth shown in Fig. 4.1b.) The remaining leaf nodes do not represent a solution and can be divided into two groups. The first group represents cul-de-sacs: G and H (cf. Fig. 4.1b) at the end of paths A-B-C-E-G, A-B-C-E-D-F-H, A-B-D-E-G, A-B-D-F-H, A-C-B-D-F-H, A-C-B-D-E-G, and A-C-E-G. The second group represents path intersections that have already been visited by us. (We have found a mark made by us with chalk.) In such cases we should go back to a previous point.[10] This is the case for paths A-B-D-E-C (from point C we can go to A or B only: both points have already been visited, so we should go back to E) and A-C-E-D-B (from point B we can go to A or C only: both points have already been visited, so we should go back to D).

[8] Again, from the "perspective of Providence".

[9] Of course, assuming we do not pass through paths we have already visited.

[10] According to the old rule of walking in a labyrinth. Otherwise, we can go round in circles.

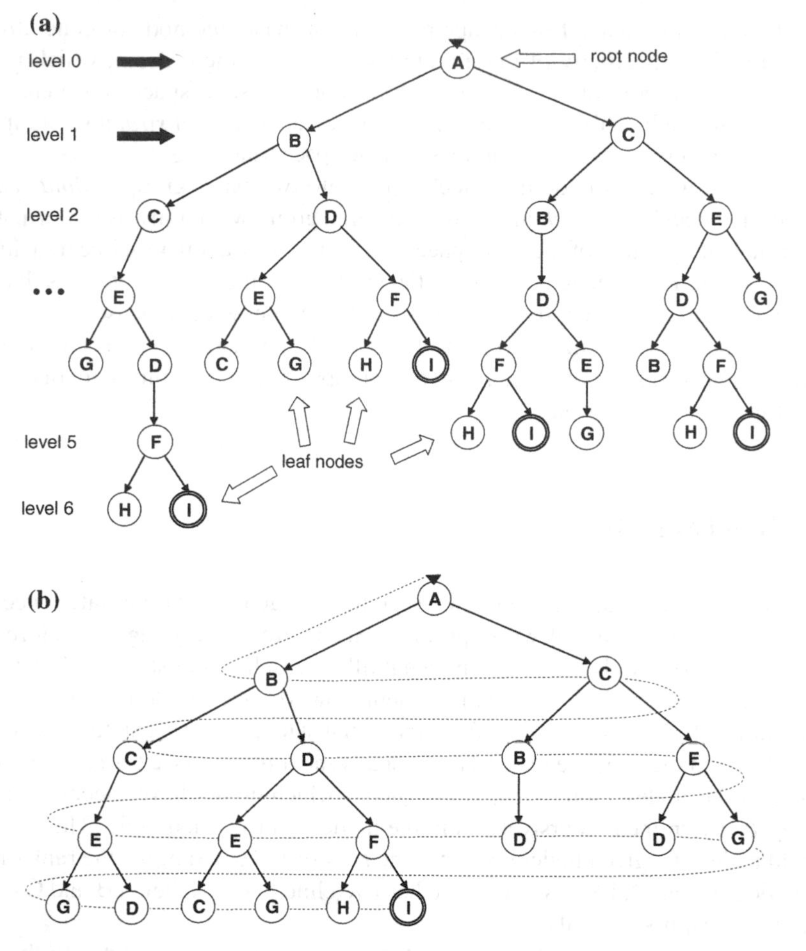

Fig. 4.3 Search trees: **a** a complete search tree for the labyrinth problem, **b** a search tree for Breadth-First Search

If we had a map in the form of such a tree, we would escape the labyrinth easily. Moreover, we could choose the shortest path to find the exit as soon as possible. However, we do not have such a map. Therefore, we have to generate the tree one node after another, i.e., wander around in the labyrinth in hope of finding the exit. Of course, we should do this in a systematic way. Blind search methods allow us to generate tree nodes in a systematic way. Now, we introduce the most important methods.

Breadth-First Search, BFS, consists of expanding tree paths by generating nodes one level after another, as shown in Fig. 4.3b. The order in which the nodes are generated is depicted with a dashed line. Let us notice that a complete search tree is not generated all at once, but generation halts at the moment of reaching the goal

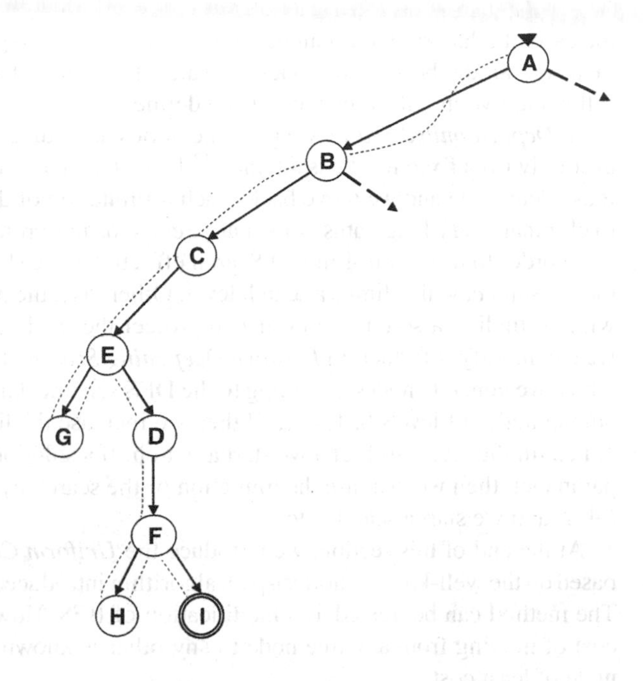

Fig. 4.4 Search tree for Depth-First Search

state, which represents the solution. In our example the path A-B-D-F-I represents the solution.

Breadth-First Search gives good results if the ranks of the nodes are not too big, i.e., a tree is not "broad" and "shallow". Otherwise, if we generate nodes across the tree in such a case, we waste a lot of time. If the nature of the problem is such that there are a lot of direct transitions from any state to other states,[11] then a search tree is just "broad" and "shallow". In such a case we should use another method, namely *Depth-First Search, DFS*, which generates only one node of the nth level for a node of the $(n - 1)$th level. Then, we generate only one node of the $(n + 1)$th level for a node of the nth level, etc. For our search tree, an application of the DFS method is shown in Fig. 4.4. Let us notice that we really explore deeply into the tree. If we reach a leaf node that is not a solution (such as the cul-de-sac node G in Fig. 4.4), then we have to move back to the nearest ancestor and continue our exploration from there.

In our example, we have found a solution making only ten steps (including steps performed back from cul-de-sac nodes) with the DFS method, whereas we had to make many more steps with the BFS method (cf. Figs. 4.3b and 4.4). Does this mean DFS has an advantage over BFS in general? Of course not. If the nature of the problem is such that a search tree is *unbalanced*, i.e., some leaf nodes have a very small level and some of them have a very big level,[12] then DFS can be not *efficient* at all. (In

[11] In the case of a labyrinth this would mean that there are a lot of alternative paths at each intersection.

[12] In case of a labyrinth it would mean that some paths are very long during a search process and some of them are very short.

the case of a labyrinth this can be dangerous. If we try to go along a very long path, then we can die because of a lack of water, food, etc.) To improve the method the following two modifications have been defined.

In *Depth-Limited Search* we generate nodes according to the DFS scheme, however only till a fixed level c is attained.[13] If we reach a node of level c, then we treat it as a leaf node and we move back. Such a limitation of the search depth allows us to eliminate very long paths, which are treated as not promising ones.

In order to use Depth-Limited Search effectively we should be able to determine the constant c, which limits a search level. Otherwise, the method can finish a search without finding a solution. In order to protect the method against such a situation we can modify it further. In *Iterative Deepening Search* [167] we fix a level l, up to which we generate nodes according to the DFS scheme, but if a solution is not found among nodes of levels 0, 1, 2, ..., l then we increase the limitation of the search by 1, i.e., till the level $l + 1$, and we start a search. If a solution is not found with such a parameter, then we increase the limitation of the search by 1 again, i.e., till the level $l + 2$, and we start a search, etc.

At the end of this section, we introduce the *Uniform Cost Search* method that is based on the well-known shortest-path algorithm introduced by Edsger W. Dijkstra.[14] The method can be treated as a modification of BFS. However, we assume that the cost of moving from any tree node to any other is known.[15] Then, we move to the node of least cost.

Summing up this section, let us notice that if we use blind search strategies, then we search for a solution by the systematic generation of new states and checking whether the solution is found (by chance).[16] Of course, such a strategy is not efficient. Therefore, if we can use knowledge concerning the nature of the problem, then we use heuristic strategies. These are discussed in the next section.

4.3 Heuristic Search

Heuristic search methods can be used if we are able to define a *heuristic function* that estimates "how far" a search tree node is from a solution. This allows us to expand those paths that seem to be the most promising. Let us assume that in our labyrinth,

---

[13]The constant c is a parameter of the method.

[14]Edsger W. Dijkstra—a professor of computer science at the Eindhoven University of Technology. His contribution to computer science includes Reverse Polish Notation, the ALGOL compiler, structured programming, and a semaphore model used in operating systems.

[15]In the case of the labyrinth the cost could be defined as the difficulty of going along a segment of the path. The difficulty could be defined as the width of the path (let us assume we are rather fat) and the slope of the path (let us assume we are not fit).

[16]In the case of Uniform Cost Search the order in which new states are generated is not accidental in sensu stricto. The method is not "blind" completely, because the order in which states are generated is determined by the cost function. However, this function does not say what the distance to a solution state is. Therefore, this method is not considered a heuristic method.

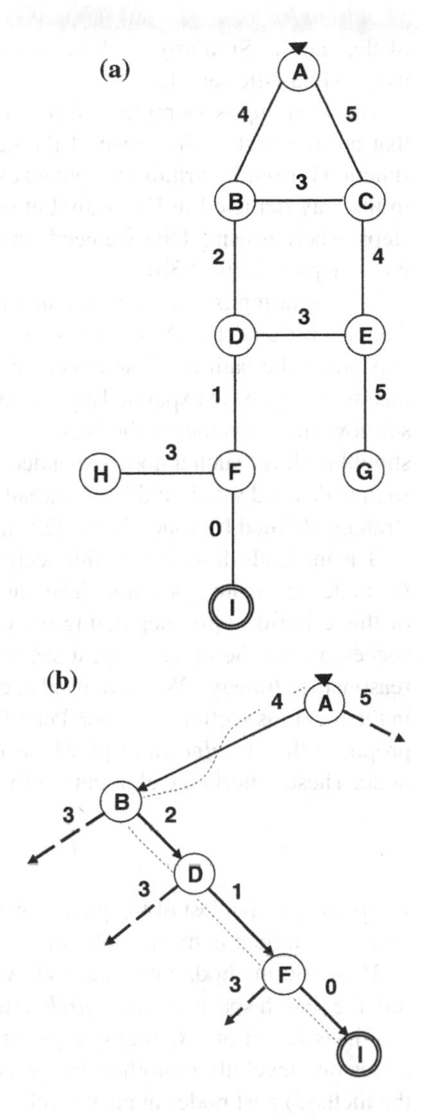

Fig. 4.5 The labyrinth problem: **a** values of the heuristic function, **b** the heuristic search

the nearer the exit we are, the more the flame of our candle flickers. Then, we can define the heuristic function h in the following way:
$h(n) = 5$—the flame is stable, $h(n) = 4$—the flame is flickering a little bit, $h(n) = 3$—the flame is flickering, $h(n) = 2$—a high frequency flicker in the flame is observed, $h(n) = 1$—a very high frequency flicker in the flame is observed, $h(n) = 0$—the candle has been blown out (we are at the exit).

The labyrinth plan, which contains values of the function h, from the "perspective of Providence", is shown in Fig. 4.5a. As we have mentioned, we do not have such a plan, so we have to look for the exit. However, in this case, contrary to a *blind search*, we can choose the path with the help of the heuristic function, i.e., observing a flame of the candle. Similarly to the case of blind methods, there are a lot of strategies using a heuristic search.

Hill climbing is a simple modification of DFS which consists of expanding nodes that are the best in the sense of the heuristic function.[17] If values of the heuristic function for our labyrinth are defined as shown in Fig. 4.5a, then the tree is expanded in the way depicted in Fig. 4.5b. Let us notice that we find a solution node in four steps, whereas using DFS we need ten steps (cf. Fig. 4.4), and for BFS we need even more steps (cf. Fig. 4.3b).

Hill climbing is a *local search* method, i.e., we can go only to neighboring states from the present state. Sometimes, the value of the heuristic function for a node after expanding the path is worse than a value for some node generated in the past. This means that we have expanded a path that seemed to be promising at first, however we see now that the values of the heuristic function are worsening. In such a situation, we should look for such a node expanded before now, which has the best value among such nodes and we should begin expanding other paths starting from this node. This strategy, defined by Judea Pearl [221], is called *Best-First Search.*

The methods described in this section use a heuristic function that estimates *how far* nodes are from a solution. The choice of succeeding node is made on the basis of this criterion only, neglecting the cost of moving from the present node to the successor. On the other hand, it seems that taking this cost into consideration is a reasonable strategy. (We have used such a cost in Uniform Cost Search, introduced in the previous section.) In 1968 Peter E. Hart, Nils J. Nilsson and Bertram Raphael proposed the *A^* algorithm* [126]. It uses both criteria for choosing the optimum node. These criteria are elements of the *evaluation function* f, given by the formula

$$f(n) = g(n) + h(n), \tag{4.1}$$

where $g(n)$ is the cost of the path from the root node to a node n, and $h(n)$ estimates how far a node n is from a solution.

Heuristic methods introduced above apply a strategy of expanding a path deep into the search space. *Beam Search*, introduced by Bruce T. Lowerre and Raj Reddy [187], is based on expanding a path breadth-wise. Similarly to BFS, we generate nodes one level after another. However, we expand only the b (b is a parameter of the method) best nodes at each level.

At the end of this section, we introduce some properties which are required for a heuristic function. We say the heuristic function h is *admissible* if the distance to a solution is never overestimated by h. For example, if finding the smallest road distance between cities is our problem, then a heuristic function, which gives a straight-line distance between cities is admissible, because it never overestimates the actual distance. This property is very important because a heuristic function,

[17] A good example of *hill climbing* is a situation when we want to reach the top of a mountain. However, we have neither a map nor a compass, we are here at night, and a dense fog hangs over the mountains. We have only an altimeter. So, according to the idea of *hill climbing*, we should go in the direction for which the altimeter shows the biggest increase in height.

which overestimates the actual distance can make finding the best path to a goal node impossible.

On the other hand the heuristic function h can underestimate the distance a little bit, i.e., its evaluation should be moderately "optimistic". At the same time, we require the function "optimism" should be more and more realistic as we approach a solution. This means that h should estimate more and more precisely. At least we require that the cost of making each succeeding step is compensated by increasing the precision of the evaluation of the remaining distance. If the heuristic function has this property, then we call it *consistent (monotone)*.

Let h_1 and h_2 be heuristic functions. If for any state v we have $h_2(v) \geq h_1(v)$, then the function h_2 *dominates* the function h_1. Let us notice that if both functions h_1 and h_2 are admissible, then they are bounded by the actual distance to a goal. This means that for any state v, $h_1(v) \leq C(v)$ and $h_2(v) \leq C(v)$, where $C(v)$ is the actual distance to a goal state. Thus, a dominating function estimates the distance better. (It is closer to $C(v)$.) Therefore, in the case of admissible heuristic functions it is better to use a dominating function.

At the end of this section, let us mention that the properties of heuristic functions introduced above are defined formally in Appendix A.

4.4 Adversarial Search

Search methods can be used for constructing artificial intelligence systems which play games. Such systems are one of the most spectacular computer applications.[18] Search techniques used for game playing belong to a group of AI methods called *adversarial search*. In Sect. 2.1 we have introduced search methods with the example of a chess game. Let us recall that states in a state space correspond to succeeding board positions resulting from the players' moves.

In the case of games we construct an *evaluation function*,[19] which ascribes a value saying how "good" for us is a given situation, i.e., a given state. Given the evaluation function, we can define a *game tree*, which differs from a search tree introduced in previous sections.[20] The *minimax method* is the basic strategy used for adversarial search. Let us assume that *maximizing* the evaluation function is our goal, whereas our opponent (adversary) wants to *minimize* it.[21] Thus, contrary to the search trees

---

[18] In May 1997 *Deep Blue*, constructed by IBM scientists under the supervision of Feng-hsiung Hsu, won a chess match against the world champion Garry Kasparov.

[19] The evaluation function estimates the expected *utility* of a game, defined by the *utility function*. This function is a basic notion of game theory formulated by John von Neumann and Oskar Morgenstern in 1944.

[20] In the case of game trees we use a specific terminology. A level is called a *ply*. Plies are indexed started with 1 (not 0, as in common search trees).

[21] Hence, a *mini-max* method.

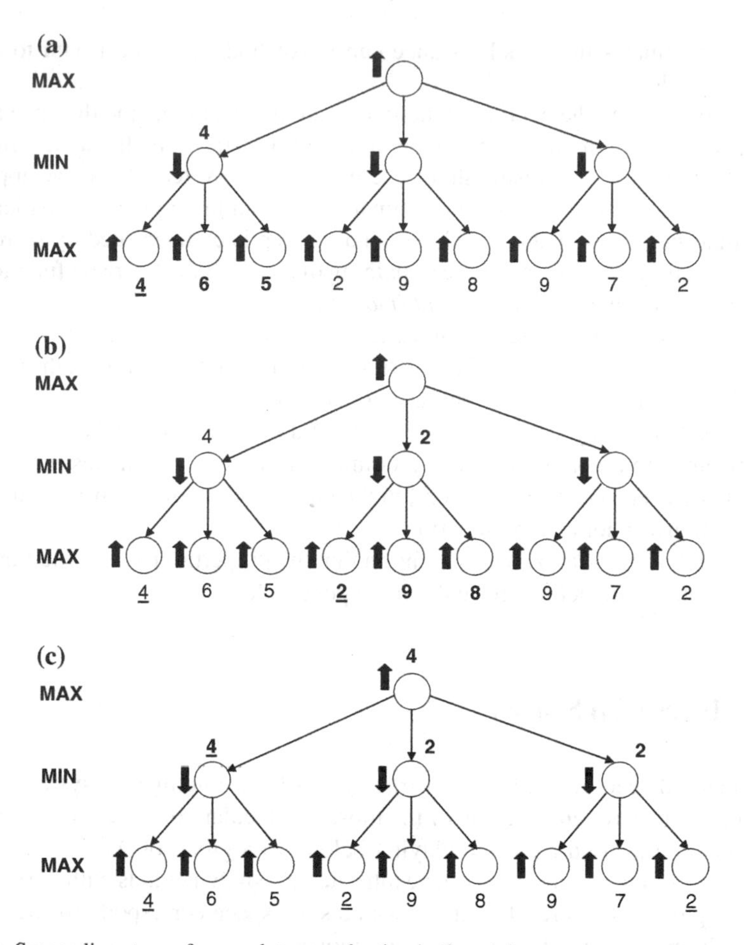

Fig. 4.6 Succeeding steps of a search tree evaluation in the minimax strategy

introduced in the previous sections, which are defined to *minimize* the (heuristic) function corresponding to the distance to a solution, in the case of a game tree we alternately maximize the function (for our moves) and minimize it (for moves of our opponent).

Let us look at Fig. 4.6a. The tree root corresponds to a state where we should make a move. Values are ascribed to leaf nodes. (It is also our turn to move at leaf nodes; it is the opponent's turn to move at nodes in the middle ply, denoted by MIN.) Now, we should *propagate* values of the evaluation function upwards.[22] Having values of leaves of the left subtree: 4, 6, 5, we propagate upwards 4, because our opponent will make a move to the smallest value for us. (He/she minimizes the evaluation function.)

[22] An arrow up means that we ascribe to a node the biggest value of its successors, an arrow down means we ascribe to a node the smallest value of its successors.

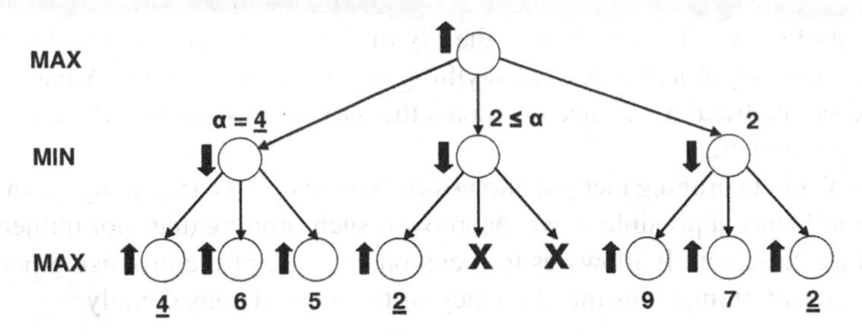

Fig. 4.7 Succeeding steps of a search tree evaluation in the α-β pruning method

The opponent will behave analogously in the middle subtree (cf. Fig. 4.6b). Having the choice of the leaves 2, 9, and 8, he will choose the "worst" path, i.e., the move to the node with the value 2. Therefore, this value is propagated upwards to the parent of these leaf nodes. After determining the values at the middle MIN ply, we should evaluate the root (which is placed at a MAX ply). Having values at the MIN ply of 4, 2, 2, we should choose the "best" path, i.e., the move to the node with the value 4. Thus, this value is ascribed to the root (cf. Fig. 4.6c).

Let us notice that such a definition of the game tree takes the opponent's behavior into account. Starting from the root, we should not go to the middle subtree or the right subtree (roots of these subtrees have the value 2), in spite of the presence of nodes having big values (9, 8, 7) in these subtrees. If we went to these subtrees, the opponent would not choose these nodes, but nodes having the value 2. Therefore, we should move to the left subtree. Then, the opponent will have the choice of nodes with the values 4, 6, and 5. He/she will choose the node with the value 4, which is better for us than a node with the value 2.

In case a game is complex, the minimax method is inefficient, because it generates huge game trees. The *alpha-beta pruning method* is a modification of minimax that decreases the number of nodes considerably. The method has been used by Arthur L. Samuel in his *Samuel Checkers-playing Program*.[23]

We explain the method with the help of an example shown in Fig. 4.7. Let us assume that we evaluate successors of the root. After evaluating the left subtree, its root has obtained the value $\alpha = 4$. α denotes the minimum score that the maximizing player is assured of. Now, we begin an evaluation of the middle subtree. Its first leaf has obtained the value 2, so temporarily its predecessor also receives this value. Let us notice that if any leaf, denoted by X, has a value greater than 2, then the value of its predecessor is still equal to 2 (we minimize). On the other hand, if the value of any leaf X is less than 2, then it does not make any difference for our evaluation, because 2 is less than α—the value of the neighbor of the predecessor. (This means that this neighbor, having the value $\alpha = 4$, will be taken into account at the level

[23] Research into a program that plays checkers (English draughts) was continued by a team led by Jonathan Schaeffer. It has resulted in the construction of the program *Chinook*. In 2007 Schaeffer's team published an article in *Science* including a proof that the best result which can be obtained playing against *Chinook* is a draw.

of the root of the whole tree, since at the ply of the root we maximize.) Thus, there is no need to analyze leaves X (and anything that is below them).[24] A parameter β plays the symmetrical role, since it denotes the maximum score that the minimizing player is assured of.

The alpha-beta pruning method allows one to give up on a tree analysis, in case it does not influence a possible move. Moreover, such pruning does not influence the result of a tree search. It allows us to focus on a search of the promising parts of a tree. As a result, it improves the efficiency of the method considerably.[25]

4.5 Search for Constraint Satisfaction Problems

Search methods can be used for solving *Constraint Satisfaction Problems, CSPs*.[26] CSPs are characterized by a set of requirements, treated as *constraints*, which should be fulfilled. However, it is very difficult to fulfill them at the same time, because they are in conflict. Drawing up a timetable in a school is a good example of a CSP. We should fulfill at the same time such constraints as the availability of classrooms (of various sizes and equipment), the availability of classes (a class of pupils cannot have two lessons at the same time), the availability of teachers, etc. There are a lot of important application areas of search methods for CSPs, e.g., shop floor control, project management, logistics, VLSI design, and molecular biology.

Constraint satisfaction problems can be described by the following elements:

- a finite set of *variables*: $x_1, x_2, \ldots, x_n$,
- for each variable x_i the set of its possible values D_{x_i}, called a *domain*,
- a finite set of *constraints* that are combinations of values of variables.

For example, let us consider the following problem. Let there be seven kingdoms K1, K2, ..., K7 (they are the variables in our CSP) on a hypothetical island shown in Fig. 4.8a. One should color in the map with three colors: blond (B), silver (S), and charcoal (C), so that no two adjacent kingdoms have the same color. Formally, the set {B, S, C} is the domain for each variable, and a set of constraints is defined in the following way: K1 $\neq$ K2, K1 $\neq$ K3, K1 $\neq$ K5, K2 $\neq$ K5, K2 $\neq$ K7, K3 $\neq$ K5, K3 $\neq$ K7, K4 $\neq$ K7, K5 $\neq$ K7, K6 $\neq$ K7.

Backtracking search is the simplest method. It consists of assigning consecutive variables (e.g., in the order B, S, C) to subsequent variables by generating a DFS tree. If an assignment conflicts with the set of constraints, then we backtrack from our

---

[24]In case we analyze a tree having more plies.

[25]It is of great importance in complex games, e.g., chess. The *Deep Blue* computer, playing against G. Kasparov, expanded some paths to 40 plies.

[26]Constraint satisfaction problems are of great importance in computer science. Therefore, there is a variety of mathematical models, mainly based on graph theory, operational research, and linear algebra, that are used for their solution. In the monograph we introduce only basic ideas of AI heuristic search methods which are used for solving these problems. We refer the reader interested in CSPs to the well-known monograph of Edward Tsang [305]. For constraint programming, the monograph by Krzysztof R. Apt [8] is recommended.

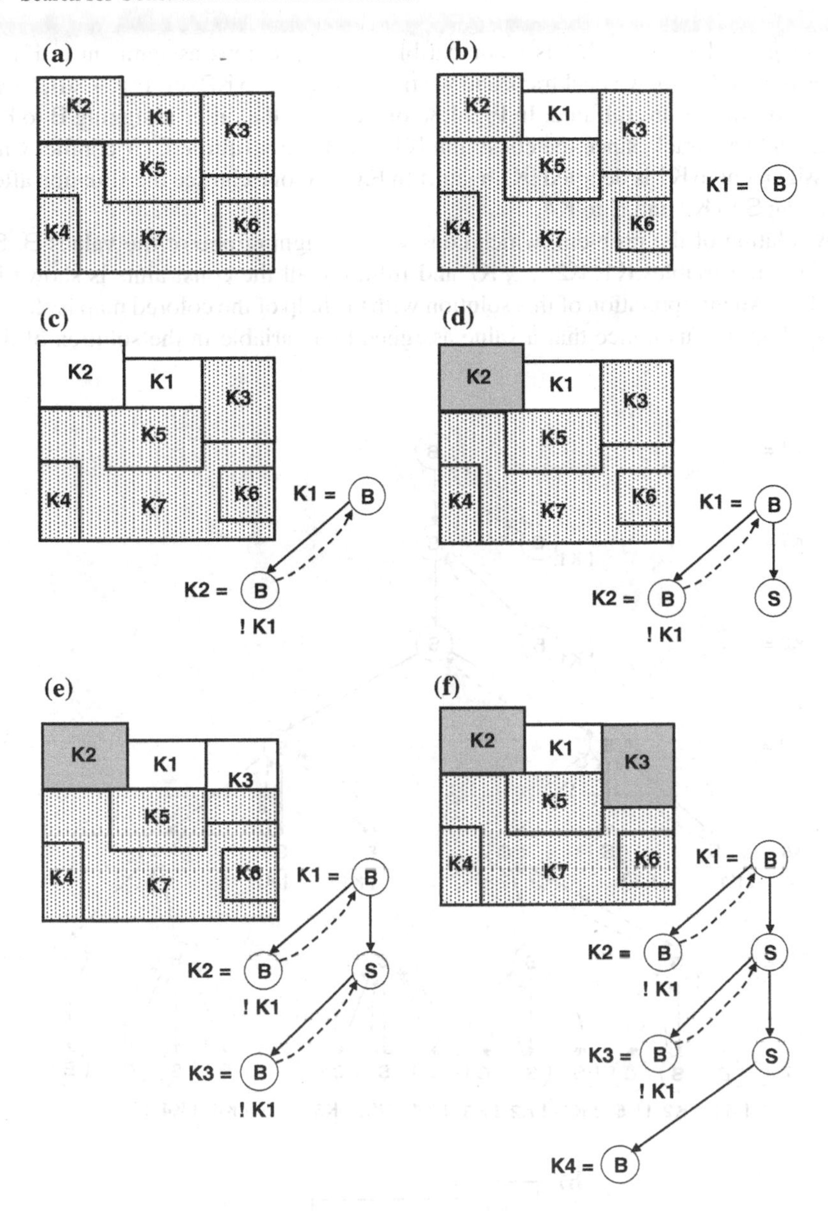

Fig. 4.8 Succeeding steps of search tree generation for a constraint satisfaction problem

current path. Let us look at Fig. 4.8b. Firstly, we generate the tree root corresponding to the variable K1 and we assign the value B to it. (We color in the kingdom K1 with blond.)[27] In the next step (cf. Fig. 4.8c), we assign B to K2 at first. However, such an assignment conflicts with the constraint K1 $\neq$ K2. (A conflict caused by

[27]Each subsequent level in the tree corresponds to a subsequent variable.

assigning a value to K2 that is incompatible with a previous assignment to K1 is denoted by !K1.) So, we backtrack and we try to assign S to K2 (cf. Fig. 4.8d). Now, all the constraints are fulfilled. In Fig. 4.8e one can see that after assigning B to K3 we should backtrack, since the constraint K1 $\neq$ K3 is not fulfilled. (A conflict with an assignment to K1 is denoted !K1 again.) In Fig. 4.8f one can see the situation after assigning S to K3 and B to K4.

A solution of the problem, which consists of assigning one of the values B, S, C to all the variables K1, K2, ..., K7 and fulfilling all the constraints is shown in Fig. 4.9a. An interpretation of this solution with the help of the colored map is shown in Fig. 4.9b. Let us notice that a value assigned to a variable in the solution of the

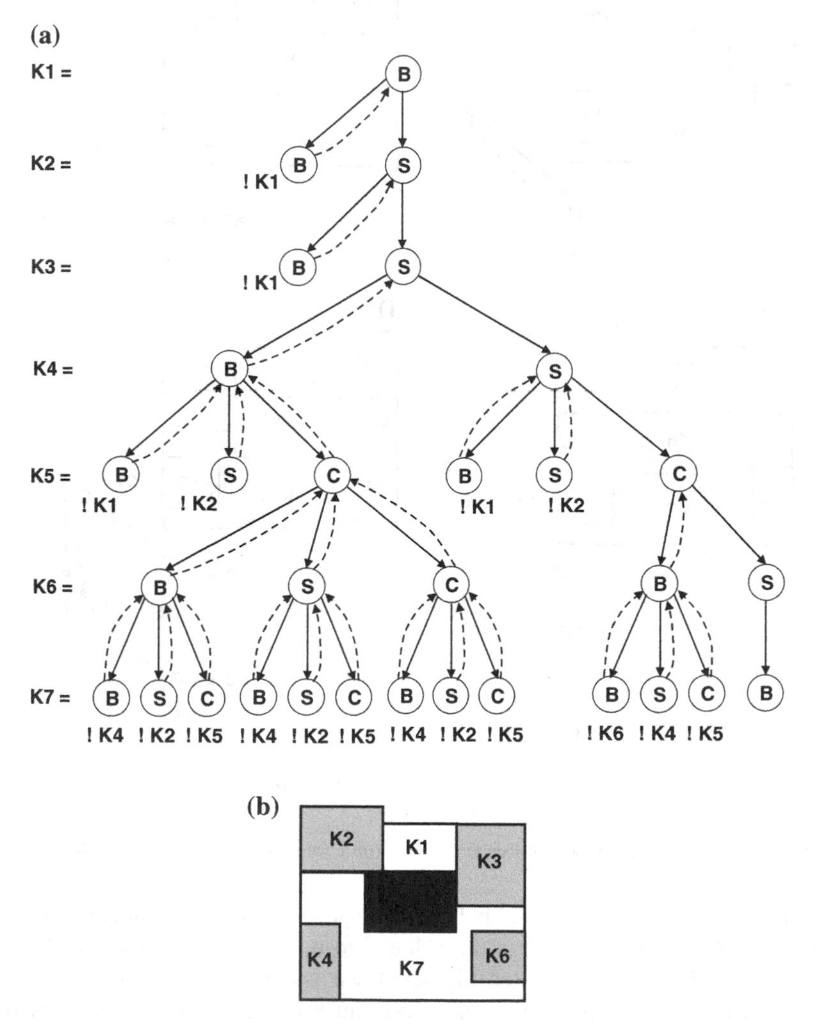

Fig. 4.9 An example of a constraint satisfaction problem: **a** a search tree generated at the moment of finding a solution—all the variables have values ascribed and all the constraints are satisfied, **b** the map colored according to a solution found

problem is represented by the last assignment to this variable in the search tree. For example, in the left subtree starting from the assignment of B to K4, we have tried to ascribe all the possible values to K7 for all combinations of colors for K6. However, we have had to backtrack from the whole subtree. This has resulted from an incorrect assignment of B to K4. (After this assignment there is no solution to the problem.) Only when we have changed this assignment to K4 = S (the right subtree) do we find a solution, completed when the variable K7 has received the value B. Thus, a solution in a backtracking search method is represented by the outermost right path in the tree. Everything that is to the left of this path represents backtracking. Let us notice that for quite a simple problem we have to generate quite a big search tree to find the solution. This means that the method is not efficient. In fact, backtracking search is a basis for defining more efficient methods. Let us introduce two of them.

We begin with the following observation. Let us look at Fig. 4.9a once more. After assigning B to K4 and C to K5 there is no "good" assignment for K7. This is illustrated by the repeated trial of assigning all the possible values to K7 after an assignment of all the possible values to K6. (The subtree below the node K5 = C is "complete".) If we had knew about this at the moment of assigning C to K5, we would not have had to generate a subtree below the node K5 = C. Checking whether assignments performed till now eliminate all the possible values for one of the remaining variables is the basic idea of *forward checking search*.

Let us analyze this method for our example with the help of Fig. 4.10a. We will track the variable K7. As we can see, after assigning S to K2, the set of values that can be ascribed to K7 is reduced to {B, C}. This results from the constraint K2 $\neq$ K7 (K2 is adjacent to K7). Then, after assigning B to K4, the set of values that are possible for K7 is reduced to {C} (the constraint K4 $\neq$ K7). Finally, after assigning C to K5, the set of values that are possible for K7 is the empty set, because of the constraint K5 $\neq$ K7. This means there is no point in expanding this path. Similarly, in the right subtree there is no sense in expanding the path after assigning B to K6. One can easily notice the better efficiency of a forward checking search than a backtracking search, comparing the search trees shown in Figs. 4.9a and 4.10a.

In order to improve the efficiency of the methods presented till now a lot of modifications have been proposed, including search with variable ordering, search with value ordering, etc. However, one of the most efficient approaches to CSP search is based on the *constraint propagation* technique. We analyze this technique for our example with the help of Fig. 4.10b. After assigning B to K1, the set of possible assignments to K5 is equal to {S, C} (because of the constraint K1 $\neq$ K5). The next assignment K2 = S causes the following sequence of subsequent restrictions for variables K5, K3, and K7:

(a) the set of possible assignments for K5 is reduced to {C}, because of the constraint K2 $\neq$ K5,
(b) the set of possible assignments for K3 is reduced to {S}, because of (a) and the constraints K1 $\neq$ K3 and K3 $\neq$ K5,
(c) the set of possible assignments for K7 is reduced to {B}, because of (a), (b), and the constraints K3 $\neq$ K7 and K5 $\neq$ K7.

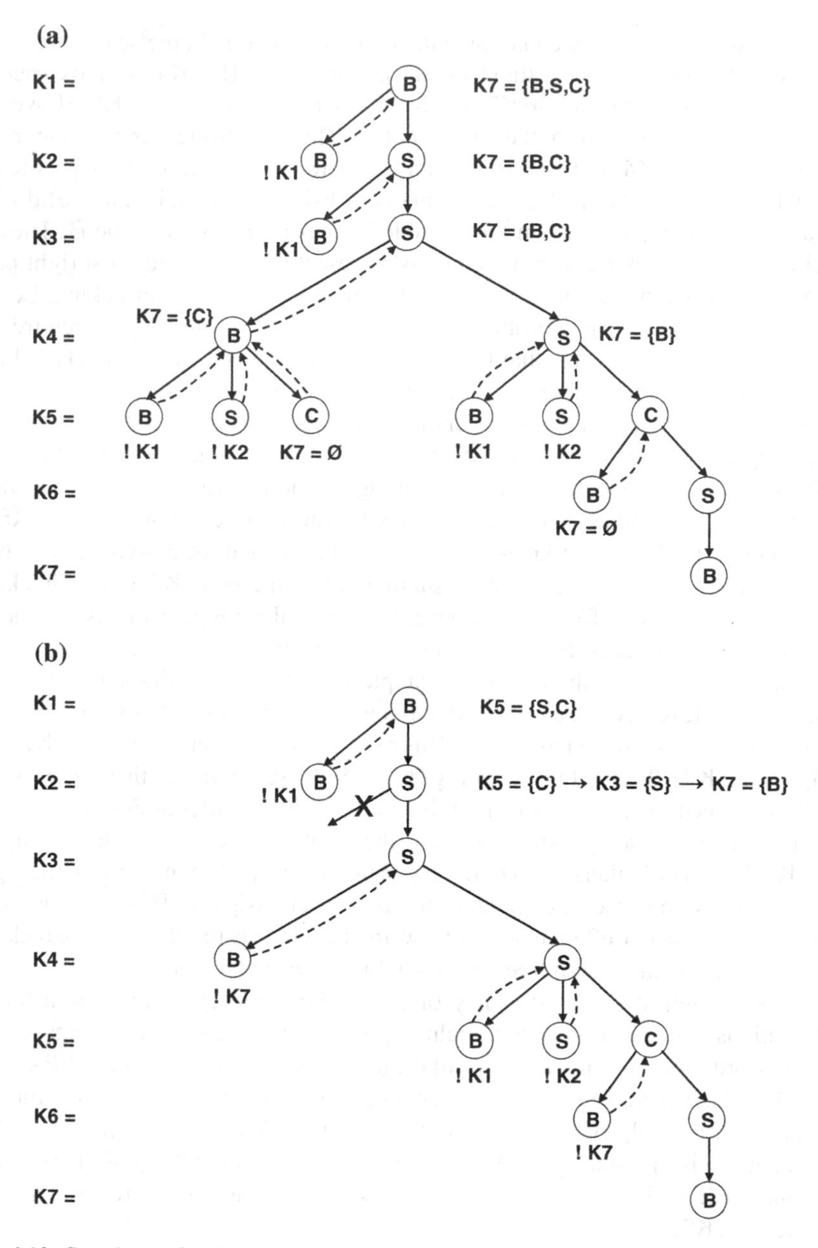

Fig. 4.10 Search tree for the constraint satisfaction problem: **a** for the forward checking search method, **b** for the constraint propagation method

As we can see in Fig. 4.10b, in the constraint propagation method, just after assigning S to K2 (i.e., as early as on the second level of the tree) we are able to determine admissible assignments for three variables that reduce the search tree considerably.

Summing up, after fixing the set of admissible values for a variable, we *propagate* consequences of the restrictions imposed on this set to the remaining variables (hence, the name of the method). A lot of efficient algorithms, such as AC-3, AC-4, PC-2, and PC-4, have been developed based on the constraint propagation approach.

Local search is the third approach which is used for solving CSP problems. It is similar to *hill climbing*, introduced in Sect. 4.3. We assign values to all the variables and we successively improve this assignment until it violates the constraints. *Min-conflicts search* is one of the most popular methods. First of all, we randomly generate initial assignments. Then, after choosing a variable with a value which conflicts with some constraints, we change the value in order to minimize the number of conflicts with other variables. This method gives good results, although it is strongly dependant on the initial assignment. Due to its iterative nature, the method is especially recommended in case constraints change over time.

4.6 Special Methods of Heuristic Search

Finishing our considerations concerning search methods, let us come back to the first heuristic method discussed in this chapter, i.e., hill climbing. The main idea of this method consists of expanding a tree in the direction of those states whose heuristic function value is most promising. Let us assume that a problem is described with the help of a *solution space* (X_1, X_2) (instead of an abstract model of a problem), which is typical for optimization problems. Thus, a solution space is the domain of a problem. Then, let us assume that for points (X_1, X_2) of this space, values of the heuristic function $h(X_1, X_2)$ are known and they are defined as shown in Fig. 4.11. Our goal is to climb to the high hill situated in the middle of the area with the hill-climbing method.[28] If we start our search at the base of the hill, we will conquer the summit, i.e., we will find a solution. However, if we start at the base of the lower hill situated in the bottom-leftmost subarea, then we will climb this hill and never leave it.[29] This means, however, that we will not find the optimum solution. We find ourselves in a similar situation if we land in a plain area (*plateau*). Then, we gain no benefit from the heuristic information and we never reach a solution.

In order to avoid a situation in which we find a local extremum (minimum/maximum) instead of the global one,[30] a lot of heuristic search methods have been constructed. Let us introduce the most important ones.

The *simulated annealing* method was introduced by Scott Kirkpatrick,[31] C. Daniel Gelatt and Mario P. Vecchi [159] in 1983. In order to avoid getting stuck in a local

[28] In a mathematical formulation, we seek a global maximum in the solution space.

[29] As we more away from the summit of this hill, values of the heuristic function decrease.

[30] In practice finding a local extremum means finding *some* solution which is not satisfactory.

[31] Scott Kirkpatrick—a professor of physics and computer science (MIT, Berkeley, Hebrew University, IBM Research, etc.). He is the author of many patents in the areas of applying statistical physics in computer science, distributed computing, and computer methods in physics.

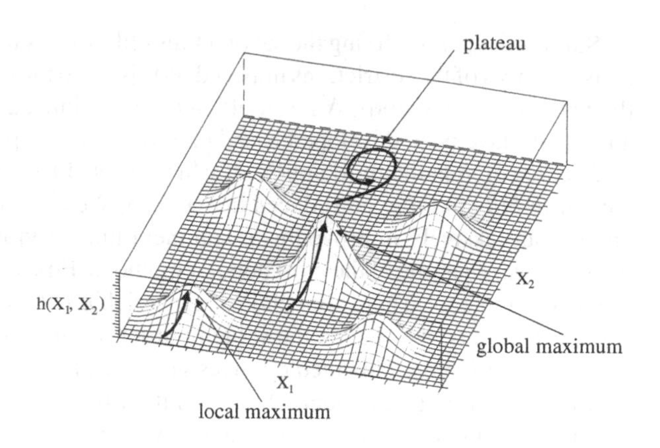

Fig. 4.11 Potential problems that can appear during "hill climbing"

extremum, the interesting physical phenomenon of metal (or glass) annealing is used. In order to improve the properties of a material (e.g., its ductility or its hardness), it is heated to above a critical temperature and then it is cooled in a controlled way (usually slowly). Heating results in "unsticking" atoms from their initial positions. (When they are in their initial positions, the whole system is in a local internal energy minimum.) After "unsticking", atoms drift in a random way through states of high energy. If we cooled the material quickly, then a microstructure would "get stuck" in a random state. (This would mean reaching a local minimum of the internal energy of the system.) However, if we cool the material in a controlled, slow way, the internal energy of the system reaches the *global minimum*.[32]

The Kirkpatrick method simulates the process described above. The internal energy of a system E corresponds to the heuristic function f, and the temperature T is a parameter used for controlling the algorithm. From a current (temporary) solution i, a "rival" solution j is generated randomly out of its neighborhood. If the value of the "rival"solution $E(j)$ is not worse than that of the current solution (i.e., $E(j) \leq E(i)$, since we look for the global minimum), then it is accepted. If not, then it can be accepted as well, however with some probability.[33] Thus, moving from a better state to a worse state is possible in this method. It allows us to leave a local extremum. As we have mentioned, a parameter T (temperature) is used for controlling the algorithm. At the beginning, T is big and the probability of accepting a "rival" worse solution is relatively big, which allows us to leave local minima. In succeeding steps of the algorithm "the system is cooled", i.e., the value of T decreases. Thus, the more stable a situation is, the less the probability of choosing a "rival" worse solution.

[32]This improves the properties of a material.

[33]This probability is determined according to the Boltzmann distribution. This defines the distribution of energy among particles in a thermal equilibrium as $P = e^{(-\Delta E_{ij}/kT)}$, where $\Delta E_{ij} = E(j) - E(i)$, and k is the Boltzmann constant.

The *tabu search* method was introduced by Fred Glover[34] in 1986 [109]. In this method the current solution is *always* replaced by *the best solution in its neighborhood* (even if it is worse). Additionally, a solution which has been already "visited" is forbidden for some time. (It receives the *tabu* status.) A *visited* solution is added to a short *tabu list*. A newly added solution replaces the oldest on the list.[35] The search process finishes after a fixed number of steps.

There are a lot of modifications of tabu search. The method is often combined with other heuristic methods. A combination of the method with evolutionary computing gives especially good results. Evolutionary computing, which can be treated as a considerable extension of heuristic methods, is discussed in the next chapter.

Bibliographical Note

Search methods are among the earliest methods of Artificial Intelligence. Therefore, they are described in fundamental monographs concerning the whole area of AI [189, 211, 241, 256, 261, 262, 273, 315].

The foundations of constructing heuristic search strategies are discussed in [221].

In the area of CSP a monograph [305] is the classic one. For constraint programming a book [8] is recommended.

[34] Fred W. Glover—a professor of computer science, mathematics, and management science at the University of Colorado. An adviser at Exxon, General Electric, General Motors, Texas Instruments, etc.

[35] The *tabu list* is a LIFO queue (Last-In-First-Out).

Chapter 5
Evolutionary Computing

Evolutionary computing is the most important group of methods within the *biology-inspired approach*, because of their well-developed theoretical foundations as well as the variety of their practical applications. As has been mentioned in Sect. 3.2, the main idea of these methods consists of simulating natural evolutionary processes. Firstly, four types of such methods are presented, namely *genetic algorithms*, *evolution strategies*, *evolutionary programming*, and *genetic programming*. In the last section other biology-inspired models, such as *swarm intelligence* and *Artificial Immune Systems*, are introduced.

5.1 Genetic Algorithms

The first papers of Alex Fraser concerning *genetic algorithms* were published in 1957 [102]. However, this approach only became popular after the publication of an excellent monograph by Holland [139] in 1975. As we have mentioned at the end of the previous chapter, genetic algorithms can be treated as a significant extension of the heuristic search approach that is used for finding the optimum solution to a problem. In order to avoid finding local extrema instead of the global one and to avoid getting stuck in a *plateau* area (cf. Fig. 4.11), a genetic algorithm goes through a *space of* (potential) *solutions* with many search points, not with one search point as in (standard) heuristic search methods. Such search points are called *individuals*.[1] Thus, each individual in a solution space can be treated as a candidate for a (better or worse) solution to the problem. A set of individuals "living" in a solution space at any phase of a computation process is called a *population*. So, a population is a

[1]The analogy is with individuals living in biological environments.

M. Flasiński, *Introduction to Artificial Intelligence*,
DOI 10.1007/978-3-319-40022-8_5

set of representations of potential solutions to the problem. Succeeding populations constructed in iterated phases of a computation are called *generations*. This means that a *state space* in this approach is defined with succeeding generations constructed by a genetic algorithm.[2]

For example, let us look at an exemplary solution space shown in Fig. 5.1. The position of each individual in this space is determined by its coordinates (X_1, X_2). These coordinates determine, in turn, the *genotype* of the individual. Usually, it is assumed that a genotype consists of one *chromosome*. Each coordinate is *binary-coded*, i.e., it is of the form of a string of *genes*: 0000, 0001, 0010, ..., 1000, etc. Thus, a genotype built of one chromosome consists of eight genes. The first four genes determine the coordinate X_1 and the second four genes determine the coordinate X_2. For example, the individual marked with a circle in the solution space in Fig. 5.1 has the genotype 01010111. As we have mentioned in the previous chapter, each search point of a space of (potential) solutions represents a set of values ascribed to parameters of the problem considered.[3] In the terminology of genetic algorithms such a set of values is called the *phenotype* of this point/individual. In order to evaluate the "quality" of an individual (i.e., a potential solution), we evaluate its phenotype with the help of a *fitness function*.[4]

Let us assume for further considerations that we look for the (global) maximum of our fitness function, which equals *11* and is marked with a dark grey color in Fig. 5.1. In the solution space there are two local maxima with a fitness function value equal to *8*, which are marked with a light grey color. Let us notice that if we searched this space with the hill-climbing method and we started with the top-leftmost point, i.e., the point $(X_1, X_2) = (0000, 1000)$ with a fitness function value equal to *3*, then we would climb a local "peak" having coordinates $(X_1, X_2) = (0001, 0111)$ with a fitness function value equal to *8*. Then, we would stay at the "peak", because the fitness function gives worse values in the neighborhood of this "peak". However, this would make it impossible to find the best solution, i.e., the global maximum. Using genetic algorithms we avoid such situations.

Now, let us introduce the scheme of a genetic algorithm, which is shown in Fig. 5.2. Firstly, the initial population is defined by random generation of a fixed number (a parameter of the method) of individuals. These individuals are our initial search points in the solution space. For each individual the value of the fitness function is computed. (This corresponds to the evaluation of a heuristic function for the search methods discussed in a previous chapter.) In the next phase the best-fitted individuals are *selected* for "breeding offsprings". Such individuals create a *parent population*.

---

[2]Thus, succeeding populations are equivalent to states of this space.

[3]Each search point corresponds to a certain *solution*. (Such a "*solution*" does not satisfy us in most cases.) If we deal with an abstract model of a problem (as in the previous chapter), not with a solution space, then such a point corresponds to a certain phase of problem solving (for our example of a labyrinth it is the path we have gone down) instead of representing values ascribed to parameters of the problem.

[4]The fitness function plays an analogous role to the heuristic function defined for the search methods presented in the previous chapter.

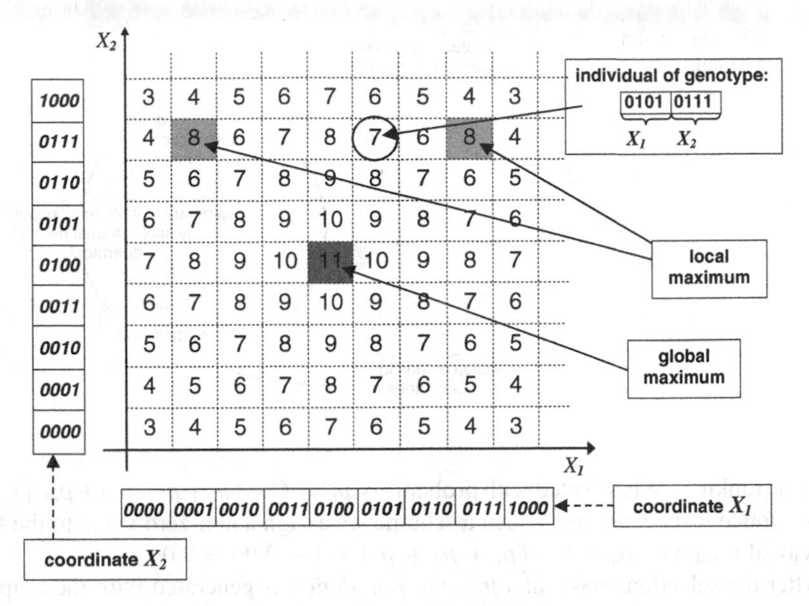

Fig. 5.1 Formulation of a search problem for a genetic algorithm

The simplest method for such a selection is called *roulette wheel selection*. In this method we assume that the roulette wheel area assigned to an individual is directly proportional to its fitness function value. For example, let the current population consist of four individuals $P = \{ind_1, ind_2, ind_3, ind_4\}$ for which the fitness function h gives the following values: $h(ind_1) = 50$, $h(ind_2) = 30$, $h(ind_3) = 20$, $h(ind_4) = 0$. Then, taking into account that the sum of the values equals 100, we assign the following roulette wheel areas to individuals: $ind_1 = 50/100\,\% = 50\,\%$, $ind_2 = 30\,\%$, $ind_3 = 20\,\%$, $ind_4 = 0\,\%$. Then, we randomly choose individuals with the help of the roulette wheel. Since no wheel area is assigned to the individual ind_4 (0 %), at least one of the remaining individuals must be chosen twice. (The individual ind_1 has the best chance, because its area comprises half of the wheel, i.e., the same area as ind_2 and ind_3 combined.)

In order to avoid a situation in which some individuals with a very small fitness function value (or even the zero value, as in the case of the individual ind_4) have no chance of being selected for "breeding offsprings", one can use *ranking selection*. In this method a ranking list which contains all the individuals, starting from the best-fitted one and ending with the worst-fitted one, is defined. Then, for each individual a *rank* is assigned. The rank is used for a random selection. For example, the rank can be defined in the following way. Let us fix a parameter for computing a rank, $p = 0.67$. Then, we choose the individual ind_1 from our previous example with probability $p_1 = p = 0.67$. The individual ind_2 is selected with probability $p_2 = (1 - p_1) \cdot p = 0.33 \cdot 0.67 = 0.22$. The succeeding individual ind_3 is chosen with probability $p_3 = (1-(p_1+p_2)) \cdot p = 0.11 \cdot 0.67 = 0.07$. Generally, the nth individual

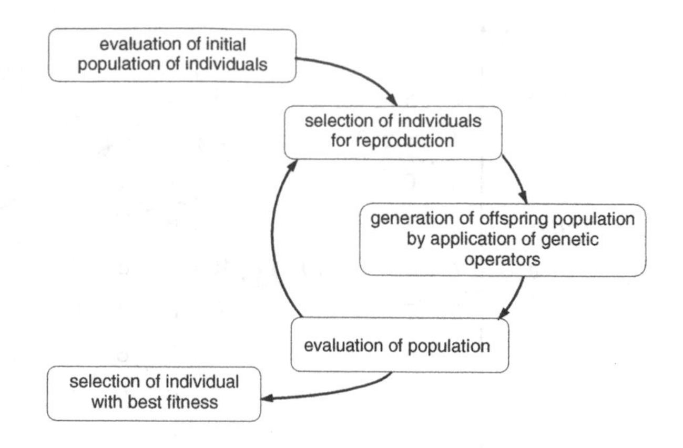

Fig. 5.2 The scheme of a genetic algorithm

from the ranking list is selected with probability $p_n = (1-(p_1+p_2+\cdots+p_{n-1}))\cdot p$. Let us notice that according to such a scheme we assign a non-zero value to the last individual ind_4, i.e., $p_4 = 1 - (p_1 + p_2 + p_3) = 1 - 0.96 = 0.04$.

After the selection phase, an *offspring population* is generated with the help of *genetic operators* (cf. Fig. 5.2). Reproduction is performed with the *crossover (recombination) operator* in the following way. Firstly, we randomly choose[5] pairs of individuals from the parent population as candidates for mating. These pairs of parents "breed" pairs of offspring individuals by a recombination of sequences of their chromosomes. For each pair of parents we randomly choose the *crossover point*, which determines the place at which the chromosome is "cut". For example, a crossover operation for two parent individuals having chromosomes 01001000 and 01110011, with a fitness function value of *7* (for both of them) is depicted in Fig. 5.3. The randomly chosen crossover point is *4*, which means that both chromosomes are cut after the fourth gene. Then, we recombine the first part of the first chromosome, i.e., 0100, with the second part of the second chromosome, i.e., 0011, which gives a new individual (offspring) with the chromosome 01000011. In the same way we obtain a second new individual having the chromosome 01111000 (by a recombination of the first part of the second chromosome and the second part of the first chromosome). Let us notice that one "child" (the one having the chromosome 01111000) is "worse fitted" (to the environment) than the "parents", because its fitness function value equals *4*. This individual corresponds to a worse solution of the problem. On the other hand, the fitness function value of the second "child" (01000011) equals *10* and this individual reaches a satisfying solution (the global maximum) in one step. Sometimes we use more than one crossover point; this technique is called *multiple-point crossover*.

[5]This random choice is made with high probability, usually of a value from the interval [0.6, 1] in order to allow many parents to take part in a reproduction process. This probability is a parameter of the algorithm.

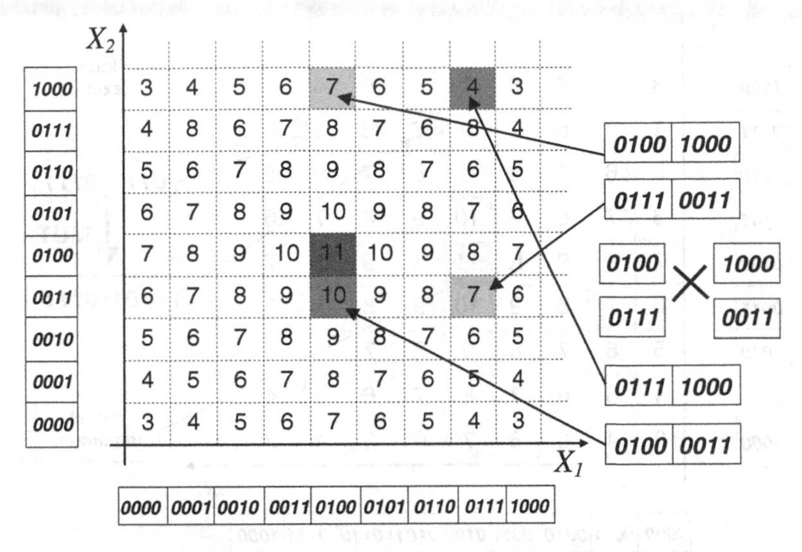

Fig. 5.3 Applying the crossover operator in a genetic algorithm

The *mutation operator* changes the value of a single gene from 0 to 1 or from 1 to 0. Contrary to a crossover, a mutation is performed very rarely.[6] For example, in Fig. 5.4 we can see an individual represented by the chromosome 01110111. Although its fitness function value (*8*) is quite big, this individual is a local maximum. If we used this individual in the hill-climbing method, then we would get stuck in this place (cf. Fig. 4.11). However, if we mutated the third gene of its chromosome from 1 to 0, then we would obtain a new individual, which has the chromosome 01010111 (cf. Fig. 5.4). This search point has a better chance to reach the global maximum.[7]

In the third phase, called *evaluation of a population*, values of the fitness function are computed for individuals of the new offspring population (cf. Fig. 5.2). After the evaluation, the termination condition of the algorithm is checked. If the condition is fulfilled, then we select the individual with the best fitness as our solution of the problem. The termination condition can be defined based on a fixed number of generations determined, the computation time, reaching a satisfying value of the fitness function for some individual, etc. If the condition is not fulfilled, then the work of the algorithm continues (cf. Fig. 5.2).

In this section we have introduced fundamental notions and ideas for genetic algorithms. As we can see, this approach is based on biological intuitions. Of course, for the purpose of constructing an AI system, we should formalize it with mathematical notions and models. The *Markov chain* model is one of the most elegant formalizations used in this case. This model is introduced in Appendix B.2.

---

[6]Since mutations occur very rarely in nature, we assume a small probability in this case, e.g., a value from the interval [0, 0.01]. This probability is a parameter of the algorithm.

[7]In genetic algorithms mutation plays a secondary role. However, as we will see in subsequent sections, mutation is a very important operator in other methods of evolutionary computing.

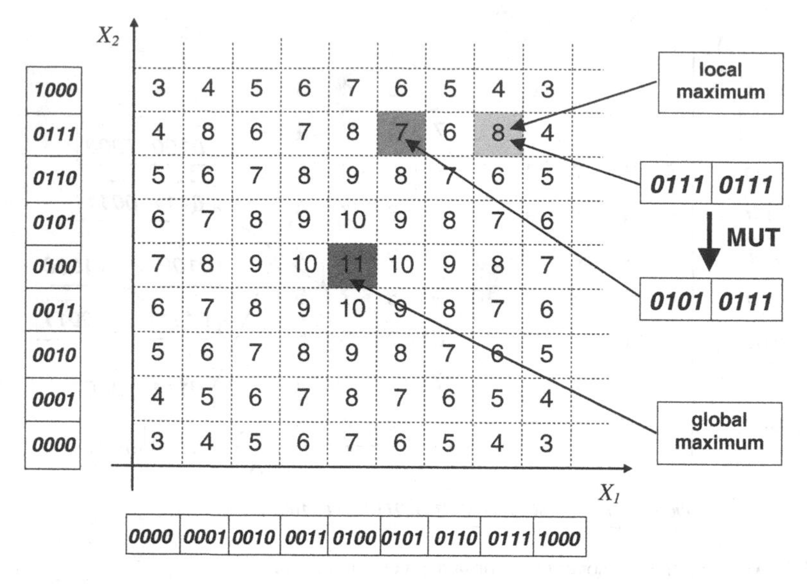

Fig. 5.4 Applying the mutation operator in a genetic algorithm

5.2 Evolution Strategies

In the genetic algorithm discussed in the previous section, an individual (a potential solution) has been coded in binary. Such a representation is convenient if we are searching a *discrete* solution space,[8] for example in the case of a discrete optimization problem. *Evolution strategies* were developed in the 1960s by Rechenberg [236] and Schwefel [267] at the Technical University of Berlin in order to support their research into *numerical optimization* problems.[9] In this approach, an individual is represented by a pair of vectors $(\mathbf{X}, \boldsymbol{\sigma})$, where $\mathbf{X} = (X_1, X_2, \ldots, X_n)$ determines the location of the individual in the n-dimensional (continuous) solution space,[10] $\boldsymbol{\sigma} = (\sigma_1, \sigma_2, \ldots, \sigma_n)$ is a string of parameters of the method.[11]

Let us discuss the general scheme of an evolution strategy shown in Fig. 5.5a. Similarly to the case of genetic algorithms, we begin with the initialization and evaluation of a μ-element parent population R. Then, we start the basic cycle of the method, which consists of three phases.

During the first phase λ-element offspring population O is generated. Each descendant is created in the following way. Firstly, ρ individuals, which will be used

[8] An example of a discrete solution space has been defined in the previous section in Fig. 5.1.

[9] The research into fluid dynamics was carried out at the Hermann Föttinger Institute for Hydrodynamics at TUB.

[10] A vector $\mathbf{X}$ represents here the chromosome of the individual, and its components $X_1, X_2, \ldots, X_n$, being real numbers, correspond to its genes.

[11] A parameter σ_i is used for mutating a gene X_i.

for the production of a given descendant, are chosen.[12] These "parents" are drawn with replacement according to the uniform distribution.[13] Then, these ρ parents produce a "preliminary version" of the descendant with the crossover operator. After recombination the element $\boldsymbol{\sigma}$ of the child chromosome, which contains parameters of the method, is mutated. Finally, a mutation of the individual, i.e., a mutation of the element **X** of its chromosome, is performed with the help of the mutated parameters of $\boldsymbol{\sigma}$.

In the second phase an evaluation of the offspring population O is made. This is performed in an analogous way to genetic algorithms, that is with the help of the fitness function.

The third phase consists of the selection of μ individuals to form a new parent population P according to the values of the fitness function. There are two main approaches to selection. In the selection of the $(\mu + \lambda)$ type a choice is made from among individuals which belong to both the (old) parent population and the offspring population. This means that the best parents and the best children create the next generation.[14] In selection of the (μ, λ) type we choose individuals to form the next generation from the offspring population.

Similarly to genetic algorithms, a termination condition is checked at the end of the cycle. If it is fulfilled, the best individual is chosen as our solution to the problem. Otherwise, a new cycle is started (cf. Fig. 5.5a).

After describing the general scheme let us introduce a way of denoting evolution strategies [23]. We assume an interpretation of parameters μ, λ, ρ as in the description above. If we use selection of the $(\mu + \lambda)$ type, then the evolution strategy is denoted by $(\mu/\rho + \lambda)$. If we use selection of the (μ, λ) type, then the evolution strategy is denoted by $(\mu/\rho, \lambda)$.

Now, we present a way of defining genetic operators for evolution strategies.

Let us assume that both parents *Father* and *Mother* are placed in a solution space according to vectors $\mathbf{X^F} = (X_1^F, X_2^F)$ and $\mathbf{X^M} = (X_1^M, X_2^M)$, respectively, as shown in Fig. 5.5b.[15] Since an individual is represented by a vector of real numbers, calculating the average of the values of corresponding genes which belong to the parents is the most natural way of defining the *crossover operator*. Therefore, the position of *Child* is determined by the vector $\mathbf{X^C} = (X_1^C, X_2^C)$, where $X_1^C = (X_1^F + X_1^M)/2$ and $X_2^C = (X_2^F + X_2^M)/2$ (cf. Fig. 5.5b).

In case of crossover by averaging, we also compute $\boldsymbol{\sigma}^C$ by taking the average values of $\boldsymbol{\sigma}^F$ and $\boldsymbol{\sigma}^M$, which represent the parameters of the method. An exchange of single genes of parents can be an alternative way of creating an offspring.

---

[12] This means that a "child" can have more than two "parents".

[13] Firstly, this means that each individual has the same chance to breed an offspring (the uniform distribution). Secondly, any individual can be used for breeding several times (drawing with replacement). This is the main difference in comparison to genetic algorithms, in which the best-fitted individuals have better chances to breed an offspring (roulette wheel selection, ranking selection).

[14] As one can see, we try to improve on the law of Nature. A "second life" is given to outstanding parents.

[15] The reader is advised to compare Fig. 5.5b with Fig. 4.11 in the previous chapter. For clarity there is no axis $h(X_1, X_2)$ corresponding to the fitness function in Fig. 5.5b.

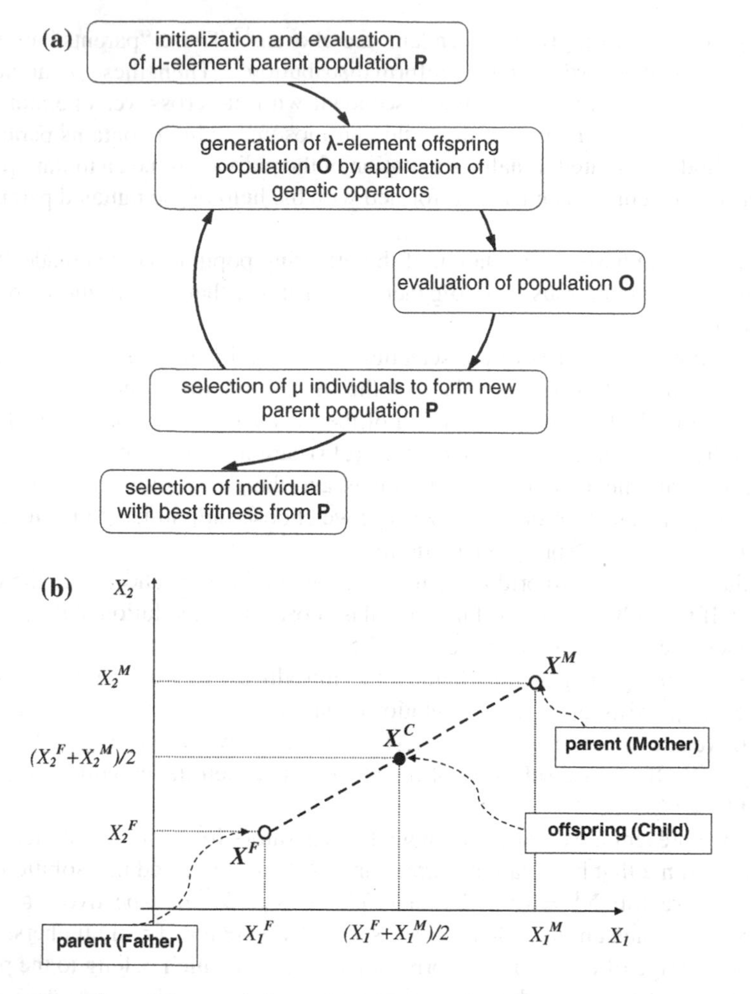

Fig. 5.5 **a** The scheme of an evolution strategy, **b** an example of crossover by averaging in an evolution strategy

As we have mentioned above, firstly, *mutation* is performed for an element $\boldsymbol{\sigma}$, secondly for an element $\mathbf{X}$. A mutation of the element $\boldsymbol{\sigma}$ consists of multiplying its every gene by a certain coefficient determined by a random number,[16] which is generated according to the normal distribution.[17] After modifying the vector $\boldsymbol{\sigma}$, we use it for a replacement of the individual in the solution space, as shown in Fig. 5.6. As we can see in the figure, the position of the individual represented by $\mathbf{X}$ is determined by its genes (coordinates) X_1 and X_2. Now, we can e.g., mutate its gene X_1 by adding

[16]Sometimes, the coefficient is determined by several random numbers.

[17]Formal definitions of notions of probability theory which are used in this chapter are contained in Appendix B.1.

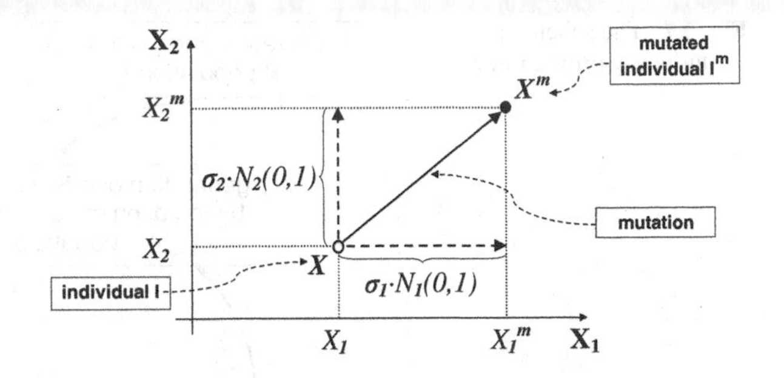

Fig. 5.6 Mutation of an individual in an evolution strategy

a number $\sigma_1 \cdot N_1(0, 1)$, where σ_1 is its corresponding gene—and the parameter of the method—and $N_1(0, 1)$ is a random number generated according to the normal distribution with an *expected value (average)* equal to 0 and a *standard deviation* equal to 1.

Let us notice that the element $\boldsymbol{\sigma} = (\sigma_1, \sigma_2, \ldots, \sigma_n)$ contains parameters which determine how big a mutation is.[18] As we have seen, these parameters are modified,[19] which means that evolution strategies are self-adapting.

5.3 Evolutionary Programming

In 1966 Lawrence J. Fogel introduced an approach to evolutionary computing which is called *evolutionary programming* [99]. The main idea of this approach differs from the methods discussed above. One difference concerns the level of abstraction at which evolution processes are simulated. In genetic algorithms and evolution strategies search points in a solution space correspond to individuals in a population. However, in the case of evolutionary programming instead of individuals we deal with a species-level abstraction.[20] This influences, of course, how the method is constructed. First of all, there is no crossover operation, since there is no crossover among species. Secondly, a mutation is defined in such a way that radical changes occur with low probability and small changes are preferred.

The second important difference with respect to the methods introduced in previous sections is the fact that in evolutionary programming we do not assume any specific form of representation of an individual.[21] A representation of an indi-

[18]Bigger values of these parameters cause a bigger change to an individual in a solution space. Strictly speaking, the bigger the value of a parameter σ_k, the more the gene X_k is mutated, which corresponds to moving the individual along the $\mathbf{X_k}$ axis.

[19]The probabilities of both a crossover and a mutation are constant parameters in genetic algorithms.

[20]Let us remember that a *population* is a set of individuals of the same species which live in the same area. Thus, in the case of evolutionary programming we should rather use the term *biocoenosis* instead of *population*, which is more correct from the point of view of biology.

[21]We have assumed a binary representation of individuals in genetic algorithms and real number vectors in evolution strategies.

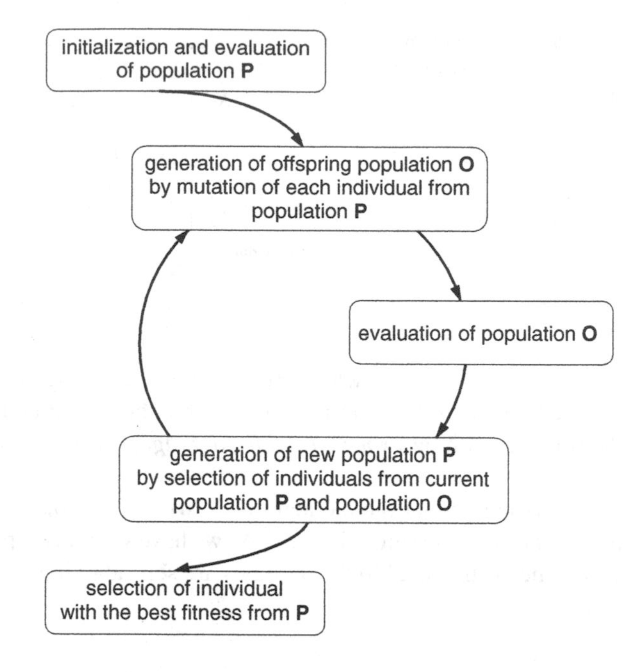

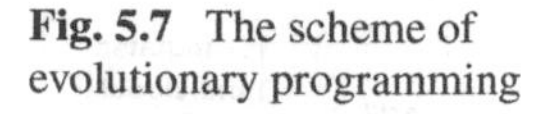
Fig. 5.7 The scheme of evolutionary programming

vidual should, simply, be adequate for a given problem. A variety of representations (variable-length vectors, matrices, etc.) are used in evolutionary programming projects for defining abstract models of problems.

Now, we present a scheme of evolutionary programming, which is shown in Fig. 5.7. The initialization and evaluation of a (parent) population P is a preliminary phase. Then, we begin the basic cycle of the method. The generation of the offspring population O by a mutation of each individual from the population P is performed in the first phase. Mutation is made randomly according to the normal distribution. The second phase consists of the evaluation of the population O. The generation of a new population P by a selection of individuals from the current population P and the population O is performed in the third phase. In a standard version of evolutionary programming a ranking selection is applied for this purpose. Then, as in previous methods, a termination condition is tested. If it is fulfilled, we choose the best individual as the solution to the problem. If not, a new cycle is begun.

At the end of the twentieth century David Fogel[22] introduced two improvements in evolutionary programming. Firstly, instead of ranking selection, a certain variant of *tournament selection* is applied. Tournament selection consists of dividing a population into groups which usually contain several individuals and selecting the best

[22]David Fogel—a researcher in the area of evolutionary computing. In his famous research project *Blondie24*, an evolutionary-computing-based AI system became an expert in English draughts (checkers). Fogel has been the President of the IEEE Computational Intelligence Society and the Editor-in-Chief of IEEE Transactions on Evolutionary Computation. He is the son of Lawrence J. Fogel.

individuals from each group separately. This method is especially useful for multi-criteria optimization problems, when we optimize more than one function. Secondly, D. Fogel has introduced self-adapting mechanisms similar to those used in evolution strategies.

Since no specific form of representation of an individual is assumed in evolutionary programming, the approach may be applied to a variety of problems, e.g., control systems, pharmaceutical design, power engineering, cancer diagnosis, and signal processing. In Artificial Intelligence the approach is used not only for solving problems, mainly optimization and combinatorial problems, but also for constructing self-learning systems.

In fact, the history of this approach began in 1966in the area of self-learning systems. L.J. Fogel in his pioneering paper [99] discussed the problem of formal grammar induction,[23] strictly speaking the problem of synthesizing a formal automaton[24] on the basis of a sample of sentences belonging to some language. A formal automaton is a system used to analyse a formal language. The synthesis of an automaton A by an AI system consists of an automatic construction of A on the basis of a sample of sentences, which belong to a formal language. L.J. Fogel showed that such a synthesis can be made via evolutionary programming. In his model a formal automaton evolves by the simulation of processes of crossover and mutation in order to be able to analyze a formal language. Let us notice a difference between problem solving by genetic algorithms/evolution strategies discussed previously and solving the problem of synthesis of an automaton by the AI system constructed by L.J. Fogel. In the first case, generation of *a problem solution* is the goal of the method, whereas in the second case we want to generate *a system* (automaton) that solves a certain class of problems (a formal language analysis). Such an idea would appear twenty years later in the work of M.L. Cramer, which concern genetic programming. This approach is introduced in the next section.

5.4 Genetic Programming

Although *genetic programming* was popularized in the 1990s by John Koza due to his well-known monograph [172], the main idea of this approach was introduced in 1985 by Cramer [61]. In genetic programming instead of searching a solution space with the help of a program, which is implemented on the basis of principles of evolution theory, a population of programs is created. Then, a space of programs is searched in order to find the one which can solve a class of problems in a satisfactory way. Of course, we have to define a function, to assess the quality (adequacy) of such programs. Thus, automatic synthesis of a computer program to solve a given problem is the objective of genetic programming. This objective has been extended to other

---

[23]This problem is discussed in Sect. 8.4.

[24]Formal automata are introduced in Sect. 8.2.

systems in the technical sciences, such as digital circuits (electronics), controllers (automatic control), etc.[25]

In order to achieve such an ambitious objective, a human designer has to deliver certain knowledge to an AI system [173]. Firstly, the system has to know which components are to be used for generating a solution. In the case of a software system synthesis, arithmetic operations, mathematical functions, and various instructions of a programming language are such components. In the case of a digital circuit synthesis AND, OR, NOT, NAND, NOR logic gates, various flip-flops, etc. are such components. Secondly, a human designer has to define the fitness function. It seems that this is the most difficult problem. In order to define the fitness function, one has to formalize the task of a synthesized system precisely. Solving problems which belong to a certain class is, of course, the main objective of a synthesized system. Thus, systems-individuals should evolve in order to solve problems in a satisfactory way. In other words, the fitness function should define how well the system solves these problems. Thirdly, a human designer should deliver control parameters such as the size of the population, the probability of applying genetic operators, the termination condition, etc.

In genetic programming, programs are usually represented by tree or graph structures. In Fig. 5.8a the expression $-b/2a$ is represented with the help of a tree structure. All expressions of a programming language can be represented with such a tree representation. The specific form of the "chromosomes" of individuals results in the specific form of the genetic operators. A mutation is shown in Figs. 5.8a and 5.8c. A part of the first tree, which is mutated—the subtree $-b$ (it is encircled by a dashed line in Fig. 5.8a) is removed. A subtree representing the expression $-b - \sqrt{\Delta}$ (encircled by a dashed line in Fig. 5.8c) is joined to the tree in place of the removed part. A tree which represents the expression $\frac{-b - \sqrt{\Delta}}{2a}$ is obtained as a result of this mutation. A crossover operator consists of exchanging subtrees of trees (individuals). Let us cross a tree which has been obtained as a result of the mutation (Fig. 5.8c) and a tree which represents the expression $\frac{a+b}{-b+\sqrt{\Delta}}$, which is shown in Fig. 5.8b. The parts of the "chromosomes" which are exchanged are encircled by a dashed line. As a result of the crossover we obtain the tree shown in Fig. 5.8d, which represents the expression $\frac{a+b}{-b-\sqrt{\Delta}}$ (it is not interesting), and the tree shown in Fig. 5.8e, which represents the expression $\frac{-b+\sqrt{\Delta}}{2a}$. This second expression has a well-known mathematical interpretation, as does the initial expression $-b/2a$ shown in Fig. 5.8a.

Let us notice that we have to formulate the *well-defined* fitness function in order to generate (with genetic operators) a system which solves a certain class of problems. The fitness function directs the actions of the genetic operators. In genetic program-

[25] If we analyze the applications of genetic programming, it seems that the synthesis of systems of electronics or automatic control is easier than the synthesis of software systems.

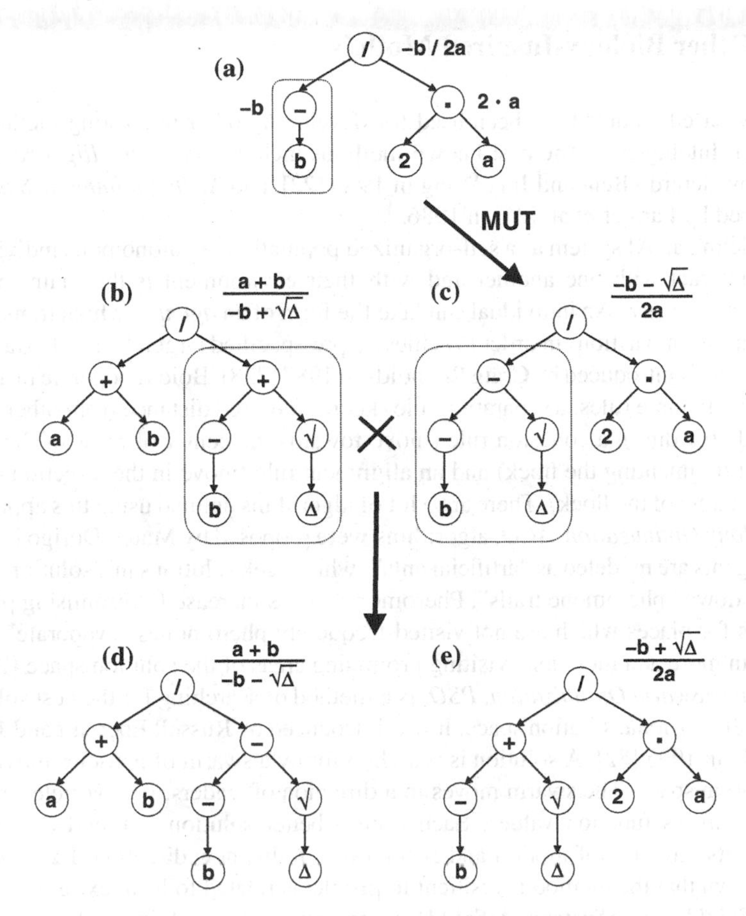

Fig. 5.8 **a** An exemplary expression of a programming language which is coded as a tree structure and its mutation, **b, c** the crossover of two structures, **d, e** descendant structures which are results of the crossover

ming crossing reasonable solutions is a basic operator. The mutation operator plays an auxiliary role.

On the basis of genetic programming a very interesting approach, called *meta-genetic programming*, was defined in 1987 by Jürgen Schmidhuber of the Technical University of Munich [265]. In this approach both chromosomes and genetic operators evolve themselves, instead of being determined by a human designer. Thus, a meta-genetic programming system evolves itself with the help of genetic programming. A similar approach was used for constructing the *Eurisko* system by Douglas Lenat[26] in 1976. The system, which is based on heuristic rules, also contains meta-rules allowing it to change these heuristic rules.

[26]Douglas Lenat—a professor of computer science at Stanford University and Carnegie-Mellon University. A president of Cycorp, Inc., which researches the construction of a common-sense knowledge base (ontology) *Cyc*.

5.5 Other Biology-Inspired Models

Biology-based models have been used for developing other interesting methods in Artificial Intelligence. The best known methods include *swarm intelligence*, introduced by Gerardo Beni and Jing Wang in 1989 [22], and *Artificial Immune Systems*, developed by Farmer et al. [86] in 1986.

Modeling an AI system as a self-organized population of autonomous individuals which interact with one another and with their environment is the main idea of *swarm intelligence*. An individual can take the form of an *agent*,[27] which transforms observations into actions in order to achieve a pre-specified target. It can also take the form of a *boid* introduced by Craig Reynolds in 1987 [238]. Boids cooperate in a flock according to three rules: a separation rule (keep a required distance from other boids to avoid crowding), a cohesion rule (move towards the center of mass of the flock to avoid fragmenting the flock) and an alignment rule (move in the direction of the average target of the flock). There are a lot of algorithms defined using this approach. *Ant Colony Optimization, ACO*, algorithms were proposed by Marco Dorigo in 1992 [72]. Agents are modeled as "artificial ants", which seek solutions in a solution space and lay down "pheromone trails". Pheromone values increase for promising places, whereas for places which are not visited frequently pheromones "evaporate". This results in more and more ants visiting promising areas of the solution space [298].

Particle Swarm Optimization, PSO, is a method of searching for the best solution in an n-dimensional solution space. It was introduced by Russell Eberhart and James Kennedy in 1995 [82]. A solution is searched for by a swarm of particles moving in the solution space. The swarm moves in a direction of leaders, i.e., particles having the best fitness function values. Each time a better solution is found the swarm changes its direction of motion and accelerates in this new direction. Experiments have shown that the method is resilient to problems related to local extrema.

Artificial Immune Systems, AISs [43], are mainly used for solving problems related to detecting anomalies. The idea of differentiating between normal/"own" cases and pathological/"alien" cases is based on the immune system of a biological organism. All the cases that are not "similar" to known cases are classified as anomalies. When an unknown case appears and its characteristics are similar to those recognized by one of the detectors of anomalies, it is assumed to be an "alien" and this detector is activated. The activated detector is "processed" with operators such as mutation and duplication. In such a way the system learns how to recognize pathological cases.

Bibliographical Note

A general introduction to the field can be found in [83, 65, 169, 201, 260]. Genetic algorithms are discussed in [111, 139], evolution strategies in [268], evolutionary programming in [100], and genetic programming in [172].

---

[27] Agent systems are discussed in Chap. 14. Therefore, we do not define them in this section.

Chapter 6
Logic-Based Reasoning

Two models of problem solving which are based on logic, reasoning as theorem proving and reasoning as symbolic computation, are discussed in this chapter. Both models are implemented in AI systems with the help of the declarative programming paradigm, which has been introduced in Sect. 2.2.

In a programming language which belongs to the declarative paradigm, *control of an execution* of a computer program follows a certain *general (standard) scheme*. *A general scheme* means here *the same scheme* for all possible programs[1] coded in this language. In practice this means that such a language should be based on a very precise formal model, which allows a reasoning system to interpret a program in an unambiguous way.[2] In a method based on theorem proving, such a precise formal model is based on the syntax of First-Order Logic (FOL) and FOL rules of inference,[3] especially the resolution rule of inference. In a method based on symbolic computation, lambda calculus is one such precise formal model.

Reasoning as theorem proving is discussed in the following sections as follows.

- *The description of the world with the help of First-Order Logic.* A way of representing some aspect of the real world with the help of the FOL language is discussed in the first section. We introduce fundamental rules of inference that are based on such a representation.
- *The resolution method of inference.* In the second section a basic method of reasoning which is used in AI systems, i.e., the resolution method, is introduced.

---

[1] Of course, we mean programs which are correct syntactically.

[2] As we have mentioned in Sect. 2.2, we do not specify *how* a program should be executed in the declarative paradigm.

[3] A rule of inference in FOL should not be mistaken for the notion of a rule in rule-based systems, which are a subclass of expert systems. In the first case rules of inference are formulas of reasoning schemes (e.g., the *modus ponendo ponens* rule), which are used in order to process *axioms*, which are stored in a knowledge base. These *axioms* are either facts, which concern the domain of an application of the AI system, or principles (rules), which are valid in this domain. In the second case (rule-based systems) rules are equivalents of such *axioms*-principles, which are valid in the domain.

M. Flasiński, *Introduction to Artificial Intelligence*,
DOI 10.1007/978-3-319-40022-8_6

- *Techniques to transform FOL formulas to standard (normal) forms.* A resolution method can be applied, if FOL formulas are expressed in special forms, called *normal forms*. In the third section techniques to transform formulas to such forms are presented.
- *Special forms of FOL formulas in reasoning systems.* If a problem is described using formulas in normal forms, we can begin the implementation phase of constructing an AI system. In the fourth section special forms of FOL formulas, which are convenient for implementing AI systems are introduced.

Reasoning as symbolic computation, which is based on *Abstract Rewriting Systems, ARSs*, and *lambda calculus* (*λ-calculus*) are discussed in the fifth section.

All formal notions concerning First-Order Logic, the resolution method, and lambda calculus are contained in Appendices C.1, C.2 and C.3, respectively.

6.1 World Description with First-Order Logic

In order to reason using theorem proving, we should describe an aspect of the world which is interesting to us with the help of First-Order Logic. For such a description the following elements, called *terms*, are used.

- *Individual constant symbols*, which correspond to objects such as human beings, animals, buildings, etc. Examples of individual constants include: *John III Sobieski* (King of Poland), *Wawel Castle in Cracow*, *Norcia* (an individual constant corresponding to an individual object, which is a dog of the Yorkshire Terrier breed and belongs to my daughter).
- *Individual variable symbols*, which range over individual objects, usually denoted by x, y, z, etc.
- *Function symbols*, which ascribe objects to other objects, e.g., square-root(), length(), father().

Additionally, *predicate symbols* are added to the FOL language. We can treat them as functions defined over terms which give one of two values: *True* or *False*. Examples of predicates include *is_less_than*(), *is_high*(), *is_brother_of*(). For example, the value of the predicate *is_less_than*(4, 9), i.e., in standard arithmetic notation $4 < 9$, is *True* and the value of the predicate *is_less_than*(3, 1), i.e., in standard arithmetic notation $3 < 1$, is *False*. A predicate with fixed arguments is called an *atomic formula*.

Finally, we add logical symbols $\neg$ (it is not true that ...), $\wedge$ (... and ...), $\vee$ (... or ...), $\Rightarrow$ (if ..., then ...), $\Leftrightarrow$ (... is equivalent to ...), and quantifiers $\forall$ (for each ...), $\exists$ (there exists ...such that). With the help of these symbols we can define *formulas*. For example, a formula which states that for every individual object x the following holds: "If x barks, then x is a dog." can be defined as follows: $\forall x\ [barks(x) \Rightarrow is\_dog(x)]$. A formula which states that there exist black cats, strictly speaking that there exists such an individual object y that y is black and y is a cat, can be defined as follows: $\exists y\ [black(y) \wedge is\_cat(y)]$.

Quantifiers *bind* variables in formulas. In the formulas defined above, a *variable* x *is bound* by a quantifier $\forall$ and a variable y is bound by a quantifier $\exists$. Variables which are not bound by quantifiers in a formula are called *free variables*. Formulas which do not contain free variables are called *sentences* of FOL.

After introducing the syntax of FOL, we will define its *semantics*. The semantics allows us to refer formulas to the world, which contains individual objects represented by individual constant symbols. A set of individual objects is called a *universe*. Relations which hold among elements of the universe are described with predicate symbols. Functions defined in the universe are represented using function symbols. An assignment of individual objects, functions, and relations to individual constant symbols, function symbols, and predicate symbols, respectively, is called an *interpretation*. In other words, an interpretation is an assignment of *meaning* to elements of the FOL language. Let us explain these notions with the following example.

Suppose we have a set of individual constant symbols $\Sigma^S = \{a, d\}$, a set of variables $X = \{x, y\}$, and a set of two-argument predicate symbols $\Sigma_2^P = \{p_L, p_R\}$. Let us assume that they are all elements used for defining atomic formulas. Let us determine a universe $\mathcal{U}$ as shown in Fig. 6.1a. As we can see, the universe consists of a certain car and a certain tree. Then, let us define an interpretation $\mathcal{I}$ as follows: $\mathcal{I}(c) =$ car, $\mathcal{I}(t) =$ tree, $\mathcal{I}(p_L) =$ left_of, $\mathcal{I}(p_R) =$ right_of (cf. Fig. 6.1b) and the following holds: tree is right_of car and car is left_of tree (cf. Fig. 6.1c). We can describe it by (tree, car) $\in$ right_of and (car, tree) $\in$ left_of. A pair $(\mathcal{U}, \mathcal{I})$ is called a *structure* and we denote it by $\mathfrak{A}$.

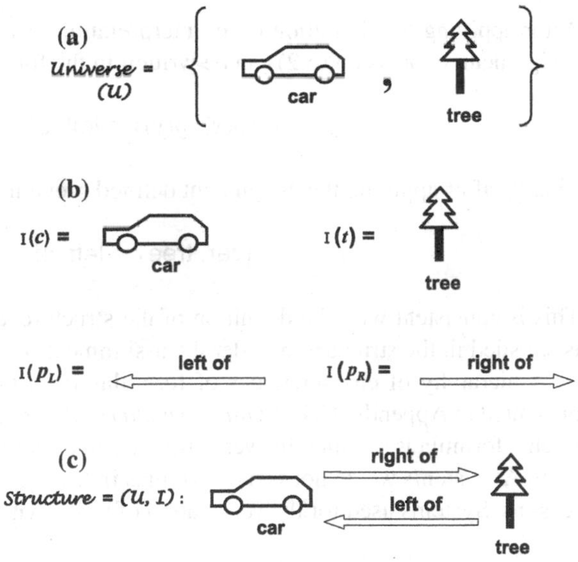

Fig. 6.1 An example of FOL semantics: **a** the universe, **b** the interpretation, **c** the structure

Before we discuss the semantics of formulas, we have to define an *interpretation* (value) of a term. For variables[4] we introduce, firstly, an *assignment* (valuation), which is denoted by ϱ. It assigns an element of the universe to a variable. Let us assume e.g., that $\varrho(x) = \mathsf{car}$, $\varrho(y) = \mathsf{tree}$. After determining the assignment ϱ, we are able to define an interpretation for a term t, which in our case is a variable, in a structure $\mathfrak{A}$. This is defined as follows: $[\![t]\!]_{\varrho}^{\mathfrak{A}} = \varrho(t)$. Thus, for our variables the following holds: $[\![x]\!]_{\varrho}^{\mathfrak{A}} = \varrho(x) = \mathsf{car}$, and $[\![y]\!]_{\varrho}^{\mathfrak{A}} = \varrho(y) = \mathsf{tree}$.

Now, we can discuss the issue of relating FOL formulas to an aspect of the world they describe. First of all, we would like to know whether a certain formula φ describes a "part" of the world in an adequate way. If yes, we say that the formula is *satisfied* in the structure $\mathfrak{A}$ under the assignment ϱ, which is denoted $(\mathfrak{A}, \varrho) \models \varphi$.

Let us consider the satisfaction of a formula, continuing our example of the universe which consists of the car and the tree. We will consider a simple atomic formula of the form: $p_L(x, y)$. Thus, we ask whether:

$$(\mathfrak{A}, \varrho) \models p_L(x, y). \tag{6.1}$$

According to the definition of satisfaction of a formula,[5] for a predicate symbol p it is assumed that: $(\mathfrak{A}, \varrho) \models p(t_1, \ldots, t_n)$ if and only if $([\![t_1]\!]_{\varrho}^{\mathfrak{A}}, \ldots, [\![t_n]\!]_{\varrho}^{\mathfrak{A}}) \in p^{\mathfrak{A}}$. Thus, we can write (6.1) as:

$$([\![x]\!]_{\varrho}^{\mathfrak{A}}, [\![y]\!]_{\varrho}^{\mathfrak{A}}) \in \mathsf{left\_of}. \tag{6.2}$$

After applying the definition of an interpretation of a variable with the help of an assignment, expression (6.2) can be written in the form:

$$(\varrho(x), \varrho(y)) \in \mathsf{left\_of}. \tag{6.3}$$

Finally, after applying the assignment defined previously we obtain:

$$(\mathsf{car}, \mathsf{tree}) \in \mathsf{left\_of}. \tag{6.4}$$

This is consistent with the definition of the structure $\mathfrak{A}$. Thus, the formula $p_L(x, y)$ is satisfied in the structure $\mathfrak{A}$ under the assignment ϱ.

A hierarchy of characteristics of formulas from the semantic point of view is presented in Appendix C.1. A *valid formula (tautology)* is at the top of the hierarchy. Such a formula is satisfied in every structure under every assignment.

In AI systems we do not verify formulas in the way presented above for practical reasons. Systems used for the verification of unknown formulas infer on the basis of

---

[4] We do not discuss here all definitions related to the FOL semantics. They are included in Appendix C.1.

[5] See Definition C.12 in Appendix C.1.

formulas which are considered to be true.[6] These systems consist of at least the two following components.

- *Axioms* are formulas which are considered to be true. They constitute the basic *knowledge base* in a reasoning system.
- *Rules of inference* are patterns which are used to derive new formulas from known formulas. A rule of inference is written in the form $\frac{\varphi_1, \varphi_2, \ldots, \varphi_n}{\psi_1, \psi_2, \ldots, \psi_k}$, where $\varphi_1, \varphi_2, \ldots, \varphi_n$ are input formulas and $\psi_1, \psi_2, \ldots, \psi_k$ are the resulting formulas.

The *modus ponendo ponens rule* is one of the fundamental rules of inference. It is formulated as follows:

$$\frac{\varphi \Rightarrow \psi, \varphi}{\psi}. \tag{6.5}$$

The rule says that if there are two formulas and the first formula is of the form of an implication and the second formula is an antecedent of this implication, then we can generate the formula which is the consequent of this implication.

The *universal instantiation rule* is also a very useful rule:

$$\frac{(\forall x \in X)\, \varphi(x)}{\varphi(a),\, a \in X}, \tag{6.6}$$

where a is an individual constant symbol. Intuitively, the rule says that if something is true for each element of a class of individuals, then it is true for a particular element of this class.

In our further considerations we will also use the *material implication rule*:

$$\frac{\varphi \Rightarrow \psi}{\neg\varphi \vee \psi}. \tag{6.7}$$

The rule allows us to replace an implication with a disjunction.

Now, we present a simple example of processing in a reasoning system. Let us assume that a knowledge base contains the axiom:

$$\forall x[barks(x) \Rightarrow is\_dog(x)]. \tag{6.8}$$

Let us assume that we can add our new axiom to the knowledge base after making some observation:

$$barks(Norcia). \tag{6.9}$$

(We have observed that an individual Norcia is barking.)

[6] Formal notions, which concern reasoning in logic are contained in Appendix F.2.

Now, we can ask the question "Is Norcia a dog?" to the reasoning system:

$$is\_dog(Norcia) \quad (???). \tag{6.10}$$

In other words, we would like to ask the system to "prove" theorem (6.10) on the basis of its axioms and with the help of rules of inference.

First of all, the system applies the universal instantiation rule (6.6) to the axiom (6.8):

$$\frac{\forall x[barks(x) \Rightarrow is\_dog(x)]}{barks(Norcia) \Rightarrow is\_dog(Norcia)}, \tag{6.11}$$

which results in generating the formula:

$$barks(Norcia) \Rightarrow is\_dog(Norcia). \tag{6.12}$$

Then, the system applies the *modus ponendo ponens* rule (6.5) to formulas (6.12) and (6.9):

$$\frac{barks(Norcia) \Rightarrow is\_dog(Norcia), barks(Norcia)}{is\_dog(Norcia)}, \tag{6.13}$$

which results in generating the formula:

$$is\_dog(Norcia). \tag{6.14}$$

Thus, the system answers "Yes" to our question (6.10).

6.2 Reasoning with the Resolution Method

In the previous section the general idea of processing in an AI system based on logical reasoning has been presented. In practice, such a way of reasoning is not convenient for designing AI systems. Such systems should be able to reason on the basis of a few rules of reasoning and facts. The *resolution method* developed by J. Alan Robinson[7] allows us to construct *efficient* logic-based reasoning systems.

The resolution method is based on *theorem proving by contradiction* (in Latin *reductio ad absurdum*).[8] In order to prove a proposition, we firstly deny it and then show that this results in a contradiction with respect to true assumptions. The method can be defined in the following way.

[7] John Alan Robinson—a philosopher, mathematician, and computer scientist, a professor of Syracuse University. His research mainly concerns automated theorem proving and logic programming. He is a founder of the prestigious *Journal of Logic Programming*.

[8] It means: *reducing to absurdity*.

- If we want to prove that a formula ψ, which is our hypothesis, results from a set of formulas $\varphi_1, \varphi_2, \ldots, \varphi_n$, which are our axioms, then
- we create the negation of the formula $\neg\psi$, we add this negation to the set of formulas $\varphi_1, \varphi_2, \ldots, \varphi_n$, and we try to derive the *empty clause*, denoted by $\Box$, which represents the logical value *False*.

If we succeed in deriving the empty clause, then this means that the formula ψ follows from the set of formulas $\varphi_1, \varphi_2, \ldots, \varphi_n$. Thus, the reasoning system proves the formula ψ from the axioms.

Now, let us introduce a rule of inference for the resolution method. The *resolution rule* can be defined in its simplest form in the following way[9]:

$$\frac{\neg\alpha \vee \beta, \alpha \vee \gamma}{\beta \vee \gamma}, \tag{6.15}$$

where the resulting formula $\beta \vee \gamma$ is called the *resolvent* of the input formulas $\neg\alpha \vee \beta$ and $\alpha \vee \gamma$; the input formulas are called *clashing formulas*.

Each formula appearing in the rule (6.15) is of the form of a disjunction, which consists of an atomic formula or a negated atomic formula. The first type of atomic formula is called a *positive literal*, the second type is called a *negative literal*. A formula which is a disjunction of finite literals (positive or negative) is called a *clause*. A single literal is a specific case of a clause.

Users of AI reasoning systems often write formulas in a knowledge base in the form of an implication, because of its intuitive character. We also have defined a formula (6.8) as an implication: "If x barks, then x is a dog." On the other hand, in order to use the resolution rule, one has to formulate a formula as a disjunction (clause). We can transform an implication into a disjunction with the help of the material implication rule (6.7). Then we obtain a clause that contains at most one positive literal,[10] i.e., a Horn clause.[11]

Returning to the form of the resolution rule, let us notice that its special cases include a rule of the form:

$$\frac{\neg\alpha, \alpha \vee \gamma}{\gamma}, \tag{6.16}$$

and the *rule of deriving the empty clause*:

$$\frac{\neg\alpha, \alpha}{\Box}. \tag{6.17}$$

[9] In the general case the resolution rule is defined for disjunctions consisting of any (finite) number of elements in the following way: $\dfrac{\neg\alpha \vee \beta_1 \vee \cdots \vee \beta_k, \alpha \vee \gamma_1 \vee \cdots \vee \gamma_n}{\beta_1 \vee \cdots \vee \beta_k \vee \gamma_1 \vee \cdots \vee \gamma_n}$.

[10] If an implication formula is of the form: "If Condition-1 and Condition-2 and ... and Condition-n, then Result", formally speaking $C_1 \wedge C_2 \wedge \cdots \wedge C_n \Rightarrow R$, then we generate a clause $\neg C_1 \vee \neg C_2 \vee \cdots \vee \neg C_n \vee R$. This clause contains at most one positive literal.

[11] Alfred Horn—a professor of mathematics at the University of California, Berkeley. He introduced this form in 1951. His research results in the areas of universal algebra and lattice theory are very important in logic programming.

Now, we consider how the resolution method can be used for our previous example of Norcia, which barks. Let us begin from the step in which the universal quantifier is removed from the axiom (6.8). Then, we have obtained a formula (6.12) in the form of an implication. Now, we have to transform it into the form of a clause. We do this with the help of the material implication rule (6.7):

$$\frac{barks(Norcia) \Rightarrow is\_dog(Norcia)}{\neg barks(Norcia) \vee is\_dog(Norcia)}, \tag{6.18}$$

and we generate an equivalent formula of the form:

$$\neg barks(Norcia) \vee is\_dog(Norcia). \tag{6.19}$$

Now, we can use the resolution method. In order to prove clause (6.10), we create its negation:

$$\neg is\_dog(Norcia) \tag{6.20}$$

and we try to derive the empty clause from a set containing axioms and the negated clause. In our case we use formula (6.16), since hypothesis (6.20) is of the (special) form of an atomic formula (not a disjunction).

There are two clauses, (6.9) and (6.19), in our set of axioms. Let us choose clause (6.19) for applying in the resolution rule. After an application of the resolution rule to clauses (6.19) and (6.20), i.e.,

$$\frac{\neg is\_dog(Norcia),\ \neg barks(Norcia) \vee is\_dog(Norcia)}{\neg barks(Norcia)}, \tag{6.21}$$

we obtain a new clause:

$$\neg barks(Norcia). \tag{6.22}$$

Let us apply the resolution rule once more, this time to the newly generated clause (6.22) and the second clause from the set of axioms (6.9). As we can see below, we obtain the empty clause according to rule (6.17):

$$\frac{\neg barks(Norcia),\ barks(Norcia)}{\Box}. \tag{6.23}$$

Summing up, negating a clause-hypothesis (6.10) and using this negation with the set of axioms, we are able to derive the empty clause with the resolution method. This means that the clause (6.10) follows from the set of axioms, so the reasoning system has completed the proof.

A single proof step with the resolution method can be represented by the tree shown in Fig. 6.2a (cf. formula (6.15)). Leaves of the tree correspond to clashing clauses. The root of the tree represents a resolvent. A complete proof is represented with the help of a *resolution tree*, which shows succeeding proof steps, as illustrated for our example in Fig. 6.2b.

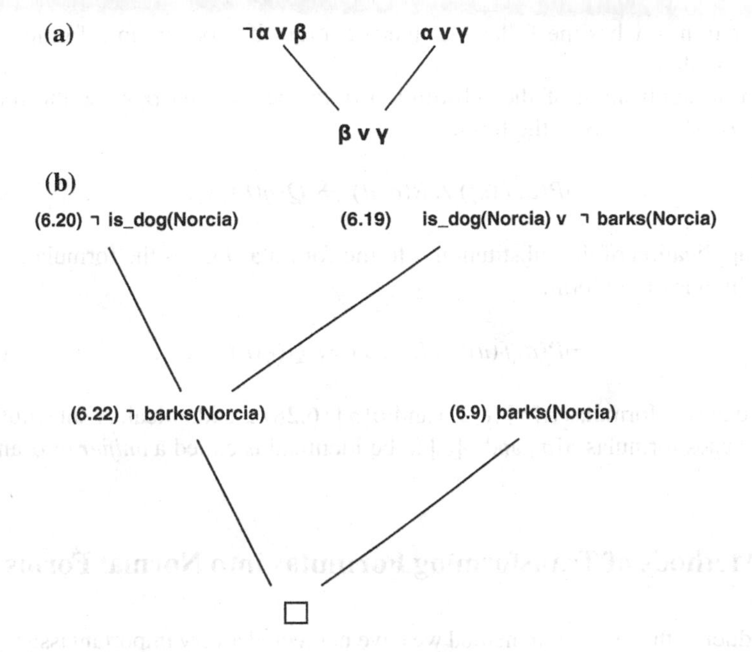

Fig. 6.2 A resolution tree

After introducing the main idea of the resolution method, we discuss a certain problem related to its practical application. It concerns *matching input formulas* during an application of rules of inference. Let us assume that we would like to use rule (6.5). Input formulas are expressed in the form: $\varphi \Rightarrow \psi, \varphi$. Of course, this does not mean that the antecedent of an implication of the first formula and the second formula should be the same *literally*. It is enough that they can be transformed into the same form with suitable *substitutions*. We can replace variables with terms.[12] The operation of transforming formulas to the same form by a substitution is called *unification* of these formulas. For example, let two formulas be given: a formula α of the form:

$$\neg P(x, f(u)) \wedge R(z, d) \Rightarrow Q(g(b), y), \tag{6.24}$$

and a formula β of the form:

$$\neg P(a, w) \wedge R(c, d) \Rightarrow Q(v, y). \tag{6.25}$$

In order to unify them we should use the substitution:

$$\sigma = \{a/x, c/z, f(u)/w, g(b)/v\}. \tag{6.26}$$

[12]Of course, not all substitutions are allowable. For example, the replacement of a variable with a term which contains this variable is not allowed.

(The notation a/x has the following interpretation. If x occurs in a formula, then replace x with a.)

Then an application of the substitution σ to the formula α gives the formula denoted $\alpha[\sigma]$, which is of the form:

$$\neg P(a, f(u)) \wedge R(c, d) \Rightarrow Q(g(b), y), \tag{6.27}$$

and an application of the substitution σ to the formula β gives the formula denoted $\beta[\sigma]$, which is of the form:

$$\neg P(a, f(u)) \wedge R(c, d) \Rightarrow Q(g(b), y). \tag{6.28}$$

As one can see, formulas $\alpha[\sigma]$ (6.27) and $\beta[\sigma]$ (6.28) are identical. A substitution σ which causes formulas $\alpha[\sigma]$ and $\beta[\sigma]$ to be identical is called a *unifier* of α and β.

6.3 Methods of Transforming Formulas into Normal Forms

In introducing the resolution method we have neglected a very important issue which concerns its practical application. In AI systems we apply the resolution method to formulas, which are expressed in special forms called *normal forms*. In this section we will transform FOL formulas in order to obtain formulas in *conjunctive normal form*.[13] Before a formula gets into such a final form, it has to be transformed into a series of temporary forms, which are the result of normalizing operations. These operations are based on rules of FOL. Now, we present these operations and the forms related to them in an intuitive way.[14]

Firstly, a formula is transformed into *negation normal form*. The transformation consists of eliminating logical operators of implication ($\Rightarrow$) and equivalence ($\Leftrightarrow$) and then moving negation operators so that they occur immediately in front of atomic formulas. For example, the formula:

$$\forall x[\neg student(x) \Rightarrow [\neg on\_leave(x) \wedge \neg\exists y(attends(x, y) \wedge course(y))]], \tag{6.29}$$

after eliminating the implication operator is transformed into the following form:

$$\forall x[\neg\neg student(x) \vee [\neg on\_leave(x) \wedge \neg\exists y(attends(x, y) \wedge course(y))]], \tag{6.30}$$

[13]In the previous section, we have made use of the fact that the starting formula (6.19) was already in conjunctive normal form.

[14]The rules of formula normalization are discussed in detail in monographs on the mathematical foundations of computer science, which are listed at the end of this chapter in a bibliographical note.

and after moving negation operators it is transformed into the following negation normal form:

$$\forall x[student(x) \vee [\neg on\_leave(x) \wedge \forall y(\neg attends(x, y) \vee \neg course(y))]]. \quad (6.31)$$

Then, formula (6.31) is transformed into *prenex normal form*, by moving all the quantifiers to the front of it, which results the following formula:

$$\forall x \forall y[student(x) \vee [\neg on\_leave(x) \wedge (\neg attends(x, y) \vee \neg course(y))]]. \quad (6.32)$$

As we can see, in the case of our formula all variables are inside scopes of universal quantifiers. Thus, the quantifiers are, somehow, redundant and we can eliminate them,[15] arriving at the following formula:

$$student(x) \vee [\neg on\_leave(x) \wedge (\neg attends(x, y) \vee \neg course(y))]. \quad (6.33)$$

A certain problem related to quantified variables can appear when we move quantifiers to the front of formulas. We discuss it with the help of the following example. Let a formula be defined as follows:

$$\forall x[\forall y[P(x, y) \vee Q(x, y)] \wedge \exists y[R(y) \vee S(y, x)]]. \quad (6.34)$$

One can easily see that the first variable y (after the quantifier $\forall$) "is different" from the second y (after the quantifier $\exists$). Therefore, before we move quantifiers to the front of the formula we have to *rename variables* so that they are distinct.[16] Thus, we transform formula (6.34) to the following formula:

$$\forall x[\forall y[P(x, y) \vee Q(x, y)] \wedge \exists z[R(z) \vee S(z, x)]]. \quad (6.35)$$

In our considerations above we have said that universal quantifiers can be eliminated after moving quantifiers to the front of the formula. And what should be done with existential quantifiers, if they occur in a formula? If existential quantifiers occur in a formula, we make use of the method developed by Skolem[17] [276], called *Skolemization*. We define this method in an intuitive way with the following example.

Let a formula be of the following form:

$$\forall x[\ \neg likes\_baroque\_music(x) \vee \\ \exists y(\ likes\_music\_of(x, y) \wedge baroque\_composer(y))]. \quad (6.36)$$

[15] We can just keep in mind that they are at the front of the formula.

[16] After moving quantifiers to the front we do not see which quantifier concerns which part of a conjunction.

[17] Thoralf Albert Skolem—a professor of mathematics at the University of Oslo. His work, which concerns mathematical logic, algebra, and set theory (Löwenheim-Skolem theorem) is of great importance in the mathematical foundations of computer science.

The second part of the disjunction says that there is such a person y that a person x likes music composed by y and y is a baroque composer. If so, then—according to Skolemization—we can define a function F, which ascribes this (favorite) baroque composer to x, i.e., $F(x) = y$. Now, if we replace y with $F(x)$, then we can eliminate the existential quantifier $\exists y$ in formula (6.36).[18] Thus, we can transform formula (6.36) into the following formula:

$$\begin{aligned}&\forall x[\ \neg likes\_baroque\_music(x) \vee \\ &\quad (\ likes\_music\_of(x, F(x)) \wedge baroque\_composer(F(x)))]. \end{aligned} \tag{6.37}$$

In the general case, we can formulate Skolemization in the following way:

- analyze successive quantifiers left-to-right,
- if an existential quantifier of the form $\exists y$ is preceded by universal quantifiers: $\forall x_1, \forall x_2, \ldots, \forall x_n$, then introduce a unique function $F(x_1, x_2, \ldots, x_n)$, replace all occurrences of the variable y with the function $F(x_1, x_2, \ldots, x_n)$, and remove $\exists y$,
- if an existential quantifier of the form $\exists y$ is not preceded by any universal quantifier, then introduce a constant a, replace all occurrences of the variable z with the constant a, and remove $\exists z$.

The function $F(x_1, x_2, \ldots, x_n)$ is called the *Skolem function*. The constant a is called the *Skolem constant*.

At the end of this section let us come back to our previous example concerning a student. We have transformed the formula into a form without quantifiers (6.33). We can transform formula (6.33) into *Conjunctive Normal Form, CNF*.[19] Our formula in the CNF form is expressed as a conjunction of disjunctions in the following way:

$$\begin{aligned}&(student(x) \vee \neg on\_leave(x)) \wedge \\ &(student(x) \vee \neg attends(x, y) \vee \neg course(y)). \end{aligned} \tag{6.38}$$

The Conjunctive Normal Form is the final result of our normalization operations.

6.4 Special Forms of FOL Formulas in Reasoning Systems

Formulas expressed in Conjunctive Normal Form are frequently simplified further in reasoning systems for better efficiency of formula-matching algorithms. Since we know that a CNF formula is a conjunction of clauses, we can eliminate conjunction symbols and divide the formula into simpler axioms. Such simpler axioms are stored

[18] After such a replacement the variable y "disappears", so the quantifier $\exists y$ does not quantify any variable. As a result, we can eliminate it.

[19] We make use of a FOL rule of distribution of disjunction over conjunction: $[\ \alpha \ \vee (\beta \wedge \gamma)] \Leftrightarrow [\ (\alpha \vee \beta) \wedge (\alpha \vee \gamma)]$.

in a knowledge base. In our case formula (6.38) can be transformed into a set which consists of two simpler clauses in the following way:

$$\begin{aligned} &student(x) \vee \neg on\_leave(x), \\ &student(x) \vee \neg attends(x, y) \vee \neg course(y). \end{aligned} \tag{6.39}$$

The form used in the *Prolog* language is especially convenient for logic programming. This language was developed by Alain Colmerauer and Phillippe Roussel in 1973 [57] on the basis of the theoretic research of Robert Kowalski, which concerned a procedural interpretation of Horn clauses. Prolog is considered a standard language used for constructing reasoning systems based on theorem proving with the help of the resolution method. In order to transform the set of clauses (6.39) into the Prolog-like form we have to transform them into CNF Horn clauses.[20] A Horn clause of the form:

$$\neg p_1 \vee \neg p_2 \vee \cdots \vee \neg p_n \vee h \tag{6.40}$$

corresponds to the following formula in the form of an implication:

$$p_1 \wedge p_2 \wedge \cdots \wedge p_n \Rightarrow h. \tag{6.41}$$

We can express it as follows:

$$h \Leftarrow p_1 \wedge p_2 \wedge \cdots \wedge p_n, \tag{6.42}$$

and in the Prolog language in the following way:

$$\mathsf{h} :\!- \mathsf{p}_1, \mathsf{p}_2, \ldots, \mathsf{p}_n. \tag{6.43}$$

Thus, our clauses from the set (6.39) can be written as:

$$\begin{aligned} &student(x) \Leftarrow on\_leave(x), \\ &student(x) \Leftarrow attends(x, y) \vee course(y), \end{aligned} \tag{6.44}$$

and in the Prolog program they can be written as the axioms-principles:

```
student(X) :- on_leave(X).     /* Axioms-principles in knowledge base */
student(X) :- attends(X,Y),course(Y).
```

Now, if we add to the Prolog program the following axioms-"facts":

```
course(Logic).                 /* Axioms-facts in knowledge base */
attends(John Smith, Logic).
```

[20] As we have mentioned above, a Horn clause contains at most one positive literal. The clauses in (6.39) are Horn clauses, because each one contains only one positive literal *student*(x).

then, after asking the following question to the system:

```
?- student(John Smith).        /* Characters "?-" are written by the system */
```

the system answers us:

```
Yes
```

In knowledge bases of reasoning systems formulas are often stored in a special form, which is called a *clause form*. It is especially convenient when using the resolution method. In such a representation a clause is replaced by the set of its literals. For example, the clause form for our set of clauses (6.39) is defined in the following way:

$$\{\{student(x), \neg on\_leave(x)\},\ \{student(x), \neg attends(x,y)\ ,\ \neg course(y)\}\}. \tag{6.45}$$

If we use such a representation, then a knowledge base can be constructed as one big set consisting of clauses in the form of sets of literals.[21]

6.5 Reasoning as Symbolic Computation

In this section we discuss an approach to reasoning treated as *symbolic computation. Abstract Rewriting Systems, ARSs*, are formal models of symbolic computation. They can be divided into *Term Rewriting Systems, TRSs*, String Rewriting Systems, and Graph Rewriting Systems. Since symbolic computation is "implemented" with the help of Term Rewriting Systems, this type of ARS is discussed in this section.[22] Here a term means an element of the world description in the form of a constant,[23] a variable, or a function.

Reasoning as symbolic computation relates to the *physical symbol system hypothesis* introduced by Newell and Simon. This hypothesis, as has been discussed in Chap. 1, reduces reasoning to automatically transforming (let us say *rewriting*) expressions, which are built out of symbols. This automatic transformation takes

[21] If we combine clauses belonging to various formulas, we should, once more, rename variables so that symbols of variables which belong to different formulas are also different.

[22] String Rewriting Systems and Graph Rewriting Systems are formal models for AI systems based on generative grammars. They are introduced in Chap. 8.

[23] In the formal definition of a term, a constant does not occur explicitly, because a zero-argument function is treated as a constant.

place without a semantic interpretation of symbols and expressions.[24] Instead, it is based on applying *term rewriting rules*.[25]

Now, we define exemplary rewriting rules for an addition operation based on the Peano axiomatic system,[26] in which addition is defined as an operation fulfilling the following conditions:

$$m + 0 = m, \qquad m + S(n) = S(m + n), \tag{6.46}$$

where S is the *successor* operation (generating the next natural number). We can interpret these rules in the following way. If we add 0 to any number, then nothing changes. The addition of m and a successor of n is equal to the result of adding m and n and taking successor of the result of this addition. If we define addition in such a way, we can define successive natural numbers as follows:

$$\begin{aligned} \mathit{0} &\equiv 0, \\ \mathit{1} &\equiv S(0), \\ \mathit{2} &\equiv S(1) \equiv S(S(0)), \\ \mathit{3} &\equiv S(2) \equiv S(S(1)) \equiv \underbrace{S(S(S}_{3\ times}(0))), \\ &\ldots \end{aligned} \tag{6.47}$$

Now, we can define the addition operation with the help of two rules corresponding to definition (6.46):

$$r_1 : A(m, 0) \rightarrow m, \qquad r_2 : A(m, S(n)) \rightarrow S(A(m, n)), \tag{6.48}$$

where $A(m, n)$ means $m + n$. The expression on the left of an "arrow" is called the *left-hand side of a rule* and an expression on the right of an "arrow" is called the *right-hand side of a rule*. An application of a rule to a term during its rewriting consists of matching the left-hand side of a rule to the term and transforming the term according to the right-hand side of the rule.

The rule r_2 can be interpreted in the following way. If a term (or its part) is of the form of an operation A for two arguments (i.e., $A(\ldots, \ldots)$) and the first one is of any form (i.e. m) and the second argument is of the form of the successor of an expression n (i.e., $S(n)$), then replace this term (or the part matched) with a term of the form: the

---

[24] This does not mean that these symbols and expressions have no meaning. Adherents of the physical symbol system hypothesis only claim that during *automated reasoning* referring to meaning (semantics) is not necessary.

[25] Applying the rules of inference in logic introduced in the previous sections is a good analogy to applying term rewriting rules.

[26] Giuseppe Peano defined arithmetic operations for natural numbers with the help of one constant—*zero*—and one operation—*successor*.

successor (i.e., $S(\ldots)$) of the result of the operation A (i.e., $S(A(\ldots, \ldots)))$, which has been applied to expressions m and n (i.e., $S(A(m, n))$).

Let us add 2 to 3, which means transforming the expression $2+3$, strictly speaking the expression: $A(S(S(0)), S(S(S(0))))$, with the help of our term rewriting rules.

$$\begin{aligned} A(S(S(0)), S(S(S(0)))) &\overset{2}{\rightarrow} S(A(S(S(0)), S(S(0)))) \\ &\overset{2}{\rightarrow} S(S(A(S(S(0)), S(0)))) \\ &\overset{2}{\rightarrow} S(S(S(A(S(S(0), 0))))) \\ &\overset{1}{\rightarrow} S(S(S(S(S(0))))). \end{aligned} \tag{6.49}$$

We have transformed the initial expression to a correct final expression, which denotes *the fifth successor of zero* and means the number 5 according to the definition (6.47). Indices above arrows are the indices of the rules applied. Let us notice that the goal of such a rewriting process can be defined as the reduction of the second element of the sum to zero by the multiple use of the second rule followed finally by an application of the first rule, which eliminates the operation A. The reduction of the second element of the sum (following rule r_2) consists of *taking away* a single character S from this element and *embedding* the character S before the symbol of the addition operation A.

We have used the words: *taking away* and *embedding* deliberately in order to describe the process of applying rules r_1 and r_2 to transform the expression $2 + 3$. These words convey the essence of *symbolic computation*, which consists of rewriting terms without analyzing what they mean. Let us imagine somebody who does not know what addition is. If we teach this person how to apply rules r_1 and r_2 and how to represent natural numbers (6.47), then this person will not know that he/she is adding two numbers when transforming the expression $2 + 3$ according to these rules.[27]

Since we do not use the meaning of an expression in symbolic computation, we should ensure the formal correctness of such reasoning. Let us return to our term rewriter, who does not know that he/she is adding. If he/she does not know this, then maybe he/she does not know when rewriting should be finished. (For example, if the term rewriter finishes after two steps, then he/she does not know whether the expression generated is the final expression, i.e., the result.) Therefore, we introduce the following principle in rewriting systems: a rewriting process should be continued as long as the latest expression can be transformed. In other words, a rewriting process is finished only if there is no rule of the rewriting system which can be applied, i.e., the system cannot match the left-hand side of any rule to any part of the latest transformed expression. An expression that cannot be transformed with any rewriting rule is called a *normal form.* Let us notice that the final expression in our example is a normal form.

The second important issue can be formulated in the following way. Does the order of rule applications influence the final result of rewriting? In our example such a problem does not appear, since the order of rule applications is determined by

[27] Such a situation is analogous to the "Chinese room" thought experiment discussed in Chap. 1.

their form.[28] However, in the general case there is more than one possible sequence of rule applications. If we constructed a rewriting system, which allowed us to add more than two addends (in one expression), then there would be more alternative sequences for obtaining the final result. Still, we should obtain the same final result of such an addition regardless of the order of term rewriting. If the final result of rewriting terms in a system does not depend on the rewriting order, then we say that the rewriting system has the *Church-Rosser property*.[29] Of course, this property of the system is required.[30]

At the end of our considerations concerning term rewriting systems, we show a way of implementing our rewriting rules (6.48) for adding numbers according to the Peano axioms (6.46) in the *Lisp* programming language. Let us define a function add (m n), which corresponds to the operation $A(m, n)$. This function checks whether the second addend equals zero. If yes, then the function gives m as the result according to the rule r_1. If no, then it gives the successor of the sum of the first addend and the predecessor of the second addend, according to rule r_2.

```
(defun  add (m n)
        (if (zerop n)
            m
            (successor (add m (predecessor n)))
        )
)
```

Now, we introduce one of the most popular term rewriting systems, i.e., *lambda calculus* (λ-*calculus*)[31], which was developed by Alonzo Church and Stephen C. Kleene in the 1930s.[32] This calculus was used for answering the so-called *Entscheidungsproblem* (decision problem) proposed by David Hilbert during the International Congress of Mathematicians in Bologna in 1928. The *Entscheidungsproblem* is the problem of the solvability of First Order Logic. It can be formulated with the following question: Is there an effective general computation procedure that can be used for deciding whether any FOL formula is a valid formula (tautology)? The problem is of great importance, because it relates to an old philosophical issue: Can any (deductive)

[28] If we perform an addition $m + n$, then always we have to apply n times the rule r_2 and at the end we have to apply once the rule r_1. This is the only possible sequence of operations.

[29] John Barkley Rosser, Sr.—a professor of mathematics, logic, and computer science at the University of Wisconsin-Madison, a Ph.D. student of Alonzo Church. An author of excellent publications in logic, number theory, and ballistics.

[30] For example, a "system of cooking" does not have the Church-Rosser property, which is known to the author of this monograph from personal experience. Any time I replace a model sequence of culinary operations (by mistake, of course) by another one, I obtain an unacceptable result.

[31] We consider here the untyped lambda calculus.

[32] The second such popular system, namely *combinatory logic*, was introduced by Moses Schönfinkel in 1920. The ideas of Schönfinkel were further developed by Haskell Brooks Curry in 1927.

reasoning be automatized? In 1936 Church showed with the help of lambda calculus that the answer to this question is negative [48].[33]

Lambda calculus is a formal system in which theorem proving has the form of *symbolic computation*. As we have discussed above, during such a computation, we transform (rewrite) expressions without analyzing what they mean.

Now, let us try to answer the question: What do we have in mind saying "rewriting expressions without analyzing what they mean"? Let us notice that if we have to transform an expression $2 * 3$ or an expression 3^2 (in the sense of obtaining a result), then we should know how to multiple numbers and how to exponentiate numbers.[34] However, expressions of the form $x * y$ or x^y do not give any information on how they should be computed. By contrast, in lambda calculus expressions contain information on how they should be computed. This concerns a numeric calculus (arithmetic operations, etc.) as well as a symbolic calculus (such as logical reasoning). Moreover, these expressions are defined in such a way that they, somehow, *compute themselves automatically*. (We show such a computation process below.)

Now, let us introduce the fundamentals of lambda calculus. For a novice reader its notions and notational conventions can seem to be a little bit peculiar, but as we see further on they are convenient for the purpose of symbolic computation. Let us begin by introducing an infinite (countable) set of *variables*[35]:

$$\mathcal{V} = \{a, b, c, \ldots, z, a_1, b_1, c_1, \ldots, z_1, a_2, b_2, \ldots\}. \tag{6.50}$$

Such *variables* are treated as functions in lambda calculus. Thus, for example, instead of using the following notation:

$$f(z), \tag{6.51}$$

we denote it as follows:

$$fz, \tag{6.52}$$

which can be interpreted in two ways: either as the process of computing fz or as a result of such a computation.[36] An operation fz is called the *application* of f to (an argument) z.

[33]Independently, A.M. Turing showed the same with the help of his *abstract machine* (now called a *Turing machine*) in 1937 [306].

[34]In our example a term rewriter did not know how to add numbers, but he/she had two rules which allowed him/her to *perform addition* (in an automatic way).

[35]We define notions in a simplified way, since we introduce lambda calculus in this section in an informal way. Therefore, we omit the subtle difference between *lambda expression* and *lambda term* (which is the equivalence class of lambda expressions), etc. Formal definitions which concern lambda calculus are contained in Appendix C.3.

[36]Let us notice that we have already met such a situation, when we considered term rewriting systems. For example, an expression $S(S(S(0)))$ can be interpreted either as the process of computing the third successor of the constant 0 or as a result of this computation, i.e., 3—see formula (6.47).

A *lambda abstraction* (λ-*abstraction*) is the second basic notational convention of lambda calculus. Let $M \equiv M[x]$ be an expression which contains x (i.e., it depends on x), for example $x + 5$. Now, if we want to define it as a function, then we use a convention of the form $x \mapsto M[x]$ and we write it as follows:

$$x \mapsto x + 5. \tag{6.53}$$

In lambda calculus, we use a special convention for defining a function, which is of the form $\lambda x.M[x]$. Therefore, we write our function as follows:

$$\lambda x.x + 5. \tag{6.54}$$

The expression $x + 5$ is called the *body* of the lambda abstraction (function).

The constructs introduced above, i.e., variables, expressions defining applications, and expressions denoting a lambda abstraction are called *lambda expressions.*

After defining our example function as a lambda abstraction, we can use it for some argument, i.e., we can perform an *application* of the function to some argument, let us say to the number 7. According to convention (6.52), we can write this as follows:

$$(\lambda x.x + 5)7, \tag{6.55}$$

where f from formula (6.52) corresponds to our function $(\lambda x.x + 5)$ and z, which occurs in formula (6.52), corresponds to the argument 7. Of course, such an application consists of replacing the parameter x which is contained in the body by the argument 7. Thus, we obtain:

$$7 + 5. \tag{6.56}$$

Let us denote the operation of replacing all occurrences of x in an expression M by N in the following way:

$$M[x := N]. \tag{6.57}$$

Then, our operation of transforming expression (6.55) into expression (6.56) can be defined with the help of the following rewriting rule:

$$(\lambda x.M)N \rightarrow_{\beta} M[x := N]. \tag{6.58}$$

This rule, called a β-*reduction*, is the basic rewriting rule in lambda calculus and the fundamental mechanism of symbolic computing in this calculus.

Now, we extend the definition of a lambda abstraction to multi-argument functions. A multi-argument function is defined with the help of successive applications (iteratively) performed for subsequent arguments.[37] For example, a function of two arguments,

[37] This technique is called *currying* (as a reference to H.B. Curry) or *Schönfinkelisation* (referring to M. Schönfinkel).

$$(x, y) \mapsto x + 2 * y, \tag{6.59}$$

is represented by the following lambda abstraction (the body of the lambda abstraction is put in square brackets for clarity of notation):

$$\lambda x.(\lambda y.[x + 2 * y]). \tag{6.60}$$

Let us compute it for a pair $(x, y) = (3, 7)$. The computation is performed by the application

$$\lambda x.(\lambda y.[x + 2 * y])\, 3\, 7 \tag{6.61}$$

in the following steps. Firstly, an application is performed for the argument x (the place in which the application is performed is underlined):

$$\lambda \underline{x}.(\lambda y.[\underline{x} + 2 * y])\, \underline{3}\, 7 \tag{6.62}$$

and we obtain:

$$\lambda y.[3 + 2 * y]\, 7. \tag{6.63}$$

(Let us notice that λx has *disappeared*, since after replacing the variable by the argument 7 there is no x in the body.) Secondly, we perform an application for y:

$$\lambda \underline{y}.[3 + 2 * \underline{y}]\, \underline{7}, \tag{6.64}$$

and we obtain the following final result:

$$3 + 2 * 7. \tag{6.65}$$

Instead of writing $\lambda x.(\lambda y.M))$, we use a simplified notation: $\lambda x.\lambda y.M$ or an even briefer convention: $\lambda xy.M$. (Of course the convention can be used for n arguments: $\lambda x_1 x_2 x_3 \ldots x_n.M$.)

Let us introduce the following notions. An operator λ *binds* variables in a similar way as quantifiers do. A variable x which is in the scope of the binding of the operator λ in an expression M is called a *bound variable* of this expression. Otherwise, the variable is called a *free variable*. For example, in expression (6.60) both variables x and y are bound. In the expression $\lambda xy.x + y + z$ variables x and y are bound and variable z is a free variable.

Our examples of term rewriting have finished with expressions of the forms (6.56) and (6.65). However, we have promised to show the reader how lambda expressions *compute themselves automatically*. At the beginning of this section we have introduced examples in a simplified form, because we have not wanted to complicate the lambda calculus notations, which are not intuitive for a beginning reader. Now, since the reader knows these notations, we can introduce more complex forms which allow us to illustrate *self-computing* of lambda expressions. For this purpose we use an example of arithmetic of natural numbers, which has been discussed for abstract rewriting systems.

Firstly, we introduce a representation of natural numbers in the form of *Church numerals*:

$$\begin{aligned}
0 &\equiv \lambda sx.x, \\
1 &\equiv \lambda sx.sx, \\
2 &\equiv \lambda sx.s(sx), \\
3 &\equiv \lambda sx.\underbrace{s(s(s}_{3\ times}x)), \\
&\dots
\end{aligned} \tag{6.66}$$

One can easily notice that such a representation is analogous to the definition of natural numbers in the Peano axiomatic system (6.47).

The operation *successor* is defined in lambda calculus in the following way:

$$S \equiv \lambda nsx.s((ns)x). \tag{6.67}$$

For example, let us apply this operation to the number 2, i.e., let us perform the symbolic computation $S\,2$. Firstly, however, let us notice that we have used the same variables s and x for defining expressions given by formulas (6.66) and (6.67). These expressions are different from one another. Thus, if we used variables s and x of both formulas in the same expression, then such an accidental combination would be improper.[38] On the other hand, replacing variables s and x by variables t and y, respectively, in the definition of the number 2 in (6.66) does not change its definition. Therefore, we perform such a replacement in the following way:

$$2 \equiv \lambda ty.t(ty). \tag{6.68}$$

Such a treatment of *bound variables* of expressions as equivalent ones (e.g., $\lambda sx.s(sx)$ and $\lambda ty.t(ty)$) is called an *alpha-conversion.*

Now, we can compute $S\,2$. Firstly, let us replace S by a lambda expression according to (6.67), and let us replace 2 by a lambda expression according to (6.68):

$$S\,2 \equiv \lambda nsx.s((ns)x)\lambda ty.t(ty)\ . \tag{6.69}$$

Now, we perform an application of the first expression to the second one (β-reduction), i.e., the variable n in the first expression is replaced by the second expression $\lambda ty.t(ty)$[39]:

[38] We have met an analogous problem when we discussed methods based on FOL in a previous section—cf. formulas (6.34) and (6.35).

[39] Both the variable which is replaced and the expression which replaces it are underlined.

$$S\,2 \equiv \lambda \underline{n}sx.s((\underline{n}s)x)\underline{\lambda ty.t(ty)} \rightarrow_\beta \lambda sx.s((\lambda ty.t(ty)s)x). \tag{6.70}$$

Let us notice that the variable n placed immediately after λ disappears, since after the replacement there are no variables n in the body of the expression. Then, the variable t in the expression $\lambda ty.t(ty)$ is replaced by the variable s, i.e., we perform the second application (β-reduction):

$$\lambda sx.s((\lambda \underline{t}y.\underline{t}(\underline{t}y)\ \underline{s})x) \rightarrow_\beta \lambda sx.s(\lambda y.s(sy)x). \tag{6.71}$$

This time, the variable t placed immediately after the internal λ disappears, since after the replacement there are no variables t in the body of the expression. Finally, the variable y in the expression $\lambda y \cdot s(sy)$ is replaced by the variable x, i.e., we perform the last application (β-reduction):

$$\lambda sx.s(\lambda \underline{y}.s(s\underline{y})\underline{x}) \rightarrow_\beta \lambda sx.s(s(sx)) \equiv 3. \tag{6.72}$$

As we can see, the starting expression has *computed itself* with the help of a sequence of β-reductions. We have obtained the number *3*, according to the Church numerals (6.66), as the successor of *2*.

An implementation of the successor operation (6.67) in a dialect of the *Lisp* language (the *Scheme* language) consists, in fact, of rewriting this expression:

```
(define  succ
          (lambda(n)
                    (lambda(s)
                              (lambda(x)
                                        (s((ns)x))))))
```

We have presented the basic ideas of the lambda calculus with the help of simple examples, because of the very formal notational conventions of this system. In fact, we can define all programming constructs by lambda expressions. Kleene formulated the *Church-Turing thesis*, which can be interpreted in the following way: every effectively computable function is computable by the lambda calculus.[40] The lambda calculus is so attractive in Artificial Intelligence, since all functions, operations, etc. can be defined in a *constructive* way, i.e., its expressions contain information on how to obtain the final result. As we have shown above, such expressions *compute themselves* really.

[40]The *(universal) Turing machine* and *recursively definable functions* are models which are equivalent to the lambda calculus, according to this thesis.

An analogy between proving theorems in (intuitionistic) logic and the lambda calculus was discovered by H.B. Curry[41] and W.A. Howard,[42] independently. The analogy is known as the *Curry-Howard isomorphism.*[43]

In the first chapter we have mentioned J. McCarthy, who in 1958–1960 developed the programming language Lisp, which is based on the lambda calculus. Lisp is one of two[44] classic languages in Artificial Intelligence. It (and its various dialects) is still a very popular tool for developing AI systems.

Bibliographical Note

Monographs [7, 184, 212] are good introductions to constructing logic-based AI systems.

Prolog programming is presented in [37, 53, 162, 288].

Monographs [20, 46, 81, 177, 181, 244, 317] are recommended for logic fundamentals used in computer science.

Basic monographs in the area of rewriting systems include [14, 24], and for lambda calculus [16, 17, 135, 200]. Lisp programming is presented in [286, 304].

---

[41] Haskell Brooks Curry—a mathematician and logician, a professor at the Pennsylvania State University. David Hilbert and Paul Bernays were his doctoral advisors. The functional programming languages *Haskell* and *Curry* are named after him.

[42] William Alvin Howard—a mathematician and logician, a professor at the University of Illinois at Chicago. Author of well-known papers in proof theory.

[43] The analogy concerns *typed lambda calculus.*

[44] Prolog, introduced in Sect. 6.4, is the second classic language in AI.

Chapter 7
Structural Models of Knowledge Representation

Constructing so-called *ontologies*[1] is one of the main goals of applying *structural models of knowledge representation*, which have been introduced in Sect. 2.4. In Artificial Intelligence and in computer science, an ontology[2] is defined as a formal specification (conceptualization) of a certain (application) domain which is defined in such a way that it can be used for solving various problems (in the scope of this domain) with the help of general reasoning methods.[3] Such a specification is of the *structural* form. It can be treated as a kind of encyclopedia for the domain which contains descriptions of notions, objects, relations between them, etc.

Automated reasoning with a general technique is possible if we separate the *domain knowledge* from this generic (for a given technique) reasoning scheme. Semantic networks, frames, and scripts are typical structural models for representing domain knowledge. We introduce them in the next three sections.

When we introduce notions concerning structural models of knowledge representation, we refer to the corresponding definitions and notations of *description logics*. These logics were introduced in the 1980s and the 1990s as formal models of ontology representations in Artificial Intelligence. They are used, as well, for constructing efficient generic reasoning schemes, which are mentioned above.[4]

[1]Although there is an analogy between the notion of ontology in computer science and the notion of ontology in philosophy, we should differentiate between the two notions. In philosophy ontology is the study of being, its essential properties and its ultimate reasons.

[2]The system *Cyc*, which is developed by D. Lenat, is one of the biggest AI systems based on an ontology-based approach.

[3]Such standard reasoning methods are analogous to a universal reasoning scheme, which is discussed in a previous chapter.

[4]Description logics are introduced formally in Appendix D.

M. Flasiński, *Introduction to Artificial Intelligence*,
DOI 10.1007/978-3-319-40022-8_7

7.1 Semantic Networks

Semantic networks were introduced by Allan M. Collins and Ross Quillian in 1969 [56] as a result of their research into (natural) language understanding. They assumed that formulating knowledge in the form of a set of *notions* which relate to one another allows us to understand this knowledge better. Therefore, knowledge systems are constructed in just such a way. For example, in mathematics we introduce successive notions referring to the notions defined already. This is shown for geometry in Fig. 7.1a. Let us notice that notions which are successively introduced (from top to bottom in the figure) are particular cases of notions which have been introduced already. In other words, a new notion has all the properties of its predecessor notions and it has also new specific properties. Thus, a notion which is introduced later constitutes a *subclass* of a notion which has been introduced earlier. For example, *Trapezoid* is a subclass of *Quadrilateral*, which in turn is a subclass of *Polygon*, etc. This relation is represented by directed edges, which are labeled by *is subclass* in semantic networks. In description logics we talk about *general concept inclusion* and we denote it as follows:

$$Trapezoid \sqsubseteq Quadrilateral,$$
$$Quadrilateral \sqsubseteq Polygon, \text{ etc.}$$

We construct taxonomies in the natural sciences to systematize our knowledge in such a way. For example, a part of such a taxonomy defined for the notion of *Animals* is shown in Fig. 7.1b. Let us notice that concept inclusion is also defined in this case, i.e., classes which are placed at lower levels are subclasses of certain classes placed above.

Sometimes we define an ontology (or its parts) in such a way that new concepts are constructed with the help of a few simple elementary notions. Such simple elementary notions are called *atomic concepts*. For example, for a "color ontology" we can assume the following atomic concepts, which correspond to primary colors[5]: *red* (*R*), *green* (*G*), *blue* (*B*). Then, we can define successive (complex) concepts: *yellow* (*Y*) ≡ *red mixed with green*[6]; *violet* (*V*) ≡ *red mixed with blue*; *white* (*W*) ≡ *red mixed with green mixed with blue*. In a description logic such a definition of complex concepts (here, colors) is expressed in the following way:

$$Y \equiv R \sqcup G,$$
$$V \equiv R \sqcup B,$$
$$W \equiv R \sqcup G \sqcup B.$$

[5]In the RGB (Red-Green-Blue) color model.

[6]We assume that secondary colors are obtained with the help of additive color mixing, i.e., by mixing visible light from various colored light sources.

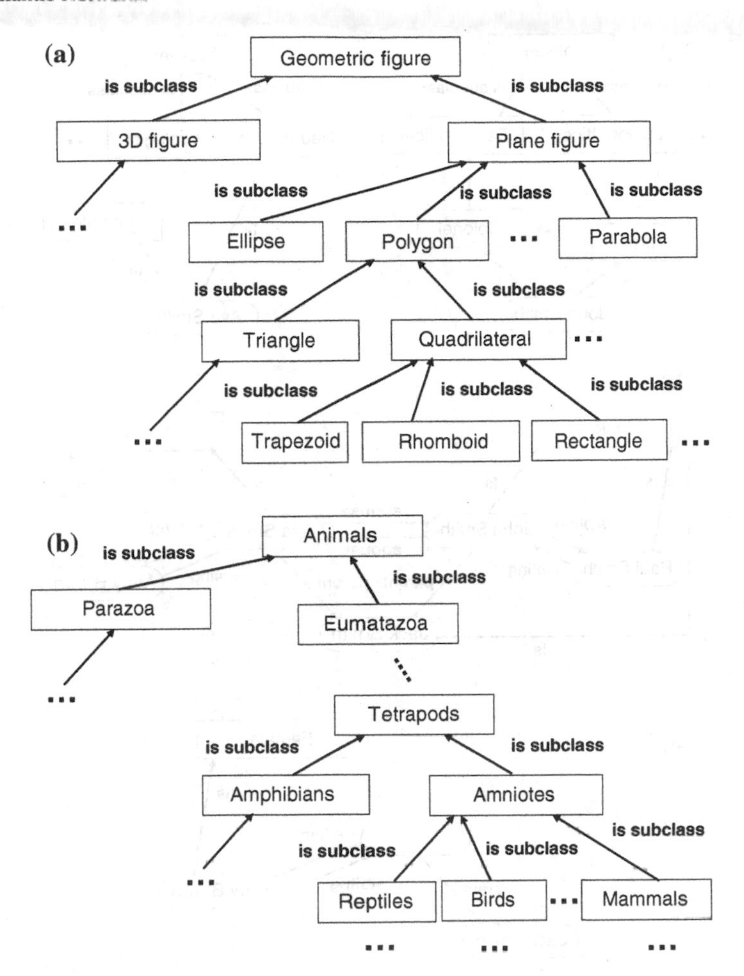

Fig. 7.1 Examples of simple semantic networks (ontologies): **a** in geometry, **b** in biology

Objects are the second generic element of semantic networks. Objects represent individuals of a certain domain. We say that objects are *instances* (examples) of a certain class (concept). For example, an object *John Smith* (a specific person having this name and this surname, who is identified in a unique way by a Social Security number) can be an instance of a class *American*. The existence of such an instance (in this case, a person) is represented by *American(John Smith)* in descriptive logics. Let us notice that a class (concept) can be treated as a set of objects. We introduce a relation *is* (*is a*) in semantic networks for denoting the fact that an object belongs to a class.

For example, a part of a semantic network which contains two objects *John Smith* and *Ava Smith*, and their characteristics is shown in Fig. 7.2a. As we can infer from this representation, the object *John Smith* is a *male* and a *colonel*. The class *colonel*

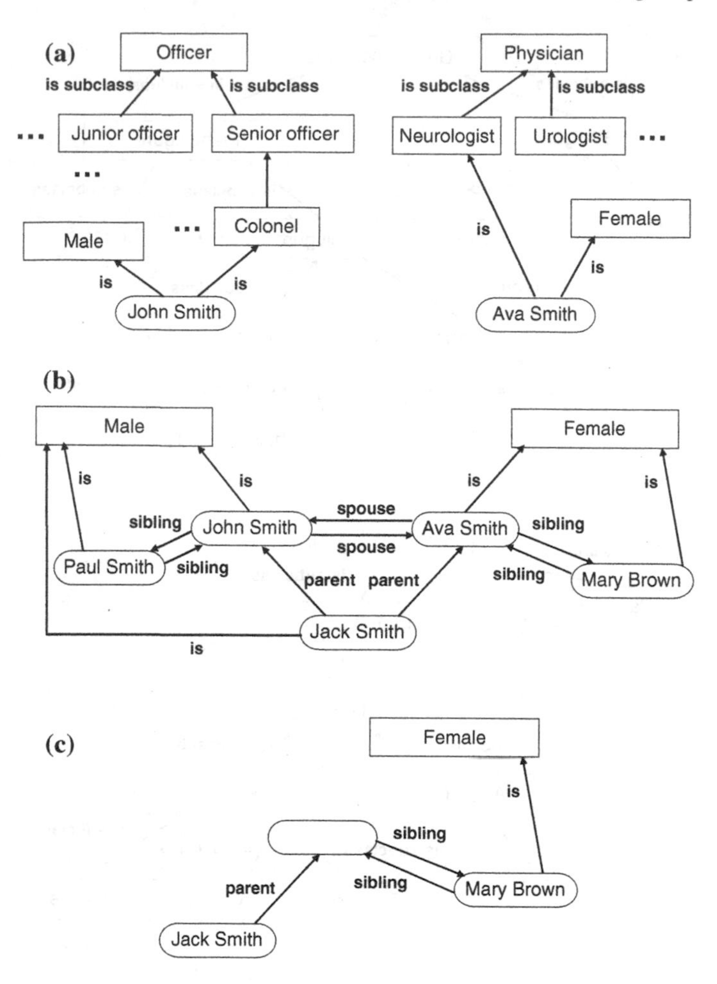

Fig. 7.2 Examples of semantic networks: **a** containing objects, **b** defining roles, and **c** a representation of a query in a system which is based on a semantic network

is a subclass of the class *senior officer* and the class *senior officer* is a subclass of the class *officer*. The object *Ava Smith* is a *female* and a *physician*, strictly speaking a *neurologist*.

Roles are the third generic element of semantic networks. Roles are used for describing relations between objects (sometimes also between classes). For example, we can introduce roles *spouse*, *parent*, and *sibling* in order to represent a genealogical knowledge. A semantic network which represents a part of a genealogy ontology is shown in Fig. 7.2b. We see that, for example, *John Smith* and *Ava Smith* are the parents of *Jack Smith*. In descriptive logics some roles represented by the semantic network shown in Fig. 7.2b can be defined in the following way:

$$parent(Jack\ Smith,\ John\ Smith),$$
$$parent(Jack\ Smith,\ Ava\ Smith),$$
$$spouse(John\ Smith,\ Ava\ Smith), \text{etc.}$$

Similarly to the case of concepts, we can define complex roles on the basis of simpler *atomic roles* (and concepts). For example, the role *grandfather* can be defined as a *parent* of a *parent* and a *male*. The role *aunt* can be defined as a *sibling* of a *parent* and a *female*.

A variety of reasoning methods have been developed in order to extract knowledge from semantic networks. One of the simplest is the *method of structural matching*. For example, if we would like to verify the validity of the following proposition:

Mary Brown is an aunt of *Jack Smith*,

then we should define a general structural pattern which represents such a proposition and then we should check whether the pattern can be matched to some part of our semantic network. A structural pattern for our proposition is shown in Fig. 7.2c. Let us notice that some elements are fixed and some elements are not fixed. The object that is a parent for *Jack Smith* is not fixed. If we denote it by *X*, then we have:

$$parent(Jack\ Smith,\ X),$$

since when we look for an aunt of *Jack Smith*, it is unimportant whether she is a sibling of his father or his mother. Of course, one can easily see that such a pattern defined for the verification of our proposition can be matched to a part of the semantic network shown in Fig. 7.2b.

The efficiency of pattern matching methods is a crucial problem of reasoning in semantic networks. A semantic network is a graph from a formal point of view. As we know from computational complexity theory, this problem, i.e., graph pattern matching, is of the non-polynomial complexity. Therefore, at the end of the twentieth century intensive research was carried out for the purpose of constructing methods of efficient graph processing. We discuss such methods in Chap. 8, in which graph grammars are introduced.

7.2 Frames

Frames were introduced by Marvin Minsky in 1975 [203]. As we have mentioned in Sect. 2.4, a frame system can be treated as an extension of a semantic network. The extension consists of replacing the nodes of a network by complex structures called *frames*, which allow us to characterize objects and classes in a detailed way. In the case of objects we talk about *object frames* and in the case of classes we talk about *class frames*.

A frame consists of *slots*, which are used for describing features and properties of an object/concept precisely. Each slot consists of *facets*. For example, if we construct a frame of an object which describes some device, then it can be characterized by certain properties such as voltage, temperature, pressure (of gas within the device), etc. Each property is represented by a slot. However, for each property various "aspects" can be defined. For example, for voltage we can define a current value, a unit (mV, V, kV, MV, etc.), an accuracy of measurement, a measuring range, etc. In a slot of a frame facets are used for storing such aspects.

Some default types of *facets* are used in frame systems. The most important ones include the following types.

- *VALUE*—the current value of the slot is stored in this facet.
- *RANGE*—this contains a measuring range or a list of values of the slot which are allowed.
- *DEFAULT*—this contains the default value of the slot. We can assume that this value is valid, if e.g., the facet of a type *VALUE* is not known at the moment.
- *belongs to a class* (for an object frame)—this contains a pointer to a class to which this object belongs.
- *is a subclass of* (for a class frame)—this contains a pointer to a class for which this class is a subclass.

Inheritance is a basic reasoning mechanism in frame systems. It is based on a fundamental property of ontologies, which is that subclasses *inherit* all features of superclasses (in the sense of *general concept inclusion*). Due to this property, if knowledge which concerns a certain class is updated/modified then it can be propagated to subclasses of this class (and to objects which belong to this class). This mechanism is enhanced, additionally, by the fact that an object can belong to more than one class. For example, the object *John Smith* belongs to two classes, *Male* and *Colonel*, in Fig. 7.2a. An analogous property concerns a class which can be a subclass of many classes. In such case, we talk about *multiple inheritance*.

Demon procedures, also called *demons*, are the second reasoning mechanism in frame systems. A *facet* of a slot can be not only of a static form (data, pointer to other frame, etc.), but also of the dynamic form of a demon. This peculiar name for these procedures comes from their idea, which is described by some authors as lying in wait to be invoked. If a demon is invoked by "jostling" its frame, e.g., by demanding some information, then it is awoken and it begins to operate. Similarly to the case of static facets, there are many types of demons. The most popular types include the following cases.

- *if-needed*—activated if a value of a facet is not known and we want to acquire it. Then, a demon tries to acquire/compute it from other frames.
- *if-added*—triggered if a new value has been added to a facet.
- *if-updated*—activated if a value in a facet has been updated.
- *if-removed*—triggered if a value has been removed from a facet.
- *if-read*—activated if a value has been read from a facet.
- *if-new*—triggered if a new frame is generated.

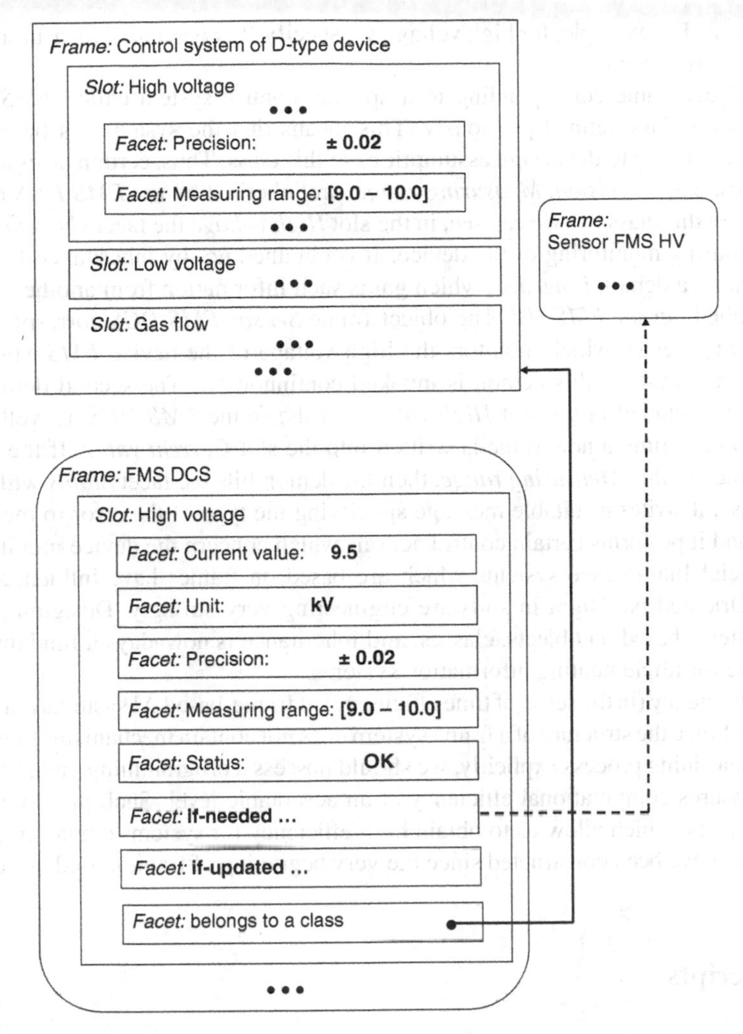

Fig. 7.3 Part of a model defined for a frame-based system for controlling complex industrial-like equipment

Let us analyze a part of a model defined for a frame-based system for controlling complex industrial-like equipment,[7] which is shown in Fig. 7.3. A class frame *Control system of D-type device* contains design characteristics of the device such as *High voltage*, *Low voltage* (a device has two types of electric power supply), *Gas flow*, etc. Each parameter relates to a slot. *Facets* of a slot specify these parameters in a

[7]This example has been defined on the basis of the documentation of the project *Generic Requirements Model for LHC Control Systems*, which was coordinated by the author and Dr. Axel Daneels at *Conseil Européen pour la Recherche Nucléaire* (*CERN*) in Geneva in 1997–1998.

precise way. For example, for high voltage we specify *Precision* (of a measurement), *Measuring range*, etc.

An object frame corresponding to a specific control system called *FMS DCS* belongs to a class defined previously. This means that the system has been constructed according to the design assumptions of this class. Thus, certain design characteristics, e.g., *Precision, Measuring range*, are inherited by the *FMS DCS* object frame from this class. As we can see, in the slot *High voltage* the facet *Current value* is filled during monitoring of the device. It is obtained not by inheritance from its class, but by a demon *if-needed*, which gains such information from another object frame called *Sensor FMS HV*. The object frame *Sensor FMS DCS* corresponds to a measuring device which monitors the high voltage of the device *FMS*. (In such a monitoring system, this demon is invoked continuously.) The second demon *if-updated* is included in the slot *High voltage* of the frame *FMS DCS* as well. It is triggered each time a new value is written into the slot *Current value*. If the value is contained within *Measuring range*, then the demon fills the facet *Status* with *OK*. Otherwise, it writes a suitable message specifying the type of the error to the facet *Status*, and it performs certain control actions which concern the device monitored.

Artificial Intelligence systems which are based on frames have influenced the Object-Oriented paradigm in software engineering very strongly. Designing software systems based on objects, classes, and inheritance is nowadays a fundamental technique for implementing information systems.

The efficiency (in the sense of time efficiency) of frame-based AI systems is a basic problem. Since the structure of a frame system does not contain mechanisms that control the reasoning process explicitly, we should possess a programming environment which ensures computational efficiency at an acceptable level. Such programming environments, which allow us to obtain high efficiency for systems containing a lot of frames, have been constructed since the very beginning of frame-based systems.[8]

7.3 Scripts

Scripts were proposed by Roger Schank and Robert P. Abelson in 1977 [264] for Natural Language Processing, NLP. The model is based on the following observation in psychology. If we want to understand a message which concerns a certain event (gossip told by a friend, coverage of a broadcast parliamentary debate, etc.), then we refer to a generalized pattern that relates to the type of this event. This pattern is constructed on the basis of earlier similar events. Then, it is stored in our memory. For example, if a child goes with her/his mom to a local clinic yet again, then

[8]For example, an AI control system containing about 100 class frames and more than 3000 object frames, which has been implemented for the high-energy physics experiment under the supervision of the author and Dr. Ulf Behrens, has processed data in real time (Flasiński M.: Further Development of the ZEUS Expert System: Computer Science Foundations of Design. *DESY Report 94-048*, Hamburg, March 1994, ISSN 0418-9833).

she/he knows from experience that this event consists of the following *sequence of elementary steps*: entering the clinic, going to the reception desk, waiting in a queue, entering the doctor's surgery, being asked "Where does it hurt?" by a doctor, having a checkup, being written a prescription by a doctor and (at last!) exiting the clinic. (When I was a child, then there was always also an obligatory visit to a toy shop.)

Such a representation defines the typical course of a certain event. Knowledge of such a form can be used in an AI system for predicting the course of events or for reasoning: What should be done in order to achieve a specific goal? In the case of Natural Language Processing problems, if some information is lacking, we can "guess" it with the help of a script. However, if a message is not *structured* in a proper way (e.g., the message is chaotic, the chronology of an event is disturbed), then matching an ambiguous description of a specific event to a pattern event (script) can be difficult.

Summing up, a *script* can be defined as a structural representation which describes an event of a certain type in a generalized/stereotyped way,[9] taking into account a particular context. The definition of a script is formalized with the help of the following elements.

- *Agents* are objects which can impact on other objects and which can be influenced by other objects. In the example of visiting a local clinic, a *child*, a *mother*, a *doctor*, etc. are agents.
- *Props* are things which occur in a script. In our example a *clinical thermometer* and a *prescription* are props.
- *Actions* are elementary events which are used for constructing the whole event. In our example *writing* a prescription and *exiting* the clinic are actions.
- *Preconditions* are propositions which have to be true at the moment of starting an inference with the help of a script, e.g., *a child is ill*, *a local clinic is open*, etc.
- *Results* are propositions which have to be true at the moment of ending an inference with the help of a script, e.g., *a prescription is written out by a doctor*.

In a standard Schank-Abelson model actions are defined in a hierarchical, two-level way. *Scenes* are defined at a higher level (e.g., *doctor is giving a child a checkup*) and are represented with the help of *conceptual dependency graphs*, which have been introduced in Sect. 2.4. *Elementary acts*, which correspond to nodes of CD graphs, are defined at a lower level. Elementary acts are constructed with *conceptual primitives* (introduced in Sect. 2.4) such as *PTRANS*, which denotes "change the physical location of an object", *SPEAK*, which denotes "produce a sound", etc.

At the end of the chapter, let us consider a (simplified) example of constructing a script. Let us assume that Paul has told the following gossip to me.

> Mark was angry with Tom. Therefore, Mark backbit Tom during a party. When Tom found out about it, he became offended at Mark. Paul decided to reconcile Mark with Tom. So, he invited them to a pub. Mark, after drinking a few beers, apologized to Tom for backbiting. As a result Tom mended fences with Mark.

---

[9] Such a stereotyped sequence of elementary steps which define an event is sometimes called a *stereotyped scenario*.

This story can be represented with the help of the object shown in the lower left-side part of Fig. 7.4.

The next day I read the following article on a dispute between Ruritania and Alvania in my favorite newspaper.

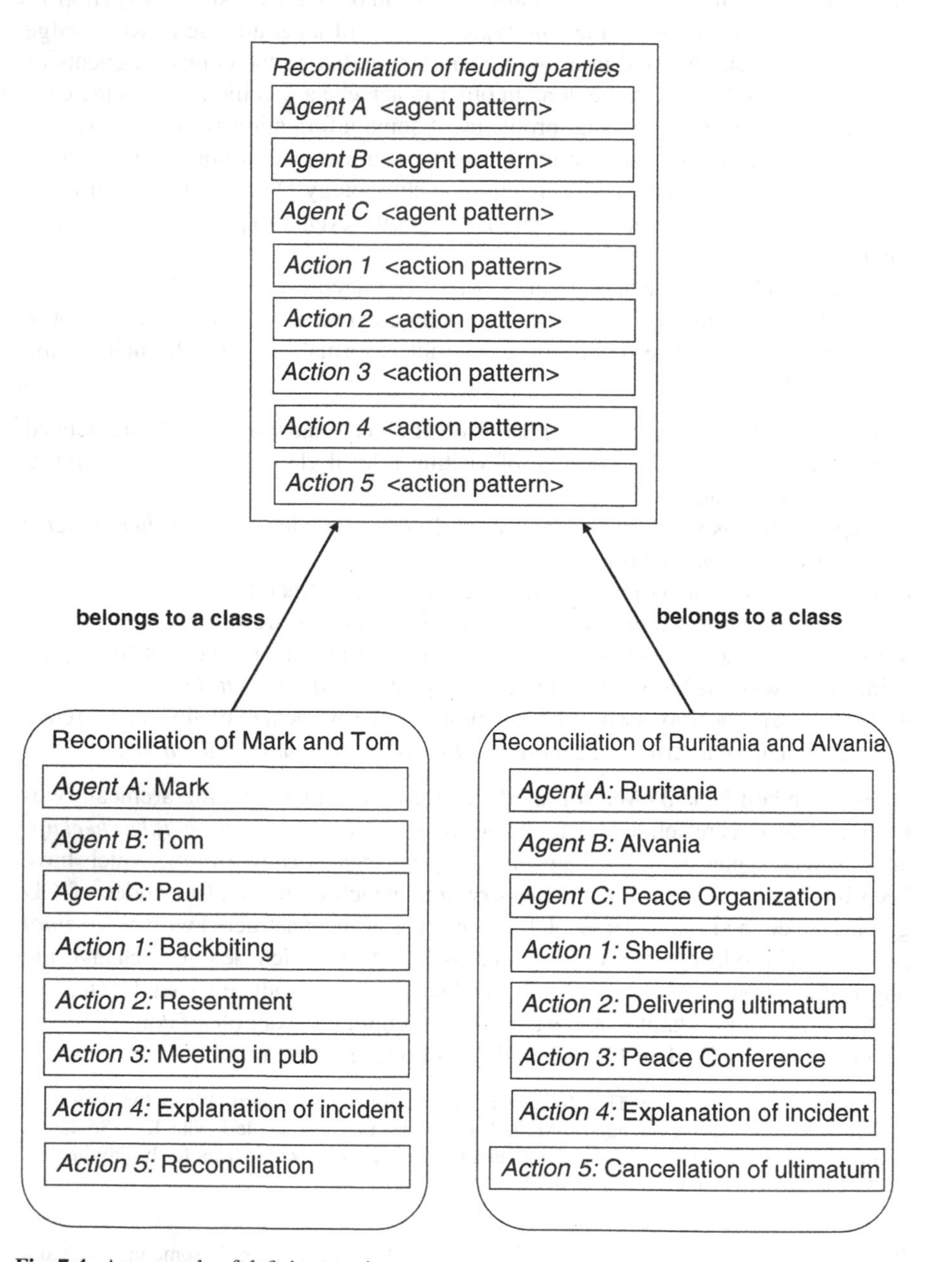

Fig. 7.4 An example of defining a script

Two weeks ago the Ruritanian troops attacked the territory of Alvania. The next day the ambassador of Alvania delivered an ultimatum to the foreign minister of Ruritania. Then, the Peace Organization decided to organize a peace conference. Representatives of Ruritania apologized to representatives of Alvania and explained it was a misunderstanding. As a result, Alvania canceled its ultimatum.

This article can be represented with the help of the object shown in the lower right-side part of Fig. 7.4. For both objects, we can define a generalized description of a reconciliation of feuding parties, which is, in fact, their class of abstraction. Such a class is, according to a definition introduced above, a *script*. This script contains a stereotyped sequence of elementary acts of the scenario, as shown in Fig. 7.4.

Bibliographical Note

Structural models of knowledge representation are introduced in classic monographs concerning Artificial Intelligence [189, 211, 315]. This area is discussed in detail in [36, 280, 281]. Monographs [130, 266] are good introductions to descriptive logics.

Chapter 8
Syntactic Pattern Analysis

In *syntactic pattern analysis*, also called *syntactic pattern recognition* [97, 104], reasoning is performed on the basis of structural representations which describe things and phenomena belonging to the world. A set of such structural representations, called (structural) patterns, constitutes the database of an AI system. This set is not represented in the database explicitly, but with the help of a formal system, which generates all its patterns. A *generative grammar*, introduced in Chap. 1 and Sect. 2.5 during a discussion of the main ideas of Chomsky's theory, is the most popular formal system used for this purpose. The grammar generates structural patterns by the application of *string rewriting rules*,[1] which are called *productions*. Thus, a (string) generative grammar constitutes a specific type of *Abstract Rewriting System, ARS*, introduced in Sect. 6.5, which is called a *String Rewriting System, SRS*. Therefore, reasoning by syntactic pattern analysis can be treated as *reasoning by symbolic computation*, which has been discussed in Sect. 6.5.

Generating structural patterns with the help of a generative grammar is introduced in the first section of this chapter. In Sects. 8.2 and 8.3 the analysis and the interpretation of structural patterns are discussed, respectively. The problem of automatic construction of a grammar on the basis of sample patterns is considered in the fourth section. In the last section graph grammars are introduced. Formal definitions of notions introduced in this chapter are contained in Appendix E.

[1] Such structural patterns can be of the form of strings or graphs. Therefore, two types of generative grammars are considered: string grammars and graph grammars. In syntactic pattern recognition tree grammars, which generate tree structures, are also defined. Since a tree is a particular case of a graph, we do not introduce tree grammars in the monograph. The reader is referred, e.g., to [104].

M. Flasiński, *Introduction to Artificial Intelligence*,
DOI 10.1007/978-3-319-40022-8_8

8.1 Generation of Structural Patterns

We begin our definition of a *generative grammar* in the Chomsky model [47] by introducing a *set of terminal symbols*. Terminal symbols are expressions which are used for constructing sentences belonging to a language generated by a grammar.[2] Let us assume that we construct a grammar for a subset of the English language consisting of sentences which contain four subjects (male names): Hector, Victor, Hyacinthus, Pacificus, two predicates: accepts, rejects, and four objects: globalism, conservatism, anarchism, pacifism. For example, the sentences Hector accepts globalism and Hyacinthus rejects conservatism belong to the language.[3] Let us denote this language by L_1. Then, a set of terminal symbols T_1 is defined as follows:

T_1 = {Hector, Victor, Hyacinthus, Pacificus, accepts, rejects, globalism, conservatism, anarchism, pacifism}.

As we have mentioned above, sentences are generated with rewriting rules called *productions*.[4] Let us consider an application of a production with the following example. Let there be given a phrase:

$$\text{Hector accepts } B\ , \tag{8.1}$$

where B is an auxiliary symbol, called a *nonterminal symbol*,[5] which denotes an object in a sentence. Let there be given a production of the form:

$$B \rightarrow \text{globalism}\ . \tag{8.2}$$

An expression placed on the left-hand side of an arrow is called the *left-hand side of a production* and an expression placed on the right-hand side of an arrow is called the *right-hand side of a production*. An application of the production to the phrase, denoted by $\Longrightarrow$, consists of replacing an expression of the phrase which is equivalent to the left-hand side of the production (in our case it is the symbol B) by the right-hand side of the production. Thus, an application of the production is denoted in the

[2]Let us remember that the notions of *word* and *sentence* are treated symbolically in formal language theory. For example, if we define a grammar which generates single words of the English language, then letters are *terminal symbols*. Then, English words which consist of these letters are called *words (sentences)* of a formal language. However, if we define a grammar which generates sentences of the English language, then English words can be treated as *terminal symbols*. Then sentences of the English language are called *words (sentences)* of a formal language. As we see later on, *words (sentences)* of a formal language generated by a grammar can represent any string structures, e.g., stock charts, charts in medicine (ECG, EEG), etc. Since in this section we use examples from a natural language, strings consisting of symbols are called sentences.

[3]We omit the terminal symbol of a full stop in all examples in order to simplify our considerations. Of course, if we use generative grammars for Natural Language Processing (NLP), we should use a full stop symbol.

[4]We call them *productions*, because they are used for generating—"producing"—sentences of a language.

[5]Nonterminal symbols are usually denoted by capital letters.

following way:

$$\textsf{Hector accepts } B \Longrightarrow \textsf{Hector accepts globalism}\,. \tag{8.3}$$

Now, we define a set of all productions which generate our language. Let us denote this set by P_1:

$$\begin{array}{ll}
(1) & S \rightarrow \textsf{Hector}\, A \\
(2) & S \rightarrow \textsf{Victor}\, A \\
(3) & S \rightarrow \textsf{Hyacinthus}\, A \\
(4) & S \rightarrow \textsf{Pacificus}\, A \\
(5) & A \rightarrow \textsf{accepts}\, B \\
(6) & A \rightarrow \textsf{rejects}\, B \\
(7) & B \rightarrow \textsf{globalism} \\
(8) & B \rightarrow \textsf{conservatism} \\
(9) & B \rightarrow \textsf{anarchism} \\
(10) & B \rightarrow \textsf{pacifism}\,.
\end{array}$$

For example, the sentence Pacificus rejects globalism is generated as follows:

$$S \overset{4}{\Longrightarrow} \textsf{Pacificus } A \overset{6}{\Longrightarrow} \textsf{Pacificus rejects } B \overset{7}{\Longrightarrow} \textsf{Pacificus rejects globalism}\,. \tag{8.4}$$

Indices of the productions applied are placed above double arrows. A sequence of production applications used for generating a sentence is called a *derivation* of this sentence. If we are not interested in presenting the sequence of *derivational steps*, then we can simply write:

$$S \overset{*}{\Longrightarrow} \textsf{Pacificus rejects globalism}\,, \tag{8.5}$$

which means that we can generate (derive) a sentence Pacificus rejects globalism with the help of grammar productions starting with a symbol S.

As we can see our *set of nonterminal symbols*, which we denote by N_1, consists of three auxiliary symbols S, A, B, which are responsible for generating a subject, a predicate, and an object, i.e.,

$$N_1 = \{S, A, B\}.$$

In a generative grammar a nonterminal symbol which is used for starting any derivation is called *the start symbol (axiom)* and it is denoted by S. Thus, our grammar G_1 can be defined as a quadruple $G_1 = (T_1, N_1, P_1, S)$. A language generated by a grammar G_1 is denoted $L(G_1)$. The language $L(G_1)$ is the set of all the sentences which can be derived with the help of productions of the grammar G_1. It can be proved that for our language L_1, which has been introduced in an informal way at the beginning of this section, the following holds: $L(G_1) = L_1$.

A generative grammar has an interesting property: it is a finite "object" (it consists of finite sets of terminal and nonterminal symbols and a finite set of productions), however it can generate an infinite language, i.e., a language which consists of an infinite number of sentences. Before we consider this property, let us introduce the following denotations. Let a be a symbol. By a^n we denote an expression which consists of an n-element sequence of symbols a, i.e.,

$$a^n = \underbrace{aaa \ldots aaa}_{n\ times} , \tag{8.6}$$

where $n \geq 0$. If $n = 0$, then it is a 0-element sequence of symbols a, called the *empty word* and denoted by λ. Thus: $a^0 = \lambda$, $a^1 = a$, $a^2 = aa$, $a^3 = aaa$, etc.

For example, let us define a language L_2 in the following way:

$$L_2 = \{a^n \ , \ n \geq 1\} . \tag{8.7}$$

The language L_2 is infinite and consists of n-element sequences of symbols a, where additionally a has to occur at least once. (The empty word does not belong to L_2.) It can be generated by the following simple grammar G_2:

$$G_2 = (T_2, N_2, P_2, S) ,$$

where $T_2 = \{a\}$, $N_2 = \{S\}$, and the set of productions P_2 contains the following productions:

$$(1)\ S \rightarrow aS$$
$$(2)\ S \rightarrow a .$$

A derivation of a sentence of a given length $r \geq 2$ (i.e., $n = r$) in G_2 is performed according to the following scheme:

$$S \xRightarrow{1} aS \xRightarrow{1} aaS \xRightarrow{1} \ldots \xRightarrow{1} a^{r-1}S \xRightarrow{2} a^r , \tag{8.8}$$

i.e., firstly the first production is applied $(r - 1)$ times, then the second production is applied once at the end. If we want to generate a sentence of a length $n = 1$, i.e., the sentence a, the we apply the second production once.

Let us notice that defining a language L_2 as an infinite set is possible due to the first production. This production is of an interesting form: $S \rightarrow aS$, which means that it *refers to itself* (a symbol S occurs on the left- and right-hand sides of the production). Such a form is called *recursive*, from Latin *recurrere*—"running back". This "running back" of the symbol S during a derivation each time after applying the first production, makes the language L_2 infinite.

Now, we discuss a very important issue concerning generative grammars. There are a lot of classes (types) of generative grammars. These classes can be arranged in a hierarchy according to the criterion of their *generative power*. We introduce this criterion with the help of the following example. Let us define the next formal language as follows:

$$L_3 = \{a^n b^m \ , \ n \geq 1 \ , \ m \geq 2\} . \tag{8.9}$$

The language L_3 consists of the subsequence of symbols a and the subsequence of symbols b. Additionally, the symbol a has to occur at least once, and the symbol b has to occur at least twice. For example, the sentences *abb*, *aabb*, *aabbb*, *aabbbb*, ... , *aaabb*, etc. belong to this language. It can be generated by the following grammar:

$$G_3 = (T_3, N_3, P_3, S) ,$$

where $T_3 = \{a , b\}$, $N_3 = \{S , A , B\}$, and the set of productions P_3 contains the following productions:

$$\begin{aligned}&(1)\ S \rightarrow aA\\&(2)\ A \rightarrow aA\\&(3)\ A \rightarrow bB\\&(4)\ B \rightarrow bB\\&(5)\ B \rightarrow b\ .\end{aligned}$$

The first production is used for generating the first symbol a. The second production generates successive symbols a in a recursive way. We generate the first symbol b with the help of the third production. The fourth production generates successive symbols b in a recursive way (analogously to the second production). The fifth production is used for generating the last symbol b of the sentence. For example, a derivation of the sentence a^3b^4 is performed as follows:

$$S \overset{1}{\Longrightarrow} aA \overset{2}{\Longrightarrow} aaA \overset{2}{\Longrightarrow} aaaA \overset{3}{\Longrightarrow} aaabB \overset{4}{\Longrightarrow}$$
$$\overset{4}{\Longrightarrow} aaabbB \overset{4}{\Longrightarrow} aaabbbB \overset{5}{\Longrightarrow} aaabbbb\ . \tag{8.10}$$

Now, let us define a language L_4 which is a modification of the language L_3, in the following way:

$$L_4 = \{a^m b^m\ ,\ m \geq 1\}\ . \tag{8.11}$$

The language L_4 differs from the language L_3 in demanding an equal number of symbols a and b. The language L_4 cannot be generated with the help of grammars having productions of the form of grammar G_3, since such productions do not ensure the condition of an equal number of symbols. It results from the fact that in such a grammar we firstly generate a certain number of symbols a and then we start to generate symbols b, but the grammar "does not remember"how many symbols a have been generated. We say that a grammar having productions in the a form of the productions of G_3 has *too weak generative power* to generate the language L_4. Now, we introduce a grammar G_4 which is able to generate the language L_4:

$$G_4 = (T_4, N_4, P_4, S) ,$$

where $T_4 = \{a , b\}$, $N_4 = \{S\}$, and the set of productions P_4 contains the following productions:

$$(1)\ S \rightarrow aSb$$
$$(2)\ S \rightarrow ab\ .$$

For example, a derivation of the sentence a^4b^4 is performed in the following way:

$$S \overset{1}{\Longrightarrow} aSb \overset{1}{\Longrightarrow} aaSbb \overset{1}{\Longrightarrow} aaaSbbb \overset{2}{\Longrightarrow} aaaabbbb\ . \qquad (8.12)$$

As we can see a solution to the problem of an equal number of symbols a and b is obtained by generating the same number of both symbols in each derivational step.

Thus, the grammar G_4 has *sufficient generative power* to generate the language L_4. The generative power of classes of formal grammars results from the form of their productions. Let us notice that the productions of grammars G_1, G_2, G_3 are of the following two forms:

<nonterminal symbol> → *<terminal symbol><nonterminal symbol>*
or *<nonterminal symbol>* → *<terminal symbol>* .

Using productions of such forms, we can only "stick" a symbol onto the end of the phrase which has been derived till now. Grammars having only productions of such a form are called *regular grammars*.[6] In the Chomsky hierarchy such grammars have the weakest generative power.

For grammars such as the grammar G_4, we do not demand any specific form of the right-hand side of productions. We require only a single nonterminal symbol at the left-hand side of a production. Such grammars are called *context-free grammars*.[7] They have greater generative power than regular grammars. However, we have to pay a certain price for increasing the generative power of grammars. We discuss this issue in the next section.

8.2 Analysis of Structural Patterns

Grammars are used for generating languages. However, in Artificial Intelligence we are interested more in the languages' analysis. For this analysis formal automata are applied. Various types of automata differ from one another in their construction (structure), depending on the corresponding classes of grammars. Let us begin by defining an automaton of the simplest type, i.e., a *finite-state automaton*. This class

[6]In fact, such grammars are *right regular grammars*. In *left regular grammars* a nonterminal symbol occurs (if it occurs) before a terminal symbol.

[7]There are also grammars which have a stronger generative power in the Chomsky hierarchy, namely *context-sensitive grammars* and *unrestricted (type-0) grammars*. Their definitions are contained in Appendix E.

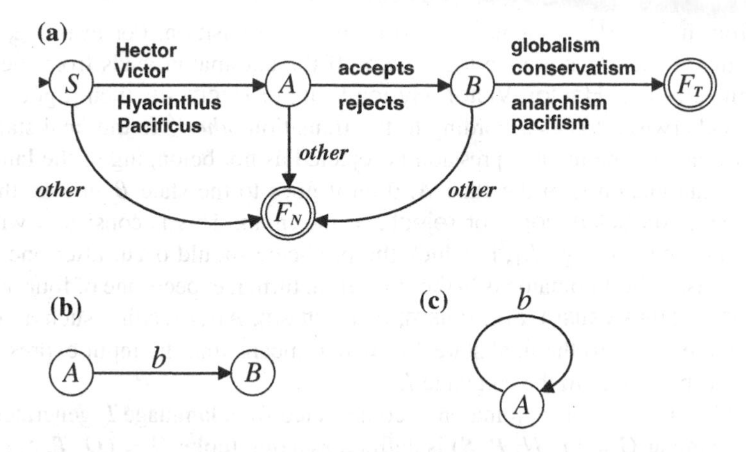

Fig. 8.1 **a** The finite-state automaton A_1, **b-c** basic constructs for defining a finite-state automaton

was introduced by Claude Elwood Shannon[8] in [272][9] in 1948, and formalized by Stephen C. Kleene in [160] in 1956 and Michael Oser Rabin[10] and Dana Stewart Scott[11] in [234] (nondeterministic automata) in 1959. A finite-state automaton is used for analysis of languages generated by regular grammars, called *regular languages*.[12]

Let us start by defining the automaton A_1 shown in Fig. 8.1a, which is constructed for the language L_1 (the grammar G_1) introduced in the previous section. Each node of the graph is labeled with a symbol (S, A, B, F_T, F_N) and represents a possible *state* of the automaton. The state in which the automaton starts working is called the *initial state* S and is marked with a small black triangle in Fig. 8.1a. States in which the automaton finishes working (F_T and F_N in Fig. 8.1a) are called *final states* and they are marked with a double border. Directed edges of the graph define *transitions* between states. A transition from one state to another takes place if the automaton

[8] Claude Elwood Shannon—a professor of the Massachusetts Institute of Technology, a mathematician and electronic engineer, the "father" of information theory and computer science.

[9] The idea of a finite-state automaton is based on the model of Markov chain which is introduced in Appendix B for genetic algorithms.

[10] Michael Oser Rabin—a professor of Harvard University and the Hebrew University of Jerusalem, a Ph.D. student of Alonzo Church. His outstanding achievements concern automata theory, computational complexity theory, cryptography (Miller-Rabin test), and pattern recognition (Rabin-Karp algorithm). In 1976 he was awarded the Turing Award (together with Dana Scott).

[11] Dana Stewart Scott—a professor of computer science, philosophy, and mathematics at Carnegie Mellon University and Oxford University, a Ph.D. student of Alonzo Church. His excellent work concerns automata theory, semantics of programming languages, modal logic, and model theory (a proof of the independence of the continuum hypothesis). In 1976 he was awarded the Turing Award.

[12] The languages L_1, L_2, and L_3 introduced in the previous section are regular languages.

reads from its input[13] an element determining this transition. For example, let us assume that the automaton is in the state S. If the automaton reads from the input one of the elements Hector, Victor, Hyacinthus, or Pacificus, then it goes to the state A. Otherwise, it goes according to the transition *other*[14] to the final state F_N, which means that the input expression is rejected as not belonging to the language L_1. If the automaton is in the state A, then it goes to the state B in case there is one of the predicates accepts or rejects at the input. This is consistent with the definition of the language L_1, in which the predicate should occur after one of the four subjects. If the automaton is in the state B, in turn, it expects one of four objects: globalism, conservatism, anarchism, or pacifism. After reading such an object the automaton goes to the final state F_T, which means that the input expression is accepted as belonging to the language L_1.

Formally, a finite-state automaton A constructed for a language L generated by a regular grammar $G = (T, N, P, S)$ is defined as a quintuple: $G = (Q, T, \delta, q_0, F)$. T is the set of terminal symbols which are used by the grammar G for generating the language L. Q is the set of states. (In our example shown in Fig. 8.1a it is the set $\{S, A, B, F_T, F_N\}$.) q_0 is the initial state, F is the set of final states. (In our example $q_0 = S$, F consists of states F_T and F_N.) δ is the *state-transition function*,[15] which determines transitions in the automaton (in Fig. 8.1a transitions are represented by directed edges of the graph). A pair (*the state of the automaton, the terminal symbol at the input*) is an argument of the function. The function *computes* the state the automaton should go into. For example, $\delta(S, \text{Hyacinthus}) = A, \delta(A, \text{accepts}) = B$ (cf. Fig. 8.1a).

A method for a generation (synthesis) of a finite-state automaton on the basis of a corresponding regular grammar has been developed. States of the automaton relate to nonterminal symbols of the grammar (the initial state relates to the start symbol, additionally we can define final states). Each production of the form $A \rightarrow bB$ is represented by a transition $\delta(A, b) = B$ (cf. Fig. 8.1b). Each recursive production of the form $A \rightarrow bA$ is represented by a recursive transition $\delta(A, b) = A$ (cf. Fig. 8.1c). Each production finishing a derivation (there is a single terminal symbol on the right-hand side of the production) corresponds to a transition to the final acceptance state. The reader can easily see that the automaton A_1 shown in Fig. 8.1a has been constructed on the basis of the grammar G_1 according to these rules. In the previous section we have said that generative grammars are arranged in a hierarchy according to their generative power. The same applies to automata. What is more, each class of grammar relates to some type of automaton. Automata which correspond to weaker grammars (in the sense of generative power) are not able to analyze languages generated by stronger classes of grammars. For example, a finite-state automaton is

[13]The *input* of the automaton is the place where the expression to be analyzed is placed. If there is some expression at the *input*, then the automaton reads the expression one element (a terminal symbol) at a time and it performs the proper transitions.

[14]The *other* transition means that the automaton has read an element which is different from those denoting transitions coming out from the current state.

[15]The state-transition function is not necessarily a function in the mathematical sense of this notion.

too weak to analyze the *context-free language* $L_4 = \{a^m b^m \ , \ m \geq 1\}$ introduced in the previous section.

A *(deterministic) pushdown automaton*[16] is strong enough to analyze languages such as L_4. This automaton uses an additional working memory, called a *stack*.[17] The transition function δ of such an automaton is defined in a different way from the finite-state automaton. A pair (*the top of the stack, the sequence of symbols at the input*[18]) is its argument. As a result, the function can "generate" various actions of the automaton. In the case of our automaton the following actions are allowed:

- accept (the automaton has analyzed the complete expression at the input and it has decided that the expression belongs to the language),
- reject (the automaton has decided, during its working, that the expression does not belong to the language),
- remove_symbol (the automaton removes a terminal symbol from the input and a symbol occurring at the top of the stack),
- apply_production_on_stack(i) (the automaton takes the left-hand side of a production i from the top of the stack and it adds the right-hand side of the production i to the top of the stack).

Before we consider the working of an automaton A_4 constructed for the language $L_4 = \{a^m b^m \ , \ m \geq 1\}$, let us define its transition function in the following way:

$$\delta(S, aa) = \text{apply\_production\_on\_stack(1)}, \tag{8.13}$$

$$\delta(S, ab) = \text{apply\_production\_on\_stack(2)}, \tag{8.14}$$

$$\delta(a, a) = \text{remove\_symbol}, \tag{8.15}$$

$$\delta(b, b) = \text{remove\_symbol}, \tag{8.16}$$

$$\delta(\lambda, \lambda) = \text{accept}, \ \ \lambda - \text{the empty word}, \tag{8.17}$$

$$\delta(v, w) = \text{reject, otherwise}. \tag{8.18}$$

The automaton A_4 tries to reconstruct a derivation of the expression which is at its input. It does this by analyzing a sequence consisting of *two* symbols[19] of the expression, because this is sufficient to decide which production of the grammar G_4 (the first or the second) is to be used at the moment of generating this sequence.

---

[16] In order to simplify our considerations we introduce here a specific case of a pushdown automaton, i.e. an *LL(k) automaton*, which analyzes languages generated by *LL(k) context-free grammars*. These grammars are defined formally in Appendix E.

[17] In computer science a *stack* is a specific structure of a data memory with certain operations, which works in the following way. Data elements can be added only to the top of the stack and they can be taken off only from the top. A stack of books put one on another is a good example of a stack. If we want to add a new book to the stack, we have to put it on the top of a stack. If we want to get some book, then we have to take off all books which are above the book we are interested in.

[18] This *sequence of symbols* has a fixed length. The length of the sequence is a parameter of the automaton. In the case of *LL(k)* automata, *k* is the length of the sequence, which is analyzed in a single working step of the automaton.

[19] The automaton A_4 is an *LL(2)* automaton.

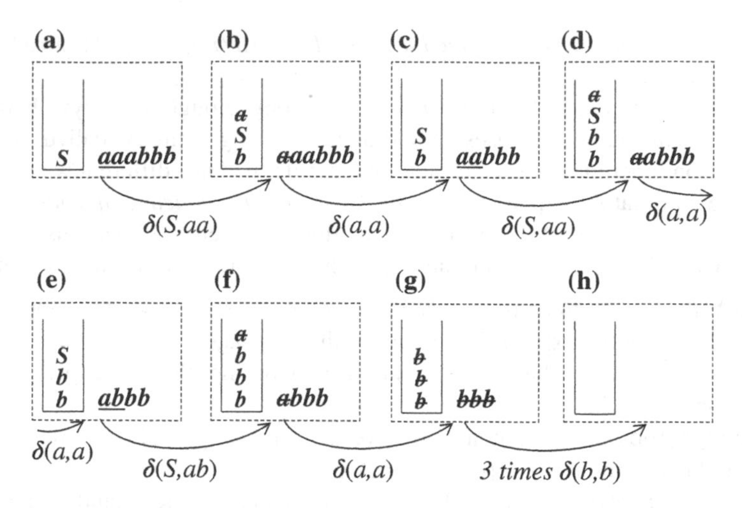

Fig. 8.2 Analysis of the sentence *aaabbb* by the automaton A_4

To be convinced that this is a proper analysis method, let us return to the derivation (8.12) in a previous section. If we want to decide how many times the first production has been applied for the generation of the sentence *aaaabbbb*, then it is enough to check two symbols forward. If we scan a sentence from left to right, then as long as we have a two-element sequence *aa* we know that production (1) has been applied, which corresponds to transition (8.13). If we meet a sequence *ab* (in the middle of the sentence), then it means that production (2) has been applied, which corresponds to transition (8.14). Now, let us analyze the sentence *aaabbb*. The automaton starts working having the start symbol S on the stack and the sentence *aaabbb* at the input as shown in Fig. 8.2a. The underlined part of the sentence means the sequence which is analyzed in a given step. Arrows mean transitions and are labeled with the transition function used in a given step. So, there is S at the top of the stack and *aa* are the first two symbols of the input. Thus, the first step is performed by the automaton according to transition (8.13). This means that the left-hand side of production (1), i.e., S, is taken off the stack and the right-hand side of this production, i.e., aSb, is put on the top of the stack[20] as shown in Fig. 8.2b. Now, a is at the top of the stack and a is at the input. Thus, the next step is performed according to transition (8.15), i.e., the symbol a is removed from the top of the stack and from the input. This is denoted by crossing out both symbols. We obtain the situation shown in Fig. 8.2c. This situation is analogous to the one before the first step. (S is at the top of the stack, *aa* is at the input.) Thus, we perform a transition according to (8.13) once more, i.e., S is taken off the stack and the right-hand side of this production, aSb, is put on the top of the stack. This results in the situation shown in Fig. 8.2d. Again a is at

[20]The right-hand side of the production, aSb, is put on the stack "from back to front", i.e., firstly (at the bottom of the stack) symbol b is put, then symbol S, then finally (at the top of the stack) symbol a.

the top of the stack and a is at the input. So, we perform the transition according to (8.15) and we get the configuration shown in Fig. 8.2e. Since S is on the top of the stack and ab are the first two symbols at the input, we should apply (8.14), which corresponds to production (2) of the grammar G_4. The automaton replaces S on the stack by the right-hand side of the second production, i.e., ab (cf. Fig. 8.2f). As we can see, the next steps consist of removing symbols from the stack according to formula (8.15) and three times formula (8.16). At the end both the stack and the input are empty, as shown in Fig. 8.2h. This corresponds to a transition according to (8.17), which means acceptance of the sentence as belonging to the language L_4. Any other final configuration of the stack and the input would result in rejecting the sentence according to formula (8.18).

One can easily notice that the working of a pushdown automaton is more complex than that of a finite-state automaton. In fact, the bigger the generative power of a generative grammar is, the bigger the computational complexity of the corresponding automaton. The analysis of regular languages is more efficient than the analysis of context-free languages. Therefore, subclasses of context-free grammars with efficient corresponding automata have been defined. The most popular efficient subclasses include: *LL(k) grammars* introduced by Philip M. Lewis[21] and Richard E. Stearns[22] in [180] in 1968, *LR(k) grammars* defined by Donald E. Knuth[23] in [163] in 1965, and *operator precedence grammars* defined by Robert W. Floyd[24] in [98] in 1963. For these types of grammars corresponding efficient automata have been defined.

The problem of syntax analysis (analyzing by automaton) becomes much more difficult if context-free grammars have too weak generative power for a certain application. As we have mentioned above, in the Chomsky hierarchy there are two remaining classes of grammars, namely *context-sensitive grammars* and *unrestricted (type-0) grammars*. *A linear bounded automaton* and *the Turing machine* are two types of

[21]Philip M. Lewis—a professor of electronic engineering and computer science at the Massachusetts Institute of Technology and the State University of New York, a scientist at General Electric Research and Development Center. His work concerns automata theory, concurrency theory, distributed systems, and compiler design.

[22]Richard Edwin Stearns—a professor of mathematics and computer science at the State University of New York, a scientist at General Electric. He was awarded the Turing Award in 1993. He has contributed to the foundations of computational complexity theory (with Juris Hartmanis). His achievements concern the theory of algorithms, automata theory, and game theory.

[23]Donald Ervin Knuth—a professor of computer science at Stanford University. The "father" of the analysis of algorithms. He is known as the author of the best-seller "The Art of Computer Programming" and the designer of the Tex computer typesetting system. Professor D. Knuth is also known for his good sense of humor (e.g., his famous statement: "Beware of bugs in the above code; I have only proved it correct, not tried it."). He was awarded the Turing Award in 1974.

[24]Robert W. Floyd—a computer scientist, physicist, and BA in liberal arts. He was 33 when he became a full professor at Stanford University (without a Ph.D. degree). His work concerns automata theory, semantics of programming languages, formal program verification, and graph theory (Floyd-Warshall algorithm).

automata which correspond to these classes of grammars, respectively. Both types of automata are inefficient computationally, so they cannot be used effectively in practical applications. Therefore, *enhanced* context-free grammars have been defined in order to solve this problem. Such grammars include *programmed grammars* defined by Daniel J. Rosenkrantz[25] in [248] in 1969, *indexed grammars* introduced by Alfred Vaino Aho[26] in [1] in 1968, and *dynamically programmed grammars* published in [95] in 1999.

8.3 Interpretation of Structural Patterns

In the previous section we have shown how to use an automaton for checking whether an expression (sentence) belongs to a language generated by a grammar. In other words, an automaton has been used to test whether an expression is built properly from the point of view of a language's syntax, which is important, e.g., in Natural Language Processing. Generally, in Artificial Intelligence we are interested not only in the syntactical correctness of expressions, but also we are interested in their semantic aspect, i.e., we want to perform a proper interpretation of expressions.[27] Let us consider once more our example of Hector, Victor, et al. introduced in Sect. 8.1. Let us assume that Hector and Victor accept globalism and conservatism, and they reject anarchism and pacifism. On the other hand, Hyacinthus and Pacificus accept anarchism and pacifism, and they reject globalism and conservatism. Let us assume that only such propositions belong to a new language L_5. Now, we can define a grammar G_5 which not only generates sentences which are correct syntactically, but also these propositions are consistent with the assumptions presented above. (That is, these propositions are true.) The set of productions P_5 of the grammar G_5 is defined as follows:

[25] Daniel J. Rosenkrantz—a professor of the State University of New York, a scientist at General Electric, the Editor-in-Chief of the prestigious *Journal of the ACM*. His achievements concern compiler design and the theory of algorithms.

[26] Alfred Vaino Aho—a physicist, an electronic engineer, and an eminent computer scientist, a professor of Columbia University and a scientist at Bell Labs. His work concerns compiler design, and the theory of algorithms. He is known as the author of the excellent books (written with J.D. Ullman and J.E. Hopcroft) *Data Structures and Algorithms* and *The Theory of Parsing, Translation, and Compiling*.

[27] Similarly to the logic-based methods discussed in Sect. 6.1.

$$
\begin{array}{lll}
(1) & S & \rightarrow \text{Hector } A_1 \\
(2) & S & \rightarrow \text{Victor } A_1 \\
(3) & S & \rightarrow \text{Hyacinthus } A_2 \\
(4) & S & \rightarrow \text{Pacificus } A_2 \\
(5) & A_1 & \rightarrow \text{accepts } B_1 \\
(6) & A_1 & \rightarrow \text{rejects } B_2 \\
(7) & A_2 & \rightarrow \text{rejects } B_1 \\
(8) & A_2 & \rightarrow \text{accepts } B_2 \\
(9) & B_1 & \rightarrow \text{globalism} \\
(10) & B_1 & \rightarrow \text{conservatism} \\
(11) & B_2 & \rightarrow \text{anarchism} \\
(12) & B_2 & \rightarrow \text{pacifism}\,.
\end{array}
$$

One can easily check that with the help of the set of productions P_5 we can generate all the valid propositions of our model of the world. On the other hand, it is impossible to generate a false proposition, e.g., Hector rejects globalism, although this sentence is correct syntactically.

Now, for the grammar G_5 we can define a finite-state automaton A_5. This automaton is shown in Fig. 8.3a. (For a simplicity we have not defined the final rejection state F_N and transitions to this state.) Let us notice that the automaton A_5 not only checks the syntactical correctness of a sentence, but it also interprets these sentences and accepts only those sentences which are valid in our model of the world.

The automata, which have been introduced till now are called *acceptors (recognizers)*. They accept (a state F_T) or do not accept (a state F_N) a sentence depending on a specific criterion such as syntax correctness or validity (truthfulness) in some model. *Transducers* are the second group of automata. During an analysis they generate expressions on their *outputs*.[28] For example, they can be used for translating expressions of a certain language into expressions of another language. The transition function of such an automaton determines a goal state and writes some expression into the output. For example, let us define a transducer A_6, which translates language L_5 into Polish. We define the transition function as follows: $\delta(A_1, \text{accepts}) = (B_1, \text{akceptuje})$, $\delta(A_1, \text{rejects}) = (B_2, \text{odrzuca})$, etc. The transducer A_6 is shown in Fig. 8.3b.

Although *Natural Language Processing, NLP,* is one the most important application areas of transducers, there are also other areas in which they are used, i.e., interpretation of the world by automata is not limited to the case of describing the world with the help of natural languages. Certain phenomena are described with the help of other representations (e.g., charts) which express their essence in a better way. Then, in syntactic pattern recognition we can ascribe a certain interpretation to terminal symbols, as shown, for example, in Fig. 8.3c. Graphical elements represented by terminal symbols (in our example: "straight arrows" and "arc arrows") are called *primitives*. Primitives play the role of elementary components used for defining charts.

[28]Therefore, transducers are called also *automata with output*.

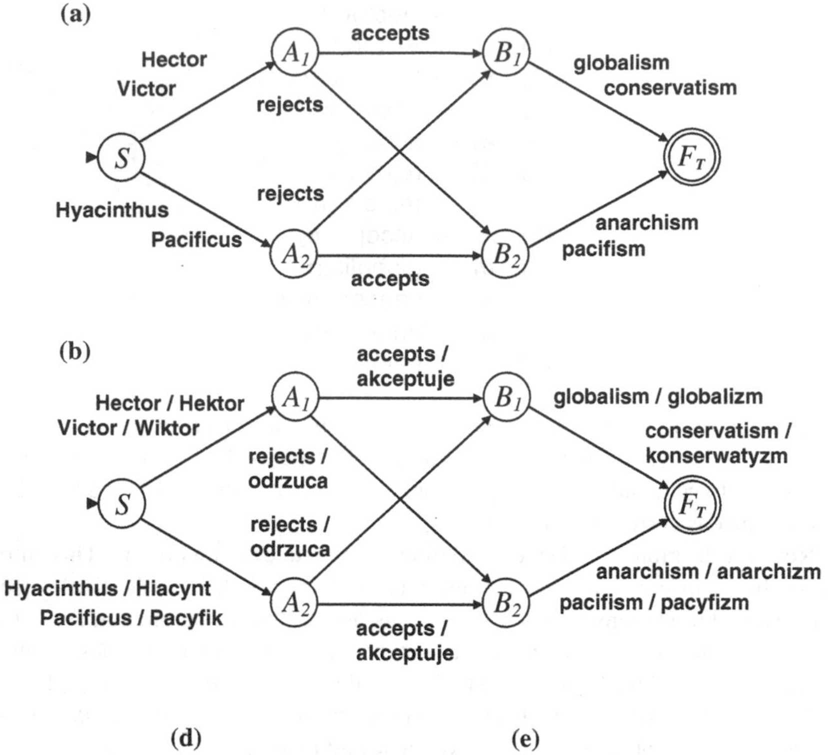

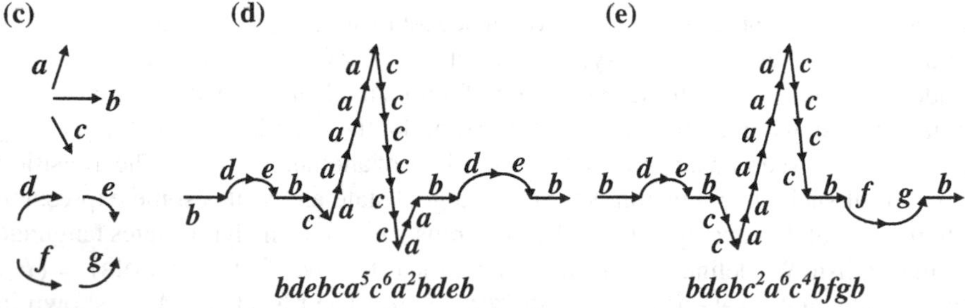

Fig. 8.3 **a** An automaton A_5 which accepts propositions that are valid in the model defined by the language L_5, **b** a transducer A_6 which translates the language L_5 into Polish, **c** an example of structural primitives, **d**-**e** structural representations of ECG patterns

For example, in medical diagnosis we use ECG charts in order to identify heart diseases. An example of a structural representation of a normal ECG is shown in Fig. 8.3d, whereas the case of a myocardial infarction is shown in Fig. 8.3e. These representations can be treated as *sentences* defined in the language of ECG patterns. Thus, on the basis of a set of such representations (sentences) we can define a grammar which generates this language. Given a grammar we can construct an automaton (transducer), which writes an interpretation of an ECG to its output.

Even if we look at an ECG casually, we notice that the primitives occurring in charts are diversified with respect to, e.g., their length or the angle of a depression. Therefore, in order to achieve a more precise structural representation, attributes can be ascribed to primitives. For example, two attributes, the length (l) and the deflection angle ($\pm\alpha$), are ascribed to the primitive a shown in Fig. 8.4a. In such a case *attributed grammars* are used for pattern generation. Automata applied to the interpretation of attributed patterns (expressions) additionally compute the *distance* between an analyzed pattern and a model pattern. This distance allows us to assess the degree of confidence of the interpretation made by the automaton.

Instead of ascribing attributes to a primitive, we can define discrete patterns of deviations of a model primitive as shown in Fig. 8.4b. Then, we can ascribe probabilities to deviations, e.g., on the basis of the frequency of their occurrence. One such model is *stochastic grammars* introduced in the 1970s and the 1980s and then developed by King-Sun Fu[29] and Taylor L. Booth[30] [34,103,104]. In such grammars the probability of the application of each production is defined. A *stochastic automaton* gives the probability that a chart represents a recognized phenomenon expressed by a corresponding structural pattern after analyzing a part of the chart. Stochastic grammars and automata are also used in Natural Language Processing, which is discussed in Chap. 16. It is interesting that *Markov chains*,[31] which have been introduced for genetic algorithms in Chap. 5 are also a mathematical model for a stochastic automaton.

In approaches to the distortion of structural patterns discussed till now we have assumed that the structure of such representations is correct. In other words, a primitive could be distorted but it has to occur in a structure in the proper place. This means that if structural representations are hand-written sentences of a natural language then vaguely hand-written letters are the only kind of errors. However, in practice we can omit some letter (e.g., if we write "gramar"), we can incorrectly add some letter (e.g., if we write "grammuar"), or we can replace a correct letter by an incorrect one (e.g., if we write "glammar"). Fortunately, in syntactic pattern recognition certain metrics are defined which can be used to compute the distance between a model pattern and its *structural* distortion. The *Levenshtein metrics*[32] [179] are some of the most popular metrics used for this purpose. They are introduced in Appendix G.

---

[29] King-Sun Fu—a professor of electrical engineering and computer science at Purdue University, Stanford University and University of California, Berkeley. The "father" of syntactic pattern recognition, the first president of the International Association for Pattern Recognition (IAPR), and the author of excellent monographs, including *Syntactic Pattern Recognition and Applications*, Prentice-Hall 1982. After his untimely death in 1985 IAPR established the biennial King-Sun Fu Prize for a contribution to pattern recognition.

[30] Taylor L. Booth—a professor of mathematics and computer science at the University of Connecticut. His research concerns Markov chains, formal language theory, and undecidability. A founder and the first President of the Computing Sciences Accreditation Board (CSAB).

[31] Markov chains are defined formally in Appendix B.2.

[32] Vladimir Iosifovich Levenshtein—a professor of computer science and mathematics at the Keldysh Institute of Applied Mathematics in Moscow and the Steklov Mathematical Institute. In 2006 he was awarded the IEEE Richard W. Hamming Medal.

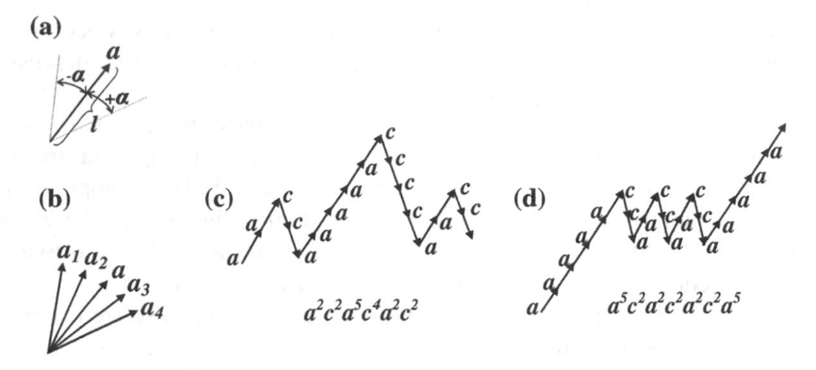

Fig. 8.4 **a** Ascribing attributes to a primitive, **b** a model primitive and patterns of its deviations, **c-d** structural representations of stock chart patterns

Syntactic-pattern-recognition-based AI systems have been used in various application areas. In medicine these include, apart from ECG, also EEG (monitoring the brain's electrical activity), PWA (pulse wave analysis), ABR (recording an auditory brainstem response for determining hearing levels), etc. Analysis of economic phenomena is another area of syntactic pattern recognition applications. For example, structural representations of stock chart patterns used for technical analysis are shown in Figs. 8.4c, d. A structural representation of the Head and Shoulders formation, which occurs when a trend is in the process of reversal, is shown in Fig. 8.4c, whereas a representation of the Flag formation, which is a trend continuation pattern, is shown in Fig. 8.4d.

In practical applications, if there are a lot of exemplary patterns (i.e., exemplary sentences of a language) then defining a grammar by hand is very difficult, sometimes impossible, because a (human) designer is not able to comprehend the whole set of sample patterns. Therefore, methods for automatic construction of a grammar on the basis of sample patterns have been developed. We discuss them in the next section.

8.4 Induction of Generative Grammars

We present the main idea of *grammar induction (grammatical inference)* with the help of a simple method of formal derivatives for regular grammars [104]. Firstly, we introduce the notion of a *formal derivative*. Let $A^{(0)}$ be a set of expressions built of terminal symbols. The formal derivative of a set $A^{(0)}$ with respect to a terminal symbol a, denoted $D_{\mathsf{a}}A^{(0)}$, is a set of expressions which are constructed from expressions of $A^{(0)}$ by removing a symbol a occurring at the beginning of these expressions. In other words, $D_{\mathsf{a}}A^{(0)}$ is the set of expressions x such that $\mathsf{a}x$ belongs to $A^{(0)}$.
For example, let there be given a set

$$A^{(0)} = \{\text{Jack cooks well, Jack runs quickly}\}.$$

Then

$$D_{\text{Jack}} A^{(0)} = \{\text{cooks well, runs quickly}\} = A^{(1)}.$$

We can continue by computing a formal derivative for the set $A^{(1)}$:

$$D_{\text{cooks}} A^{(1)} = \{\text{well}\} = A^{(2)}.$$

Now, if we compute a formal derivative once more, this time for the set $A^{(2)}$, then we obtain a set containing the empty word λ only:

$$D_{\text{well}} A^{(2)} = \{\lambda\} = A^{(3)}.$$

In fact, if a symbol well is attached to the empty word, then we obtain an expression well, which belongs to the set $A^{(2)}$.

Let us notice that computing a formal derivative can give the empty set as a result. For example, if we compute a formal derivative of the set $A^{(3)}$ with respect to any symbol, e.g., with respect to the symbol quickly, then we obtain:

$$D_{\text{quickly}} A^{(3)} = \emptyset,$$

because there is no such expression, which gives the empty word after attaching the symbol quickly (or any other symbol).

In this method we have to compute all the formal derivatives. Thus, let us do so:

$$D_{\text{runs}} A^{(1)} = \{\text{quickly}\} = A^{(4)},$$
$$D_{\text{quickly}} A^{(4)} = \{\lambda\} = A^{(5)}.$$

Computing any derivative of the set $A^{(5)}$ gives the empty set.

After computing all formal derivatives we can define the productions of a regular grammar which generates expressions belonging to the set $A^{(0)}$. A symbol $A^{(0)}$ is the start symbol of the grammar. Productions are defined according to the following two rules.

1. If the formal derivative of a set $A^{(n)}$ with respect to a symbol a is equal to a set $A^{(k)}$, i.e., $D_{\text{a}} A^{(n)} = A^{(k)}$, and the set $A^{(k)}$ does not consist of the empty word, then add a production $A^{(n)} \rightarrow \text{a}A^{(k)}$ to the set of productions of the grammar.
2. If the formal derivative of a set $A^{(n)}$ with respect to a symbol a is equal to a set $A^{(k)}$, i.e., $D_{\text{a}} A^{(n)} = A^{(k)}$, and the set $A^{(k)}$ consists of the empty word, then add a production $A^{(n)} \rightarrow \text{a}$ to the set of productions of the grammar.

As one can easily check, after applying these rules we obtain the following set of productions.

$$(1)\ A^{(0)} \rightarrow \text{Jack}\ A^{(1)}$$
$$(2)\ A^{(1)} \rightarrow \text{cooks}\ A^{(2)}$$
$$(3)\ A^{(2)} \rightarrow \text{well}$$
$$(4)\ A^{(1)} \rightarrow \text{runs}\ A^{(4)}$$
$$(5)\ A^{(4)} \rightarrow \text{quickly}\ .$$

The method introduced above is used for the induction of a grammar, which generates only a given sample of a language. Methods which try to *generalize* a sample to the whole language are defined in syntactic pattern recognition, as well. Let us notice that such an induction of a grammar corresponds to inductive reasoning (see Appendix F.2). In fact, we go from individual cases (a sample of sentences) to their generalization of the form of a grammar.

In case of the Chomsky generative grammars a lot of induction methods have been defined for regular languages. Research results in the case of context-free languages are still unsatisfactory. However, the induction of graph grammars, which are introduced in the next section, is a real challenge.

8.5 Graph Grammars

As we have mentioned in Chap. 6, reasoning as symbolic computation is based on Abstract Rewriting Systems, ARSs, which can be divided into Term Rewriting Systems, TRSs (e.g., lambda calculus introduced in Sect. 6.5), String Rewriting Systems, SRSs, which have been discussed in previous sections with the help of the example of the Chomsky generative grammars, and *Graph Rewriting Systems, GRSs*. The last ones are used for rewriting (transforming) structures in the form of graphs. *Graph grammars*, which are introduced in this section, are the most popular kind of Graph Rewriting Systems.

Graphs are widely used in Artificial Intelligence (and in general, in computer science), because they are the most general structures used for representing aspects of the world. AI representations such as semantic networks, frames, scripts, structures used for semantic interpretation in First Order Logic, Bayesian networks, structures used in model-based reasoning—all of them are graphs. Therefore, graph grammars are an important formalism for generating (in general, transforming) such representations. First of all, we show how they can be applied for modeling (describing) processes (phenomena) of the world. We consider the example of an intelligent system for integrating areas of *Computer-Aided Design* and *Computer-Aided Manufacturing*.[33] The definition of such a representation of a mechanical part which can be *translated* automatically into the *language* of technological operations performed by manufacturing equipment is a crucial problem in this area.

[33]The example is based on a model introduced in: Flasiński M.: Use of graph grammars for the description of mechanical parts. *Computer-Aided Design 27* (1995), pp. 403–433, Elsevier.

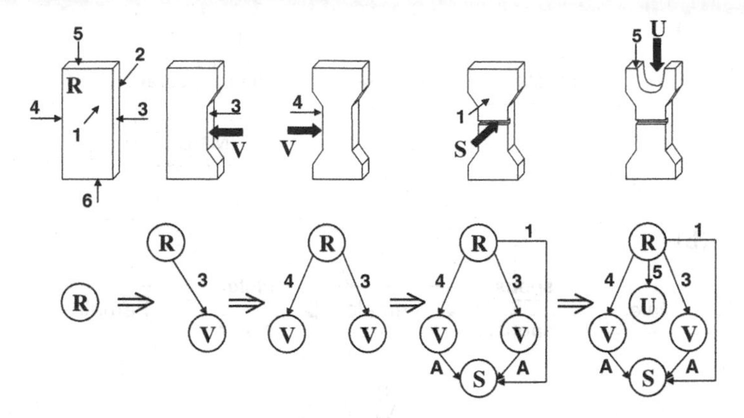

Fig. 8.5 An example of a graph grammar derivation, which represents modeling a part in a CAD system and its manufacturing controlled by a CAM system

The derivation of a graph with a graph grammar which corresponds to both a design process and a technological process is shown in Fig. 8.5. Raw material in the form of a rectangular cuboid is represented by a graph node labeled by *R*. The faces of the cuboid are indexed as shown in the figure. An application of the first production results in replacing the node *R* by a graph that consists of nodes *R* and *V*, which are connected with an edge labeled with *3*. This production corresponds to embedding a feature called a *V*-slot in the face indexed with *3* of the solid *R*.[34] In the second step of the derivation, a *V*-slot is embedded in the face indexed with *4* of the solid *R*. Then, a *S*lot is embedded in the face indexed with *1* of the solid *R*. Let us notice that this *S*lot is adjacent to both *V*-slots, which is represented by edges labeled with *A*. Finally, a *U*-slot is embedded in the face indexed with *5* of the solid *R*.

Defining a way of replacing a graph of the left-hand side of a production by a graph of the right-hand side of the production is a fundamental problem of graph grammars. (In the example above we see only the result of such a replacement.) This operation is performed with the help of the *embedding transformation*. On one hand, the embedding transformation complicates a derivation. On the other hand, it is the source of the very great descriptive power of graph grammars. It is so important that a taxonomy of graph grammars is defined on its basis. We present the embedding transformation, which was introduced by the research team of Grzegorz Rozenberg[35] for *edNLC graph grammars* in the 1980s [149].

[34] During a technological process this corresponds to milling a V-slot in the raw material.

[35] Grzegorz Rozenberg—a professor of Leiden University, the University of Colorado at Boulder, and the Polish Academy of Sciences in Warsaw, an eminent computer scientist and mathematician. His research concerns formal language theory, concurrent systems, and natural computing. Prof. G. Rozenberg was the president of the European Association for Theoretical Computer Science for 11 years.

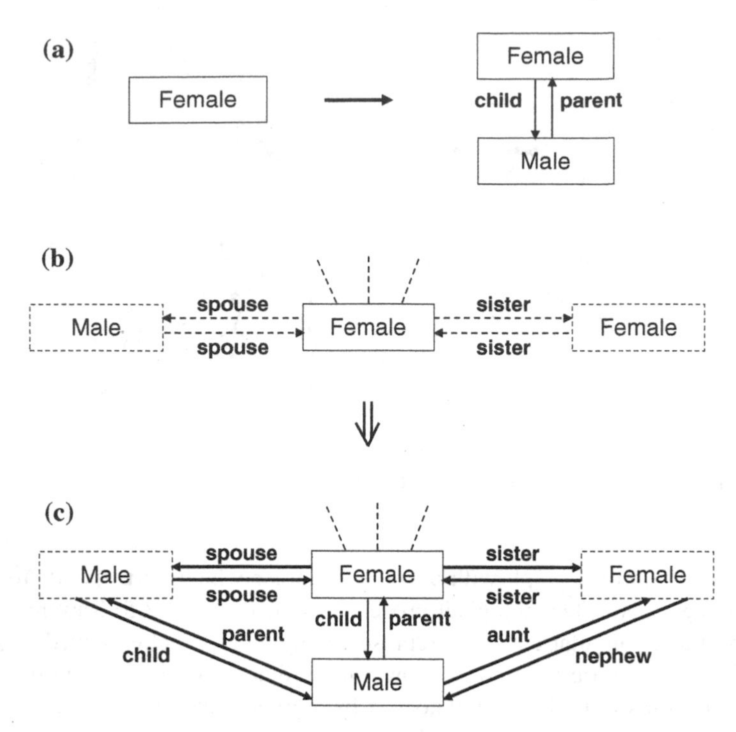

Fig. 8.6 **a** An example of a graph grammar production which is used for transforming a semantic network, **b-c** an application of a graph grammar production for transforming a semantic network

Let us transform a semantic network which represents family relations[36] with an edNLC graph grammar. Graphs of the left-hand side and the right-hand side of a production which represents the birth of a male child are shown in Fig. 8.6a. An object of the class Female is replaced by itself with an object of the class Male attached with the help of relations child-parent. A part of a semantic network before applying this production (i.e., before the birth) is shown in Fig. 8.6b and a part of the network after applying the production (i.e. after the birth) is shown in Fig. 8.6c. Let us notice that firstly, the production has to *reconstruct* (horizontal) edges connecting a (happy) mother with her husband and with her sister, because we have destroyed these edges in removing the node Female (mother) corresponding to the left-hand side of the production. Secondly, the production has to establish new edges between the child and his father as well as between the child and his aunt. All the reconstructed edges in Fig. 8.6c are bold. This reconstruction is performed by the production with the help of the embedding transformation, which is defined in the following way.

[36]A similar semantic network has been introduced in Sect. 7.1.

$$C(\textsf{spouse}, out) = \{(\textsf{Female}, \textsf{Male}, \textsf{spouse}, out), \tag{8.19}$$
$$(\textsf{Male}, \textsf{Male}, \textsf{parent}, out), \tag{8.20}$$
$$(\textsf{Male}, \textsf{Male}, \textsf{child}, in)\} \tag{8.21}$$
$$C(\textsf{sister}, out) = \{(\textsf{Female}, \textsf{Female}, \textsf{sister}, out), \tag{8.22}$$
$$(\textsf{Male}, \textsf{Female}, \textsf{aunt}, out), \tag{8.23}$$
$$(\textsf{Male}, \textsf{Female}, \textsf{nephew}, in)\} \tag{8.24}$$
$$C(\textsf{spouse}, in) = \{(\textsf{Female}, \textsf{Male}, \textsf{spouse}, in)\} \tag{8.25}$$
$$C(\textsf{sister}, in) = \{(\textsf{Female}, \textsf{Female}, \textsf{sister}, in)\}. \tag{8.26}$$

For example, formula (8.24) is interpreted in the following way.

- Each edge before the production application, which:
  - has been labeled by sister—C(sister, ...) and
  - has gone out (out) from the left-hand side of the production—C(....., out)

  should be replaced by
- the new edge, which:
  - connects a node of the right-hand side graph labeled by Male—(Male, ..., ..., ...),
  - with a node of the context of the production, which has been pointed out by the old edge[37] and which has been labeled by Female—(....., Female, ..., ...),
  - is labeled with nephew—(....., ..., nephew, ...)
  - and comes into (in) this node of the right-hand side graph—(....., ..., ..., in).

One can easily notice that formulas (8.19), (8.22), (8.25), and (8.26) reconstruct only the old edges, i.e., the edges, which previously existed in the semantic network. On the other hand, the remaining formulas establish new relations between the child and his father as well as between the child and his aunt.

In the case of the use of graph languages in AI we are interested in their analysis more than in their generation. Unfortunately, the construction of an efficient graph automaton is very difficult.[38] At the end of the twentieth century the *ETPL(k)* subclass of edNLC grammars with efficient automata was defined [93, 94]. ETPL(k) graph grammars have been applied for various AI areas such as transforming semantic networks in real-time expert systems, scene analysis in robotic systems, reasoning in multi-agent systems, intelligent integrators for CAD/CAM/CAPP, sign language recognition, model-based reasoning in diagnostic expert systems, etc. The problem of grammar induction, introduced in the previous section, has been solved for these grammars as well [96].

---

[37]The old edge has pointed out an *aunt*—C(sister, out).

[38]This was shown in the 1980s during research into the *membership problem* for graph languages, which was led (independently) by G. Turan and F.J. Brandenburg.

Bibliographical Note

Monographs [41, 104, 113, 215] are good introductions to syntactic pattern recognition.

Chapter 9
Rule-Based Systems

The main idea of reasoning in rule-based systems is, in fact, the same as in the case of logic-based reasoning introduced in Chap. 6.[1] Both models are based on deductive reasoning.[2] As a matter of fact, the form of expressions which are used for knowledge representation is the main difference between these models. In the case of logic-based reasoning expressions are formalized considerably (formulas of First Order Logic, lambda expressions), whereas in rule-based systems the expressions in the form of the so-called *rules* are represented in the following intuitive way: "*If a certain* condition *is fulfilled, then perform a certain* action". Additionally, the way of formulating both a condition and an action is much easier to comprehend than in the case of FOL terms or expressions used in symbolic computing. This is of great importance for designing knowledge bases, which are usually developed not only by IT specialists, but also by experts in the field. Therefore, *clarity* of expressions in a knowledge base is recommended.

The main components of rule-based systems and a reasoning cycle are introduced in the first section. In Sect. 9.2 two reasoning strategies which are based on progressive deduction and regressive deduction are defined. Fundamental issues of a conflict resolution and rule matching are discussed in Sect. 9.3. The relationship of rule-based systems and expert systems is discussed in the last section. We present typical classes of expert systems such as Case-Based Reasoning systems and Model-Based Reasoning systems in this section as well.

9.1 Model of Rule-Based Systems

The generic scheme of a *rule-based system* is shown in Fig. 9.1a. The *working memory* primarily contains representations of *facts* on a certain aspect of the world, which are used in a reasoning process over this aspect. Other information required in the

[1]The reader is recommended to recall the discussion in Sect. 6.1.

[2]Basic notions concerning deductive reasoning are contained in Appendix F.2.

M. Flasiński, *Introduction to Artificial Intelligence*,
DOI 10.1007/978-3-319-40022-8_9

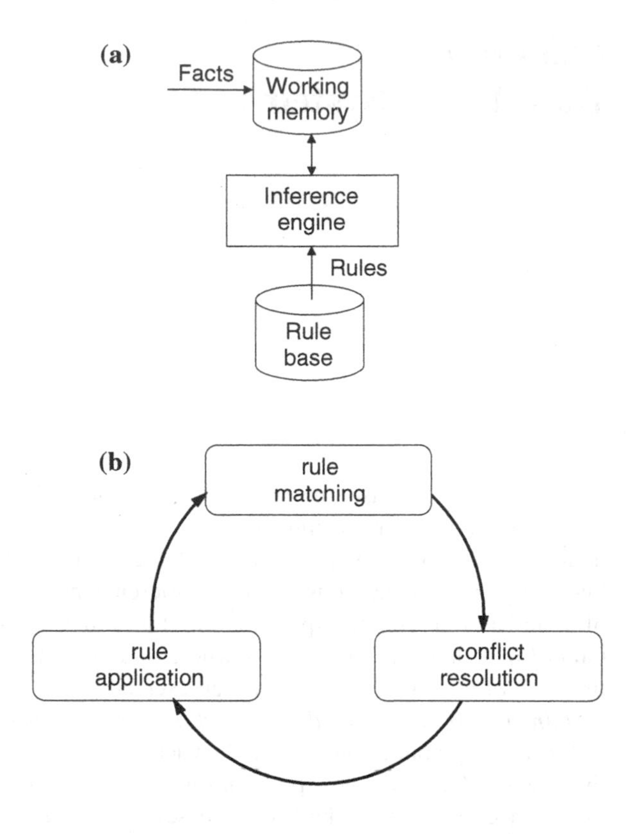

Fig. 9.1 A rule-based system: **a** a module scheme, **b** a reasoning cycle

reasoning process is stored in the working memory, e.g., *working hypotheses*, structures representing variable bindings, etc.

As we have mentioned already, *rules* are of the following form:

$$R: \quad \text{IF } COND \text{ THEN } ACT,$$

where *COND* is a *condition* (*antecedent*) of the rule and *ACT* is an *action* (*consequent*) of the rule.

The antecedent is usually defined as a logical proposition. It is of the form of a conjunction of *elementary conditions* (simple predicates). Thus, *COND* can be defined in the following way:

$$elementary\_condition\_1 \wedge elementary\_condition\_2 \wedge \cdots \wedge elementary\_condition\_k.$$

The fulfillment of the condition of the rule means that there are facts in a working memory such that variables of this condition can be substituted by these facts and after the substitution the condition is valid.

In the case of *declarative rules*, an action *ACT* is usually a logical consequent (conclusion) resulting from the rule condition. Drawing a conclusion results in modifying the content of the working memory by, e.g., adding a new fact, changing the value of some variable, etc. In the case of *reactive rules* an action can be of the form of calling a certain procedure influencing the external environment of the system, e.g., switching off a device. Rules are stored in the *rule base*.

Reasoning on the basis of the rule base and the working memory is controlled by an *inference engine*. The working cycle of an inference engine is shown in Fig. 9.1b. It consists of the following three phases.

- During the *rule-matching* phase the inference engine looks for rules which match[3] facts stored in the working memory. The set of rules which are selected preliminarily is called the *conflict set*.[4]
- In the *conflict resolution* phase the inference engine chooses one rule (sometimes a sequence of rules) to be executed. This is done according to a method of conflict resolution. We discuss such methods in Sect. 9.3.
- In the last phase a chosen rule is *applied*. (We also say that a rule is *fired*.) After a rule application the system comes back to the first phase. The whole reasoning process finishes if no rule matches the facts stored in the working memory.

The formal model of rule-based systems is introduced in Appendix F.1.

9.2 Reasoning Strategies in Rule-Based Systems

Reasoning in rule-based systems is *deductive*, i.e., it is based on the *modus ponendo ponens* rule of inference (introduced in Chap. 6). Reasoning based on this rule can be performed according to two basic strategies[5]:

- Forward Chaining, FC, which is based on progressive deduction[6] or
- Backward Chaining, BC, which is based on regressive deduction.

Now, we discuss both strategies.

A scheme of *Forward Chaining* reasoning is shown in Fig. 9.2. The inference engine tries to match the *condition* of some rule to any fact stored in the working memory. If no fact matches the condition of any rule, then the system does not proceed (cf. Fig. 9.2a). However, if a new fact appears in the working memory (the system can monitor its environment with interfaces, e.g., cameras, sensors), then the system tries to match again.[7] If this new fact matches the *condition* of some rule,

[3]The issue of rule matching is discussed in detail in the next section.

[4]It is called a set of *conflicting* rules because the system has to choose one *reasoning path*, i.e., it has to choose one of several matched rules, which *compete*. (They are in *conflict*.).

[5]In fact, we can combine both strategies into the so-called *mixed strategy* of reasoning in rule-based systems. We do not discuss this in the monograph.

[6]The reader should recall notions of progressive/regressive deduction. They are defined in Appendix F.2.

[7]In our example we do not consider the issue of a conflict situation. This issue is discussed in the next section.

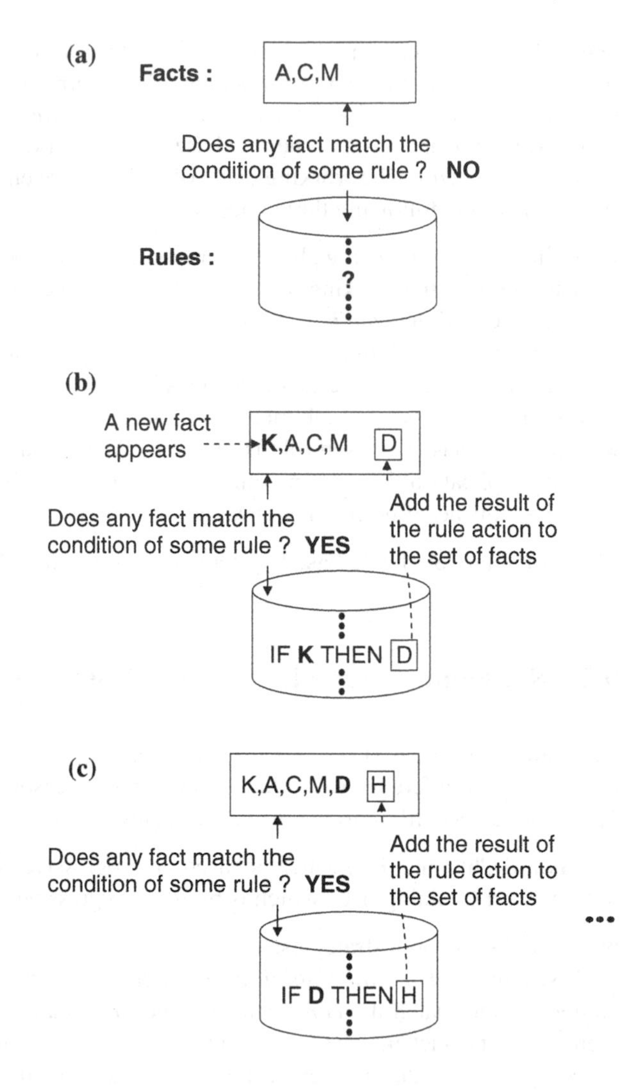

Fig. 9.2 A scheme of forward chaining

then this rule is executed, i.e., the result of performing an *action* of this rule is stored in the working memory (cf. Fig. 9.2b). Thus, changing the content of the working memory can result in matching the new fact (facts) to the condition of some rule, as shown in Fig. 9.2c. This causes the next cycle of the reasoning process.

Let us consider Forward Chaining reasoning with the help of the following example. Let there be defined the following four rules of the system, which controls a piece of industrial equipment.

R^1 : IF $temp(D)$ = high $\wedge$ $work(D)$ = unstable
THEN $status(D)$:= failure ,

R^2 : IF $temp\_sensor(D)$ > 150
THEN $temp(D)$:= high ,

R^3 : IF $status(D)$ = failure $\wedge$ $danger(D)$ = yes
THEN $power\_supply(D)$:= OFF ,

R^4 : IF $temp(D)$ = high $\wedge$ $cooling(D)$ = not_working
THEN $danger(D)$:= yes .

As we can see, the system can perform the following reasoning for a device of a class (type) D:

- the system sets the status to failure if the device is working in an unstable way and it is overheated (the first rule),
- the system considers the device to be overheated if its temperature is greater than 150 °C (the second rule),
- the system switches off the device if it is in failure mode and its environment is endangered (the third rule),
- the system considers the device to be dangerous for its environment if it is overheated and its cooling system is not working (the fourth rule).

Let us assume that at first the rule-based system "knows" two facts about the device *Device_64* of type D: the cooling system is not working and the device is working in an unstable way (see Fig. 9.3a). The inference engine cannot perform reasoning on the basis of these facts, because no rule can match them. Although rules R^1 and R^4 contain expressions *work*(D) and *cooling*(D), the inference engine has to match all the elementary conditions of the antecedent of a rule. Let us assume that at some moment a temperature sensor has measured the temperature of the device, which equals 170 °C. A message concerning this new fact has been sent to the working memory (cf. Fig. 9.3b). Now, the inference engine can match the rule R^2 and as a result of reasoning the temperature status of the device is set to high (*temp*(*Device_64*) = high), as shown in Fig. 9.3c. Now, the inference engine matches two rules, i.e., the conflict set contains rules R^1 and R^4. Let us assume that according to the conflict resolution method the system chooses rule R^1. After its application a new fact appears in the working memory, namely, *status*(*Device_64*) = failure (see Fig. 9.3d). In the next cycle the rule R^4 is matched and fired. As a result, a new fact appears in the working memory, namely *danger*(*Device_64*) = yes (cf. Fig. 9.3e). Finally rule R^3 is fired, which switches off the power supply of the device (see Fig. 9.3f).

If the inference engine tried to match subsequent rules continuously, then it would not work effectively (especially in the case of a big set of rules). Therefore, it "remembers" facts which have already matched a rule and as long as they do not change the

Working memory

(a) Device_64: D		(b)	(c)
cooling	not working	not working	not working
temp_sensor	?	**170**	170
work	unstable	unstable	unstable
status	?	?	?
temp	?	?	**high**
danger	?	?	?
power_supply	?	?	?
		↑ R^2	↑ R^1

Working memory

(d) Device_64: D		(e)	(f)
cooling	not working	not working	not working
temp_sensor	170	170	170
work	unstable	unstable	unstable
status	**failure**	failure	failure
temp	high	high	high
danger	?	**yes**	yes
power_supply	?	?	**OFF**
	↑ R^4	↑ R^3	

Fig. 9.3 An example of forward chaining

rule is not matched again. For example, now rule R^2 is not matched as long as the value sent from the temperature sensor is equal to 170 °C. Of course, for rules which have a complex condition, if one of the elementary conditions changes, then a matching process starts.

Now, we consider *Backward Chaining, BC*. Since it is more complex than Forward Chaining, we begin by analyzing an example. Let us assume that the following three rules are stored in the rule base of a system, which reasons about genealogy relations[8]:

[8]Let us remember that an *agnatic grandfather* is the father of somebody's father.

R^1 : IF $A = child(B) \wedge gender(B) = \text{male}$
THEN $B = father(A)$,

R^2 : IF $D = father(C) \wedge E = father(D)$
THEN $E = agnatic\_grandfather(C)$,

R^3 : IF $F = agnatic\_grandfather(K) \wedge H = father(K) \wedge F = father(G) \wedge gender(G) = \text{male} \wedge G \neq H$
THEN $G = paternal\_uncle(K)$.

The first rule says that B is the father of A, if A is a child of B and B is male. The second rule says that E is the agnatic grandfather of C, if there exists D who is the father of C and E is the father of D. The third rule says that G is a paternal uncle of K, if there exists F who is the agnatic grandfather of K and F is the father of G and G (a paternal uncle) is the male and G is another person than H (the father of K).[9]

The following five facts are stored in the working memory.

F^1 : Raul = *child*(Ian)
F^2 : *gender*(Ian) = male
F^3 : Karl = *father*(Ian)
F^4 : Karl = *father*(Earl)
F^5 : *gender*(Earl) = male

Now, we can begin reasoning. A Backward Chaining strategy is used for verifying hypotheses. Thus, if we want to start reasoning we have to ask the system a question. The system treats the question as a hypothesis, which should be verified on the basis of available facts and knowledge which is formalized in the form of rules.[10]

For example, let us ask the system whether Earl is a paternal uncle of Raul, that is:

$$\text{Earl} = paternal\_uncle(\text{Raul}) \quad (???).$$

Firstly, the inference engine puts this hypothesis on the top of a *stack of hypotheses*.[11] This stack is used for storing working hypotheses (see Fig. 9.4a). The engine tries to prove all the hypotheses stored in the stack by rule matching. If none of rules can be matched and the stack is not empty, then it means that the original hypothesis

[9] If the reader is confused, then she/he is advised to draw a part of a genealogy tree for the third rule.

[10] In an analogous way we have started the reasoning process in a FOL-based system in order to verify a hypothesis in Chap. 6. The way of reasoning is the difference between the two methods. In Chap. 6 we have used the *resolution method*, which is based on *theorem proving by contradiction* (in Latin: *reductio ad absurdum*). In the case of a rule-based system we use Backward Chaining, which is based on *regressive deduction*.

[11] The reader who does not know the notion of a *stack* (in the computer science sense) should read Footnote 17 about pushdown automata in Sect. 8.2.

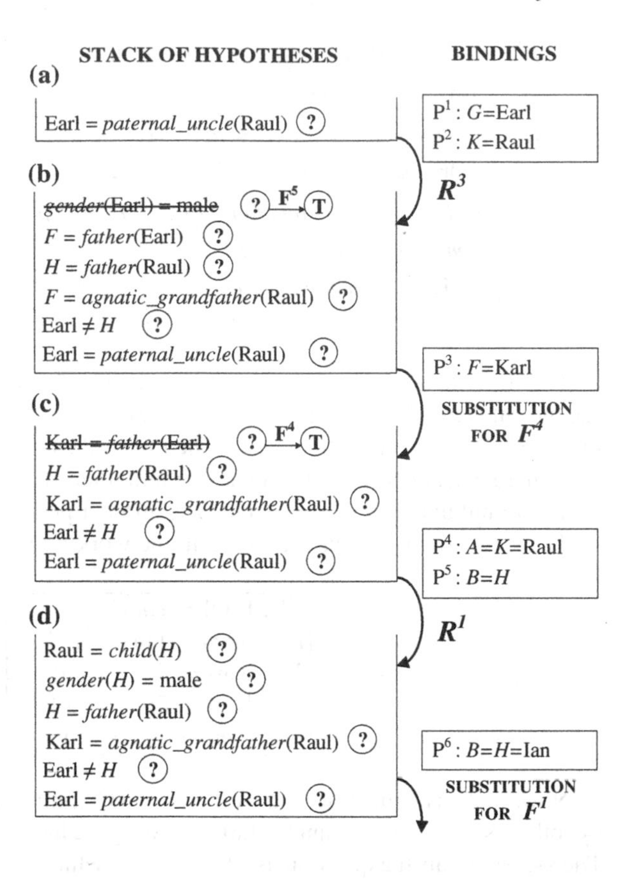

Fig. 9.4 An example of backward chaining

is not valid, i.e., the question asked to the system has the answer No. If the stack is empty, then the original hypothesis is valid.

The inference engine checks whether the hypothesis on the top of the stack belongs to the set of facts stored in the working memory. As we see, it does not belong to the set. Thus, the engine searches the rule base in order to check whether some rule matches the hypothesis. In the case of Backward Chaining this means matching the *action* of the rule[12] to the hypothesis. As we can see, the action of the rule R^3: G = *paternal_uncle*(K) matches Earl = *paternal_uncle*(Raul). Of course, it matches if the engine substitutes the variable G by the constant Earl, denoted: $G \leftarrow$ Earl, and it substitutes the variable K by the constant Raul ($K \leftarrow$ Raul).[13]

After matching the rule R^3 the engine goes to the rule application phase, which in Backward Chaining consists of adding all elementary conditions belonging to the *antecedent* of the rule to the stack of hypotheses. Of course the system has to apply

[12]This is the main difference between BC and FC strategies. In the case of the FC strategy the system matches a *condition* of a rule to facts.

[13]In Figs. 9.4 and 9.5 in the column *BINDINGS* we define bindings after performing successive substitutions.

the substitution defined, i.e., $G \leftarrow$ Earl, $K \leftarrow$ Raul. Thus, in the case of the rule R^3 the engine puts the following elementary conditions[14] on the stack: Earl $\neq H$; $F = $ *agnatic_grandfather*(Raul); $H =$ *father*(Raul); $F =$ *father*(Earl); *gender*(Earl) = male. The situation of the stack of hypotheses after adding elementary conditions of the rule R^3 is shown in Fig. 9.4b. All these elementary conditions are working hypotheses. Therefore, they should be verified. Then, the engine checks, firstly, whether they can be verified on the basis of the facts. The hypothesis on the top is equivalent to the fact F^5. Thus, it can be removed from the stack. A hypothesis which has been positively verified and removed from the stack is marked by crossing out in the figures. The change of its status from unverified (*?*) to valid (true) (*T*) is additionally denoted with the symbol of the fact or rule used for the validation. (In this case the fact $\mathbf{F^5}$ has been used—cf. Fig. 9.4b.)

Now, let us notice that if the substitution: $F \leftarrow$ Karl is made then the hypothesis at the top: $F =$ *father*(Earl) can be verified with the help of the fact F^4. The engine performs this operation as shown in Fig. 9.4c.

After removing the hypothesis Karl = *father*(Earl), the hypothesis $H =$ *father*(Raul) is on the top. No fact matches this hypothesis. Thus, the engine searches the rule base[15] and it finds out that the action of the rule R^1 matches the hypothesis after the following substitutions: P^4: $A \leftarrow$ Raul, and P^5: $B \leftarrow H$.[16] The application of this rule causes its elementary conditions: *gender*(*H*) = male; Raul = *child*(*H*) to be added to the stack (after substitutions P^4 and P^5). This is shown in Fig. 9.4d.

Now, matching the hypothesis on the top to the fact F^1 after performing the substitution: $H \leftarrow$ Ian[17] causes the following sequence of operations on the stack (cf. Fig. 9.5a).

1. The hypothesis Raul = *child*(Ian) is removed according to the fact F^1.
2. The hypothesis *gender*(Ian) = male is removed according to the fact F^2.
3. The hypothesis Ian = *father*(Raul) is recognized as a fact on the basis of the rule R^1 and facts F^1, F^2. As a result it is stored as the fact F^6 in the working memory:

$$\boxed{F^6: \quad \text{Ian} = \textit{father}(\text{Raul})}$$

and it is removed from the stack of hypotheses.

4. The hypothesis Earl $\neq$ Ian is (obviously) recognized as a fact and it is removed from the stack.[18]

Summing up, we obtain the situation shown in Fig. 9.5b.

---

[14]The engine adds elementary conditions in any order, since a conjunction is commutative.

[15]One can easily notice that the inference engine, firstly, tries to match the facts to the hypothesis on the top of the stack. Only if no fact matches the hypothesis, the engine tries to match rules. Let us notice that the fact F^k can be treated as a rule of the form: IF *TRUE* THEN F^k.

[16]Let us notice that after the substitution P^4: $A \leftarrow$ Raul, the following binding of variables and the constant holds: $A = K =$ Raul, cf. Fig. 9.4c.

[17]Let us remember that $B = H$ according to the substitution P^5. Thus, now the following binding of variables and the constant holds: $B = H =$ Ian, cf. Fig. 9.4d.

[18]In principle, this hypothesis should be removed later. However, we verify this obvious fact now.

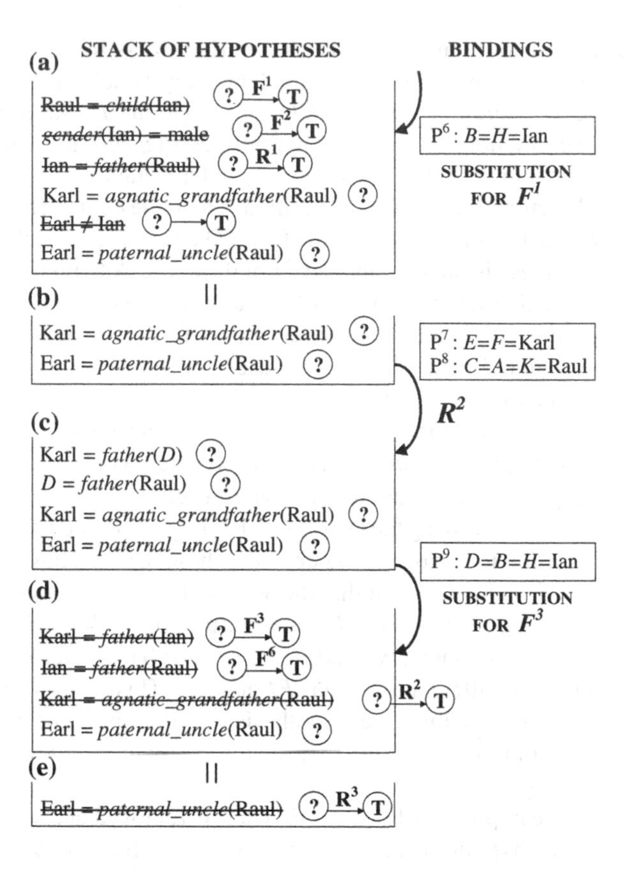

Fig. 9.5 An example of backward chaining—continued

The hypothesis on the top: Karl = *agnatic_grandfather*(Raul) can be matched to the rule R^2 after performing the following substitutions: $E \leftarrow$ Karl and $C \leftarrow$ Raul (cf. Fig. 9.5b). Its application results in the situation shown in Fig. 9.5c.

Now, matching of the hypothesis on the top to the fact F^3 after performing the substitution: $D \leftarrow$ Ian causes the following sequence of operations on the stack (cf. Fig. 9.5d).

1. The hypothesis Karl = *father*(Ian) is removed according to the fact F^3.
2. The hypothesis Ian = *father*(Raul) is removed according to the fact F^6.
3. The hypothesis Karl = *agnatic_grandfather*(Raul) is recognized as a fact on the basis of the rule R^2 and facts F^3, F^6. As a result it is stored as the fact F^7 in the working memory:

$$F^7 : \quad \text{Karl} = \mathit{agnatic\_grandfather}(\text{Raul})$$

and it is removed from the stack of hypotheses.

Finally the hypothesis Earl = *paternal_uncle*(Raul) is recognized as a fact on the basis of the rule R^3 and facts (we list them according to their occurrence in the rule)

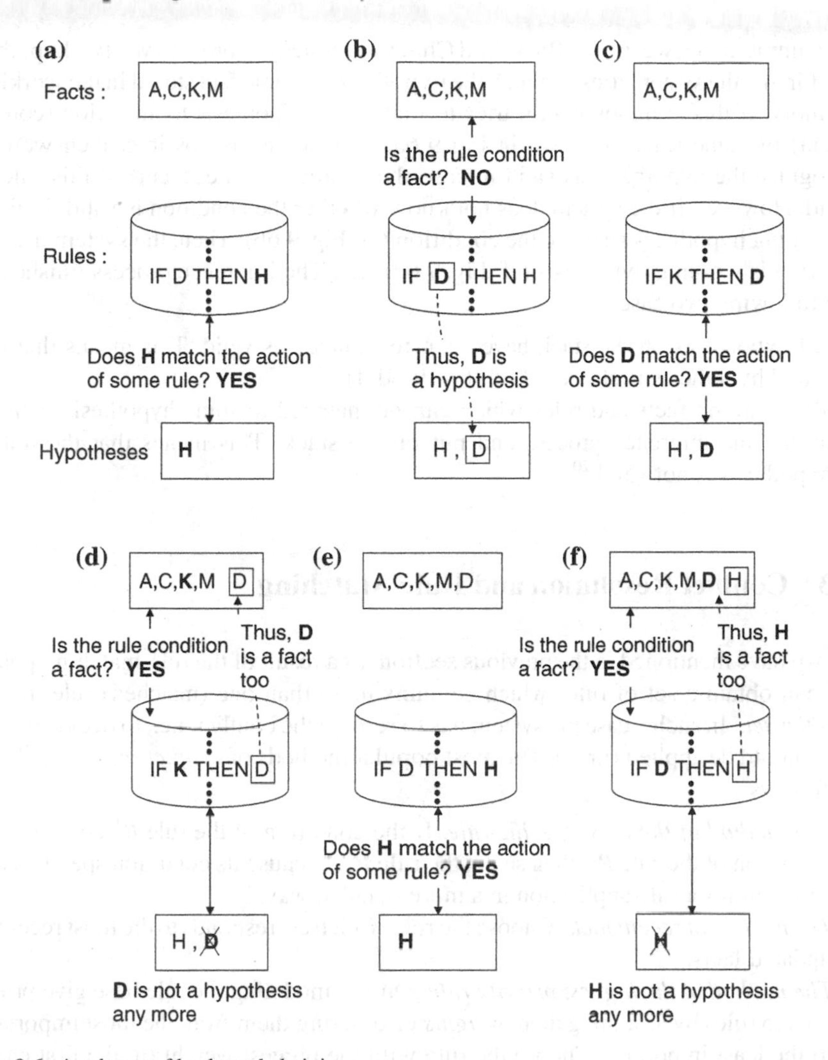

Fig. 9.6 A scheme of backward chaining

F^7, F^6, F^4, F^5, and Earl $\neq$ Ian.[19] As a result it is stored as the fact F^8 in the working memory:

$$F^8: \quad \text{Earl} = \textit{paternal\_uncle}(\text{Raul})$$

The removal of this (initial) hypothesis from the stack finishes the Backward Chaining process. The inference engine has shown that the hypothesis is true.

[19] Of course, taking into account the following bindings determined during the inference process: F = Karl, K = Raul, H = Ian, G = Earl.

Summarizing, we use the Backward Chaining strategy in order to verify a hypothesis. Firstly, the system tries to match the hypothesis to some fact stored in the working memory. If this is impossible, it tries to match the hypothesis to an action (consequent) of some rule, as shown in Fig. 9.6a. If matching is possible, then we can recognize the hypothesis as valid in case the condition (antecedent) of this rule is valid. However, if the system does not know whether the condition is valid, it gives a working hypothesis status to the condition (cf. Fig. 9.6b). Then, the system tries to verify this working hypothesis (cf. Fig. 9.6c), etc. The inference process finishes in the following two cases.

- All hypotheses on the stack have been recognized as valid. This means that the initial hypothesis is also valid (cf. Fig. 9.6d–f).
- There are no facts and rules which can be matched to some hypothesis defined during the inference process and put on the stack. This means that the initial hypothesis is not valid.[20]

9.3 Conflict Resolution and Rule Matching

As we have mentioned in the previous section, as a result of the rule-matching phase we can obtain a set of rules which contains more than one (matched) rule, i.e., a *conflict set*. In such a case the system has to resolve the conflict, i.e., to decide which rule should be applied (fired). The most popular methods of *conflict resolution*[21] are as follows.

- *The method of the most specific rule.* If the condition of the rule R^1 contains the condition of the rule R^2, then select the rule R^1, because its condition specifies the situation for a rule application in a more detailed way.
- *The method of recent facts.* Choose the rule, which corresponds to the most recently updated facts.
- *The method of the highest priority rule.* During modeling the rule base give priorities to rules by assigning them *weights* or ordering them from the most important to the least important. Choose the rule with the biggest weight (in the first case) or the rule which is found first (in the second case).
- *The method of contexts.* Divide rules into subsets related to application contexts. For example, contexts might be defined as *normal work of the equipment monitored, unstable work of the equipment monitored, failure of the equipment monitored*, etc. The inference engine checks the current context before launching the rule-matching phase and it takes into account only rules which belong to the subset relating to this context.

---

[20] In fact, it means that we are not able to recognize the initial hypothesis as valid on the basis of the facts and rules stored in the system.

[21] Apart from the methods listed, we also use the *principle of blocking a rule recently applied* in case its corresponding facts do not change. We have used this rule in the previous section for the FC strategy.

Rule matching is the second crucial issue which influences the efficiency of a reasoning process. Even in a simple rule-based system consisting of a few rules and facts, finding substitutions and creating multiple bindings are time-consuming. In fact, if we have k rules consisting of m conditions on average and n (simple) facts, then there are $k \times m \times n$ matching operations in every cycle of the system. Since in practical applications we define hundreds of rules and there are thousands of facts,[22] the efficiency of rule matching can limit the possibility of the use of a rule-based system.[23]

In 1974 Charles Forgy[24] defined the *Rete algorithm*[25] [101], which speeds up the rule-matching process considerably. The algorithm builds a net which stores relations among antecedents of rules. Variable and constant bindings which result from substitutions are also stored. On this basis all changes which concern facts are propagated in such a net.

9.4 Expert Systems Versus Rule-Based Systems

At the end of our consideration of rule-based systems, we discuss their relation to expert systems. AI systems which solve problems in some domain on the basis of knowledge of human experts that is stored in a *knowledge base* are called *expert systems* or *knowledge-based systems*. Such knowledge can be in the form of a structural representation (e.g., semantic networks, frames, scripts), a logic-based representation (e.g., FOL-like), a generative grammar, rules, etc. Thus, *expert rule-based systems* are a subclass of expert systems. On the other hand, the rule-based paradigm is also used for constructing AI systems which are based on the *cognitive simulation* approach. Soar, ACT*, and ACT-R are good examples of such systems. In these systems rules are not used for codifying specific domain knowledge, but they are applied for modeling the generic behavior of the system. Such systems are called *production systems*[26].

Returning to expert systems, we introduce two important classes of such systems, namely Case-Based Reasoning systems and Model-Based Reasoning systems.

Case-Based Reasoning, CBR, systems are often constructed when defining rules is troublesome. They are also based on domain knowledge. We model them if for similar

---

[22]For example, an AI system designed under the supervision of the author and Dr. Ulf Behrens for a particle physics experiment at the *Deutsches Elektronen Synchrotron* in Hamburg contained more than 1,300 rules and approximately 12,000 facts (see Behrens U., Flasiński M., et al.: Recent developments of the ZEUS expert system. *IEEE Trans. Nuclear Science 43* (1996), pp. 65–68).

[23]The rule-matching phase consumes about 90 % of time of a single cycle.

[24]Charles L. Forgy—a researcher in the area of rule-based systems, a Ph.D. student of Allen Newell. He designed OPS5, which was the first language used for constructing rule-based systems applied in practice.

[25]In Latin *rete* means *net*.

[26]*Rules* play a role, which is analogous to the *productions* of generative grammars.

problems in the application domain similar solutions can be applied. Descriptions of problems which have already been solved by the system and the corresponding solutions are kept in a knowledge base. If a new problem appears, the system looks for similar problems in the knowledge base and for their solutions. A suggested solution is then verified. Verification consists of trying the solution in practice or an evaluation of a computer simulation. If the result of the verification is not satisfactory, then the system proposes a modification of the solution and verifies it once more, etc. If the result of the verification is satisfactory, the system writes the *case* into its knowledge base in the form of a pair *(case, solution)* for the purpose of future use.

Case-Based Reasoning systems are sometimes equated with *systems reasoning by analogy* [123]. However, there is an essential difference between these two classes of AI systems. On one hand, in both cases we use the idea of analogy (in the sense of similarity) between problems solved in the past and the problem to be solved. On the other hand, in the case of Case-Based Reasoning the system solves problems which belong to the same domain and in the case of reasoning by analogy problems need not belong to the same domain. In fact, in typical systems which reason by analogy, discovering analogies between problems belonging to various domains is the key issue. Defining a *script* is a good example of reasoning by analogy. Let us notice that in defining the script *Reconciliation of feuding parties* in Sect. 7.3, we have analyzed two different domains: the domain of relations between friends and the domain of foreign affairs.

In *Model-Based Reasoning, MBR*, systems we use a different approach from reasoning by analogy. Instead of *direct* application of an experience resulting from solving particular cases in the past, a model of a process (a device, a phenomenon, etc.), which will be the object of future reasoning is constructed. A *model* means here an abstract representation of some aspects of the world which are defined in order to perform forecasting, diagnosing, explaining, etc. Models are often of the structural form. For example, a model of a complex device can be a representation which describes its components, functional relations among these components, *cause-effect* relations, etc. The model allows the system to perform computer simulations which can be used, e.g., for forecasting what can happen if some component breaks down.

Apart from system modules introduced in Sect. 9.1 (cf. Fig. 9.1a) expert systems include additional components which make cooperation with them easier. The following *supporting modules* are the most typical.

- An *explanation module* is used for justifying the results of reasoning. For example, returning to our "genealogical example" in Sect. 9.2, the explanation module could explain why it considers Earl to be a paternal uncle of Raul. The FC-based system considered in this section could explain why it has switched off the device *Device_64*.
- A *knowledge acquisition module* usually consists of two independent parts. The first one is used for acquiring facts about the environment of the system. Interfaces to sensors allowing the system to read signals, data, etc. are good examples of such modules. For example, our FC-based system in Sect. 9.2 acquires data concerning the device temperature. Such modules are typical for real-time control expert

systems. The second part allows a knowledge engineer to write rules to a rule base in a convenient, sometimes semi-automatic or automatic, way. Knowledge can be represented in forms different from the rule-like one. For example, it can be represented in the form of decision trees (we introduce them in the next chapter). Then, transforming such knowledge into a rule representation is the task of this submodule.

- A *graphical user interface* is used for communicating to the user effects of reasoning in a way, which is easy to understand. An example of real-time control expert systems is multi-level *cockpits*, which aid in navigation.

In practical applications we often have to operate on *imperfect knowledge* (i.e., knowledge which can be uncertain, imprecise, or incomplete) or we have to define facts with the help of *vague notions*. Then, the methods introduced in this chapter as well as logic-based methods are inadequate and we have to extend our reasoning strategies. Reasoning models used for imperfect knowledge (Bayes networks, Dempster-Shafer theory, and non-monotonic reasoning) are introduced in Chap. 12. Reasoning models used in the case of vague notions (fuzzy sets and rough sets) are considered in Chap. 13.

Bibliographical Note

Rule-based systems are discussed in classic monographs [131, 147, 182, 231, 315].

Case-Based Reasoning is presented in [166, 242] and Model-Based Reasoning in [125].

Chapter 10
Pattern Recognition and Cluster Analysis

Let us begin with a terminological remark, which concerns the notion of a pattern. In pattern recognition and cluster analysis various objects, phenomena, processes, structures, etc. can be considered as *patterns*. The notion is not limited to images, which can be perceived by our sight. There are three basic approaches in the area of pattern recognition. In the approach based on a feature space a pattern is represented by a feature vector. If patterns are of a structural nature, then syntactic pattern recognition (introduced in Chap. 8) or the structural approach[1] is used. In the third approach (artificial) neural networks are applied. (This approach is introduced in the next chapter.)

In general, *pattern recognition* consists of classifying an unknown pattern into one of several predefined categories, called *classes*.[2] *Cluster analysis* can be considered a complementary problem to pattern recognition. Grouping a set of patterns into classes (categories) is its main task.[3]

The task of pattern recognition and its basic notions are formulated in the first section. The next five sections concern various methods of pattern recognition. Cluster analysis is introduced in the last section.

[1]In structural pattern recognition patterns are represented by structural representations, similarly to syntactic pattern recognition. However, their recognition is done with the help of *pattern matching* methods, not, as in the syntactic approach, by applying formal grammars and automata.

[2]For example, patients can be considered as patterns and then *pattern recognition* can consist of classifying them into one of several disease entities.

[3]For example in the area of *Business Intelligence* we can try to group customers on the basis of their features such as the date of their last purchase, the total value of their purchases for the last two months, etc. into categories which determine a sales strategy (e.g., cross-selling, additional free services/products).

M. Flasiński, *Introduction to Artificial Intelligence*,
DOI 10.1007/978-3-319-40022-8_10

10.1 Problem of Pattern Recognition

The problem of *pattern recognition* can be described formally in the following way. Let us assume that there are C categories (classes):

$$\omega^1, \omega^2, \ldots, \omega^C, \tag{10.1}$$

into which patterns can be classified. Let us assume also that each pattern is represented by an n-dimensional *feature vector* $\mathbf{X} = (X_1, X_2, \ldots, X_n)$, where $X_i, i = 1, \ldots, n$ is called the ith *component*[4] *of the vector* $\mathbf{X}$.

In order to perform a pattern recognition task we should have a *learning (training) set*, which is defined in the following way:

$$U = (\,(\mathbf{X}^1, u^1), (\mathbf{X}^2, u^2), \ldots, (\mathbf{X}^M, u^M)\,), \tag{10.2}$$

where $\mathbf{X}^j = (X_1^j, X_2^j, \ldots, X_n^j)$, $j = 1, \ldots, M$, is the jth vector of the learning set and $u^j = \omega^k, k \in \{1, \ldots, C\}$, is the correct classification of the pattern represented by the vector $\mathbf{X}^j$. (This means that the pattern represented by the vector $\mathbf{X}^j$ belongs to the class ω^k.)

In this chapter we focus on the phase of classification, assuming that the representation of patterns in the form of a learning set has been defined in a correct way. However, before we introduce classification methods in the following sections, we consider a few issues which concern defining a correct representation of patterns. In a pattern recognition system, the following three phases precede classification:

- preprocessing,
- feature extraction,
- feature selection.

During *preprocessing* the following operations are performed: noise removal, smoothing, and normalization. Noise removal is usually done with the help of signal filtering methods.[5] Normalization consists of scaling pattern features so that they belong to comparable ranges.

For the classification phase we require the number of pattern features to be as small as possible, i.e., the dimensionality of feature vectors should be as small as possible. If the dimensionality is big, then it results in a high cost of feature measuring, a (time) inefficiency of classification algorithms, and, interestingly, often more errors in the classification phase.[6] Reduction of this dimensionality is the main task of

[4] A component represents some feature of the pattern.

[5] If patterns are images, then noise filtering, smoothing/sharpening, enhancement, and restoration are typical preprocessing operations. Then, features such as edges, characteristic points, etc. are identified. Finally, image segmentation and object identification are performed.

[6] This interesting phenomenon is discussed, e.g., in [235].

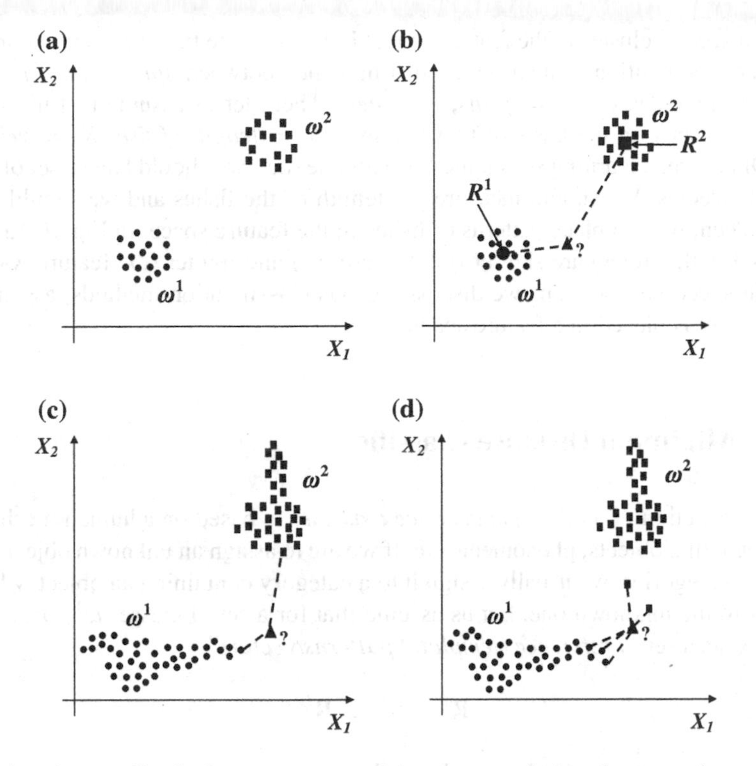

Fig. 10.1 **a** An example of a feature space containing elements of a learning set, and examples of methods: **b** minimum distance classification, **c** nearest neighbor (NN), **d** k-NN

the *feature extraction* phase. The reduction is done by combining and transforming original features into new ones.[7]

Feature selection is the next phase. It consists of selecting those features which have the biggest discriminative power. In other words, we identify those features that lead to the smallest error during the classification phase.[8]

A space containing vectors representing patterns such that their components have been extracted and selected during the phases described above is called a *feature space*. An example feature space is shown in Fig. 10.1a. It is a two-dimensional space, i.e., patterns belonging to it are represented by two features: X_1 and X_2. There are patterns belonging to a class ω^1, marked with circles, in the space. As we can see, these patterns are concentrated, i.e., they are close each to other. We say they create a *cluster*. Similarly, patterns belonging to a class ω^2, marked with rectangles,

[7]The most popular feature extraction methods include *Principal Component Analysis (PCA)*, *Independent Component Analysis*, and *Linear Discriminant Analysis*. The issues related to feature extraction methods are out of the scope of Artificial Intelligence. Therefore, they are not discussed in the book. The reader can find a good introduction to this area in monographs cited at the end of this chapter.

[8]This can be done, for example, with the help of the *search methods* introduced in Chap. 4.

create a second cluster in the feature space. Let us assume that we want to construct a pattern recognition system which distinguishes between *sprats* and *eels*. Thus, there are two classes: ω^1 = *sprats*, ω^2 = *eels*. Then, let us assume that the system classifies fishes on the basis of two features: X_1 = *length of fish*, X_2 = *weight of fish*. Of course, a learning set should be available, i.e., we should have a set of fishes of both species. We should measure the length of the fishes and we should weigh them. Then, we can place patterns of fishes in the feature space. In Fig. 10.1a sprats (marked with circles) are shorter (the feature X_1) and lighter (the feature X_2) than eels. In successive sections we discuss various classification methods, assuming a learning set is placed in a feature space.

10.2 Minimum Distance Classifier

The construction of a *minimum distance classifier* is based on a human mechanism of recognizing objects, phenomena, etc. If we are to assign an unknown object to one of a few categories, we usually assign it to a category containing an object, which is similar to the unknown one. Let us assume that for a set of classes $\omega^1, \omega^2, \ldots, \omega^C$ there exists a set of *reference (template) patterns/vectors*[9]:

$$\mathbf{R}^1, \mathbf{R}^2, \ldots, \mathbf{R}^C. \tag{10.3}$$

In case clusters corresponding to these classes are *regular*, we can assume that a vector computed as the mean (median, mode) vector of the cluster is the reference pattern (cf. Fig. 10.1b).[10]

Now, we can begin classification. If an unknown pattern $\mathbf{X}$ appears, then we should measure its features and place the corresponding feature vector in the feature space (cf. Fig. 10.1b—an unknown pattern is marked by a triangle with a question mark). Then, a minimum distance classifier computes the distances between the unknown pattern and the reference patterns, i.e.,

$$\rho(\mathbf{X}, \mathbf{R}^1),\ \rho(\mathbf{X}, \mathbf{R}^2),\ \ldots\ ,\ \rho(\mathbf{X}, \mathbf{R}^C), \tag{10.4}$$

where $\rho(\mathbf{X}, \mathbf{R}^j)$, $j \in \{1, 2, \ldots, C\}$, is the distance between the pattern $\mathbf{X}$ and the reference pattern $\mathbf{R}^j$. Finally, the classifier assigns the pattern $\mathbf{X}$ to the class ω^L containing the reference pattern $\mathbf{R}^L$ which is the nearest to the pattern $\mathbf{X}$, i.e.,

$$\rho(\mathbf{X}, \mathbf{R}^L) = \min\{\rho(\mathbf{X}, \mathbf{R}^1), \rho(\mathbf{X}, \mathbf{R}^2), \ldots, \rho(\mathbf{X}, \mathbf{R}^C)\}, \tag{10.5}$$

where the function min selects the smallest element from a set.

---

[9]Later we equate a pattern with its representation in the form of a feature vector.

[10]In our "fish example" a reference pattern corresponds to a fish of the mean length and the mean weight in a given class.

According to rule (10.5), a classifier assigns the unknown pattern to class ω^1 in Fig. 10.1b, because the distance between this pattern and the reference pattern $\mathbf{R}^1$ is smaller than the distance between this pattern and the reference pattern $\mathbf{R}^2$, which represents the second class.[11] (Distances are marked with a dashed line.)

If we use a minimum distance classifier, or other methods which compute distances, then the choice of an adequate metric is a crucial issue. For various problems the way of computing a distance influences the accuracy of the method. This issue is discussed in Appendix G.2.

10.3 Nearest Neighbor Method

Sometimes determining reference patterns which represent clusters in an adequate way is troublesome. It is very difficult if clusters are not *regular*, for example if they are *dispersed* and *scattered* in some directions. In such a case we can apply the *Nearest Neighbor, NN*, method. The main idea of the method was introduced by Evelyn Fix[12] and Joseph L. Hodges, Jr[13] in 1951 [91], then it was characterized by Thomas M. Cover[14] and Peter E. Hart[15] in 1967 [60]. In this method we compute the distances between an unknown pattern $\mathbf{X}$ and all the vectors of a learning set U. Then, we select the class containing the pattern, which is the nearest to the pattern $\mathbf{X}$. In the example shown in Fig. 10.1c we assign the unknown pattern $\mathbf{X}$ to the class ω^1, because this class contains the nearest neighbor of $\mathbf{X}$. (In Fig. 10.1c the distance between $\mathbf{X}$ and the nearest pattern belonging to the class ω^2 is also marked.) The NN method has an intuitive interpretation. If we meet an unknown object (event, phenomenon) and we want to classify it, then we can look for a resemblance to a similar object (event, phenomenon) and assign the unknown object to a class including this similar object.

The NN rule can be defined formally in the following way. Let U^k denotes the subset of the learning set including only those patterns that belong to the class ω^k, i.e.,

$$U^k = \{\mathbf{X}^j \ : \ (\mathbf{X}^j, u^j) \in U \text{ and } u^j = \omega^k\}. \tag{10.6}$$

[11] In our fish example this means that an unknown fish corresponding to an unknown pattern in Fig. 10.1b is classified as a sprat (ω^1), because it resembles the "reference sprat" $\mathbf{R}^1$ more than the "reference eel" $\mathbf{R}^2$. (That is, it is nearer to the "reference sprat" in the feature space.).

[12] Evelyn Fix—a professor of statistics of the University of California, Berkeley, a Ph.D. student and then a principal collaborator of the eminent Polish-American mathematician and statistician Jerzy Spława-Neyman, who introduced the notion of a confidence interval (also, the Neyman-Pearson lemma).

[13] Joseph Lawson Hodges, Jr—an eminent statistician (Hodges-Lahmann estimator, Hodges' estimator) at the University of California, Berkeley, a Ph.D. student of Jerzy Spława-Neyman.

[14] Thomas M. Cover—a professor at Stanford University, an author of excellent papers concerning models based on statistics and information theory.

[15] Peter E. Hart—a professor of the Stanford University, a computer scientists (a co-author of a heuristic search method A^* and the model based on the Hough transform).

Then, $\mathbf{X}$ is assigned to the class ω^L, if

$$\rho(\mathbf{X}, \mathbf{X}^r) = \min_{j=1,\ldots,M} \{\rho(\mathbf{X}, \mathbf{X}^j)\} \text{ and } \mathbf{X}^r \in U^L. \tag{10.7}$$

In practice a learning set can contain patterns characterized by values of features which have been measured in an erroneous way. Then, the NN method could give an invalid classification if a pattern with erroneous values of features is the nearest neighbor. In order to eliminate such an effect, we apply the *k-Nearest Neighbor, k-NN*, method. In this method we identify not one nearest neighbor, but the k nearest neighbors. (k is a small odd number.) Then, we check which class possesses the biggest number of representatives in the group of the nearest neighbors. An unknown pattern is assigned to this class. Let us consider the example shown in Fig. 10.1d. Two patterns belonging to the class ω^2 are placed near the class ω^1. (Their features have likely been measured in an erroneous way.) If we apply the NN method, then we assign the unknown pattern to the class ω^2 incorrectly. However, if we use the 5-NN method, i.e., $k = 5$, then in the group of the five nearest neighbors three of them belong to the class ω^1, and the unknown pattern is assigned to this class correctly.

10.4 Decision-Boundary-Based Classifiers

Instead of using reference patterns or elements of a learning set, we can try to construct boundaries which divide the feature space into subspaces corresponding to classes. Such an idea was originally introduced by Ronald A. Fisher in 1936 [90]. It is used for defining *decision-boundary-based classifiers.*

Let us begin by considering the case of two classes in a feature space which can be separated in a linear way, i.e. they can be separated by a boundary which is defined with the help of a linear function called a *linear discriminant function.* Let clusters containing elements of a learning set and representing classes ω^1 and ω^2 be placed in a feature space as shown in Fig. 10.2a. These clusters can be separated by a boundary. The points $\mathbf{X} = (X_1, X_2)$ belonging to the boundary fulfill the following equation:

$$d(\mathbf{X}) = 2X_1 - X_2 - 4 = 0. \tag{10.8}$$

Let us notice that all patterns of the learning set $\mathbf{X}$ which belong to the class ω^1 fulfill the following condition:

$$d(\mathbf{X}) > 0, \tag{10.9}$$

and all patterns of the learning set $\mathbf{X}$ which belong to the class ω^2 fulfill the following condition:

$$d(\mathbf{X}) < 0. \tag{10.10}$$

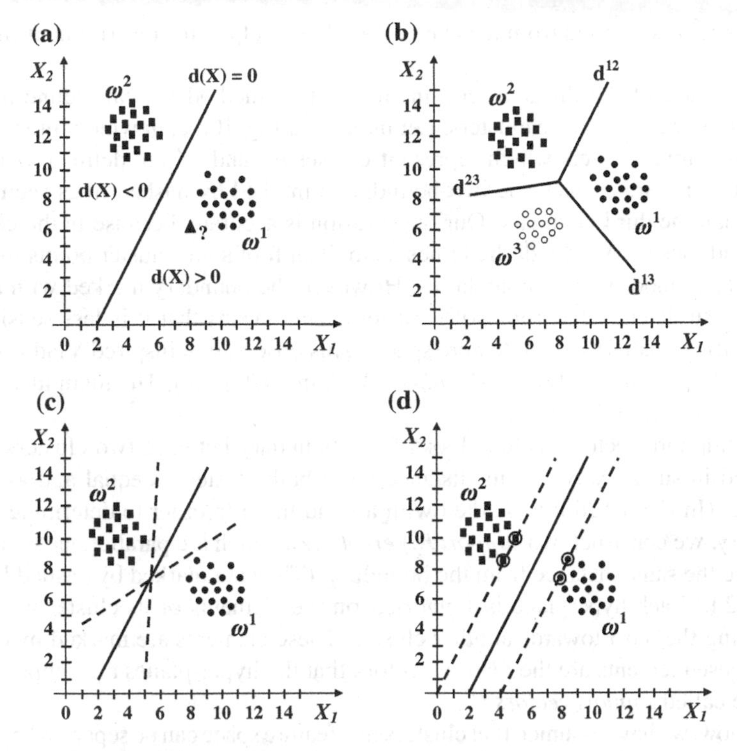

Fig. 10.2 **a** An example of a decision boundary which separates two classes, **b** a separation of three classes by three decision boundaries, **c** possible boundaries between two classes, **d** an example of the Support Vector Machine method

The function d is used for classifying unknown patterns. For example, the unknown pattern marked with a triangle in Fig. 10.2a, which has coordinates $X_1 = 8$, $X_2 = 6$ is assigned to the class ω^1, because the value of the function d for this pattern is greater than zero, according to formula (10.8).

In the general case of an n-dimensional feature space, a linear discriminant function is of the following form:

$$d(\mathbf{X}) = \sum_{i=1}^{n} W_i X_i + W_0, \qquad (10.11)$$

where $\mathbf{W} = (W_1, \ldots, W_n)$ is called the *weight vector* and W_0 is the *threshold weight*. This function corresponds to a boundary which is a hyperplane.

If there are more than two classes, then we can partition the feature space with the help of many boundaries in such a way that classes are separated *pairwise*. An example of such a dichotomous approach is shown in Fig. 10.2b. The class ω^1 is separated from the class ω^2 with the help of the discriminant function d^{12}, the class ω^1 is separated from the class ω^3 with the help of the discriminant function d^{13}, and the class ω^2 is separated from the class ω^3 with the help of the discriminant function d^{23}.

In the case of the linear discriminant function method we can define a lot of boundaries which separate clusters. Let us look at Fig. 10.2c, in which three boundaries separating clusters which represent classes ω^1 and ω^2 are defined. Although all boundaries are correct, the two boundaries marked by dashed lines seem to be *worse* than the third boundary. Our observation is accurate, because in the cases of the boundaries marked by dashed lines a small shift of some cluster points makes it necessary to modify these boundaries. However, the boundary marked with a solid line runs sufficiently far from both clusters. This means that it is less sensitive to some shifts of patterns in the feature space. This observation inspired Vladimir Vapnik[16] to define *Support Vector Machine, SVM*, in 1979 [308]. The main idea of this method is shown in Fig. 10.2d.

The Support Vector Machine looks for a boundary between two clusters which is placed in such a way that its distance from both clusters is equal and as big as possible. (In Fig. 10.2d it is marked with a solid line.) In order to determine such a boundary, we construct two *support hyper-planes*, which are parallel one to another and have the same distance from the boundary. (They are marked by dashed lines in Fig. 10.2d.) Each hyperplane is *supported* on the elements of its cluster which are protruding the most towards another cluster. These elements are marked by circles. Since these elements are the feature vectors that the hyper-planes are supported on, they are called *support vectors*.[17]

Till now, we have assumed that clusters in a feature space can be separated by linear discriminant functions. If they cannot be separated by linear functions, we have to use non-linear discriminant functions and define *non-linear classifiers*. However, constructing efficient non-linear classifiers is very difficult. In the 1990s some efficient non-linear classifiers were defined [35, 92, 259]. An alternative approach consists of using *splines*, i.e., piecewise-defined functions of smaller order.

10.5 Statistical Pattern Recognition

In *statistical pattern recognition (Bayesian approach)*, presented by Richard O. Duda[18] and Peter E. Hart in [78], the probability[19] of assigning a pattern to a class is taken into consideration. These methods are based on the Bayesian model.[20]

---

[16]Vladimir Naumovich Vapnik—a professor at the Institute of Control Sciences, Moscow from 1961 to 1990, then at AT&T Bell Labs and NEC Laboratories, Princeton. His work concerns mainly statistics and Artificial Intelligence (Vapnik–Chervonenkis theory).

[17]This is why the method is called *Support Vector Machines*.

[18]Richard O. Duda—a professor of electrical engineering at San Jose State University. His achievements concern pattern recognition. He defined the Hough transform. He is a co-author of the excellent monograph "Pattern Classification and Scene Analysis".

[19]The basic notions of probability theory are introduced in Appendices I.1, B.1 and I.2.

[20]Thomas Bayes—an eminent English mathematician and a Presbyterian minister. The "father" of statistics.

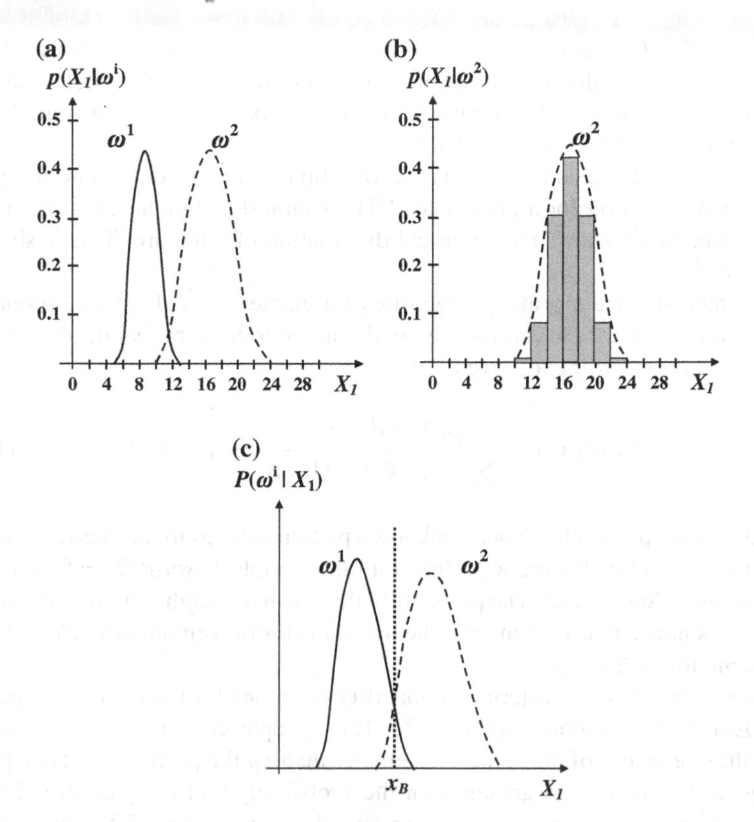

Fig. 10.3 **a** Examples of probability density functions for two classes, **b** a reconstruction of a probability density function on a basis of a histogram, **c** an example of a decision rule for the Bayesian classifier

In order to simplify our considerations, let us assume that all patterns belong to one of two classes in a one-dimensional feature space. Let us assume that for classes ω^1 and ω^2 we know the a priori probabilities $P(\omega^1)$ and $P(\omega^2)$. The a priori probability $P(\omega^1)$ gives the probability that a pattern belongs to the class ω^1 in general.[21]

Then, let us assume that for this (single) feature X_1 characterizing our patterns we know probability density functions $p(X_1|\omega^1)$ and $p(X_1|\omega^2)$ for both classes. The probability density function $p(X_1|\omega^1)$ is a function which for a given value of the variable X_1 assigns the probability of its occurrence, assuming that we say about a pattern which belongs to the class ω^1. Let us return to our "fish example". Now, ω^1 means the class of *sprats*, ω^2 means the class of *anchovy*, and the feature X_1 means the *length of a fish*. Then, $p(9|sprat) = 0.42$ means that the probability that a fish is 9 cm long, assuming that it is a *sprat*, is equal to 0.42. For example, probability density functions for two classes are shown in Fig. 10.3a. As we can see, values of

[21] We may know, for example, that there are four times more patterns belonging to the class ω^1 than patterns belonging to the class ω^2 in nature. Then, $P(\omega^1) = 4/5$ and $P(\omega^2) = 1/5$.

the feature X_1 of patterns belonging to the class ω^1 belong to the interval [5, 13]. Values around 9 are the most probable (the probability more than 0.4). Values of the feature X_1 of patterns belonging to the class ω^2 belong to the interval [10, 24]. Values around 17 are the most probable.

If we use the Bayesian approach, the question is how to determine the probability density function for a given class? The simplest technique consists of using a histogram, which shows the empirical distribution of a feature. This is shown in Fig. 10.3b.[22]

If we have defined a priori probabilities for classes ω^1 and ω^2, and probability density functions for these classes, we can define a posteriori probabilities $P(\omega^1|X_1)$ and $P(\omega^2|X_1)$, according to Bayes' rule:

$$P(\omega^j|X_1) = \frac{p(X_1|\omega^j)P(\omega^j)}{\sum_{k=1}^{2} p(X_1|\omega^k)P(\omega^k)}, \quad j = 1, 2. \tag{10.12}$$

$P(\omega^1|X_1)$ is the probability that an unknown pattern belongs to the class ω^1, depending on the value of its feature X_1. Thus, for our example $P(sprat|7) = 0.35$ is interpreted in the following way. The probability that a fish of length 7 cm is a *sprat* equals 0.35. Let us notice that we can omit the denominator of formula (10.12), because it is the same for both classes.

Example graphs of a posteriori probability functions for both classes depending on the feature X_1 are shown in Fig. 10.3c. These graphs cross for the value x_B. This means that for values of the feature X_1 greater than x_B the probability that a pattern belongs to the class ω^2 is greater than the probability that the pattern belongs to the class ω^1 (and vice versa). As we can see, there are values of X_1 for which the probabilities of belonging to both classes are non-zero. There are also values of X_1 for which the probability of belonging to a given class is equal to zero.

We can generalize our considerations to the case of more than two classes. For classes $\omega^1, \omega^2, \ldots, \omega^C$ formula (10.12) is of the following form:

$$P(\omega^j|X_1) = \frac{p(X_1|\omega^j)P(\omega^j)}{\sum_{k=1}^{C} p(X_1|\omega^k)P(\omega^k)}, \quad j = 1, 2, \ldots, C. \tag{10.13}$$

Now, we can formulate a rule for recognizing an unknown pattern characterized by one feature $\mathbf{X} = (X_1)$ with the help of the *Bayes classifier*. The classifier assigns the pattern to that class for which the a posteriori probability is the biggest. Thus, $\mathbf{X}$ is assigned to the class ω^L, if

$$P(\omega^L|X_1) > P(\omega^j|X_1) \text{ for each } j \in \{1, 2, \ldots, C\}, \; j \neq L. \tag{10.14}$$

[22]The height of the bar for an interval $[a, b]$ should be $h = p/w$, where p is the number of elements of the learning set which belong to the given class and are in the interval $[a, b]$, and w is the number of all elements of the learning set which belong to the given class.

In case we would like to recognize patterns in an n-dimensional feature space, i.e., $\mathbf{X} = (X_1, X_2, \ldots, X_n)$, we can use the so-called *naive Bayes classifier*. We make the assumption that all the features are independent. Then, the probability density function for an n-dimensional feature vector is defined in the following way:

$$p(\mathbf{X}|\omega^j) = \prod_{i=1}^{n} p(X_i|\omega^i). \quad (10.15)$$

We can generalize the Bayes classifier even more. We can assume that erroneous decisions concerning the recognition of an unknown pattern can have various costs, i.e., they have various consequences. Then, we introduce a function of the cost (of errors). This function and the a posteriori probability are the basis for defining the risk function for the classifier. In this approach we try to minimize the risk function.

10.6 Decision Tree Classifier

The pattern recognition methods introduced in the previous sections belong to a *one-stage approach*, in which we make the classification decision taking into account all classes and all features in one step. However, the classification process can be decomposed into a sequence of steps. In subsequent steps we can analyze successive features with respect to various subsets of classes. Such an approach is called a *multistage (sequential) approach*. The *decision tree classifier* introduced by J. Ross Quinlan[23] in 1979 [233] is one of the most popular methods belonging to this approach. Let us introduce it with the help of the following example.

Let us assume that we construct a classifier recognizing the creditworthiness of a customer in a bank. We take into account two features of a customer: X_1 = *Income* (yearly) and X_2 = *Debt* (of the customer to the bank). We assume two classes: ω^1 = *creditworthy customers* and ω^2 = *non-creditworthy customers*. Grouping of elements of the learning set into two clusters is shown in Fig. 10.4. After analyzing these clusters, i.e., analyzing the behavior of customers belonging to the corresponding classes, we decide to divide the feature space with the help of a boundary which separates customers with a yearly income more than 50,000 € from those who have a lower income. Defining this boundary corresponds to constructing the part of the decision tree shown in Fig. 10.4a. The condition "*Income* > 50,000" which defines the threshold is written into a node of the tree. If the condition is fulfilled, then an unknown pattern belongs to the class ω^1 (marked with a circle in the decision tree and in the feature space). Further analysis allows us to divide the feature space according to the feature *Debt* and to set a threshold of 100,000 €. Customers whose debt is greater than this threshold belong to the class ω^2 (marked with a rectangle

[23] John Ross Quinlan—an Australian computer scientist, a researcher at the University of Sydney and the RAND Corporation. His research concerns machine learning, decision theory, and data exploration.

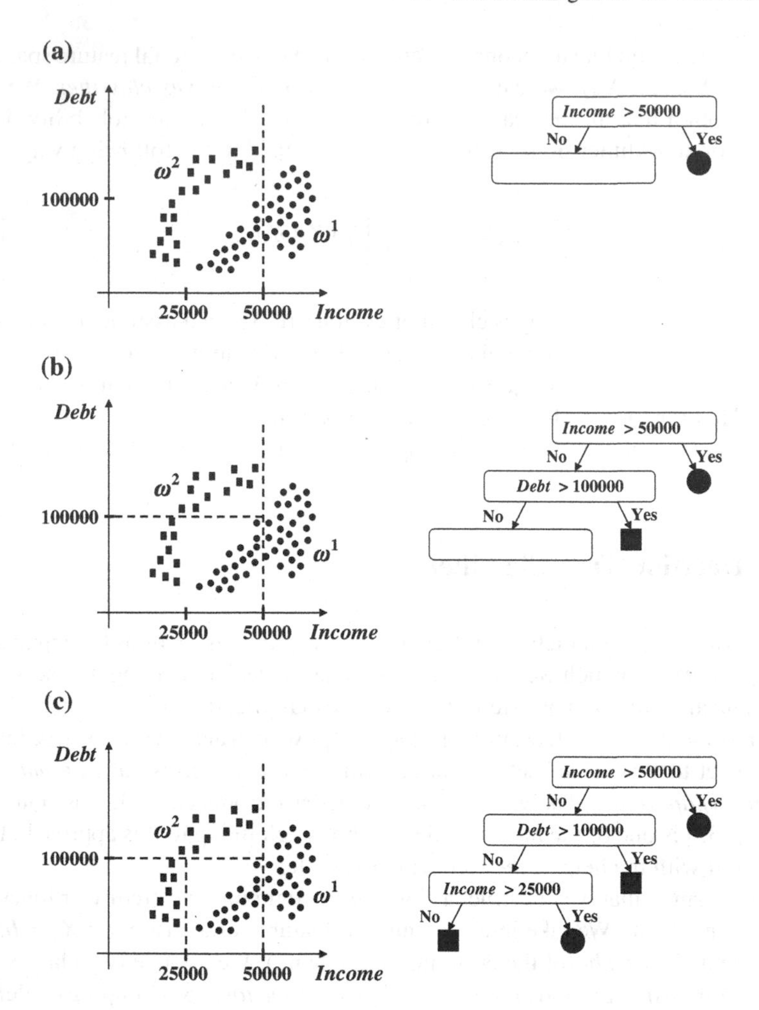

Fig. 10.4 Successive steps constructing a decision tree and partitioning a feature space

in the decision trees and in the feature space). Such a division of the feature space corresponds to developing the decision tree as shown in Fig. 10.4b. Finally, in order to separate customers of the two classes in the lower left subspace we should set a threshold equal to 25,000 € for the feature *Income*. As we can see in Fig. 10.4c this threshold separates creditworthy customers from the non-creditworthy ones in this subspace. We obtain the decision tree shown in Fig. 10.4c as a result.

Summing up, a classifier based on a decision tree divides a feature space with the help of boundaries which are parallel to the axes of the coordinate system that define the feature space. These boundaries are constructed by the sequential identification of thresholds for specific features.

10.7 Cluster Analysis

When we have formulated a pattern recognition task in previous sections, we have assumed that we have a learning set, which consists of patterns with their assignments to proper classes. We have introduced various classifiers, which assign an unknown pattern to a proper class. *Cluster analysis* is a problem which can be considered complementary to pattern recognition. We assume here that we have a set of sample patterns, however we do not know their classification. A cluster analysis task consists of grouping these patterns into clusters, which represent classes. Grouping should be done in such a way that patterns belonging to the same cluster are *similar* to one another. At the same time, patterns which belong to distinct clusters should be *different* from one another.

Thus, the notion of *similarity* is crucial in cluster analysis. Since patterns are placed in a feature space, as in pattern recognition, we use the notion of metric[24] also in this case. We compute distances between patterns of a sample set with the help of a given metric. If the distance between two patterns is small, then we treat them as similar (and vice versa).

In general, cluster analysis methods are divided into the following two groups:

- *methods based on partitioning*, where we assume that we know how many clusters should be defined, and
- *hierarchical methods*, where the number of clusters is not predefined.

K-means clustering is one of the most popular methods based on partitioning. The idea of the method was introduced by Hugo Steinhaus[25] in 1956 [287] and the algorithm was defined by James B. MacQueen[26] in 1967 [190]. Firstly, let us introduce the notion of the *centroid of a cluster*. The centroid is the mean of the positions of all patterns which belong to a given cluster. Let us assume that we want to group patterns into k clusters. The method can be defined in the following way.

1. Select k initial centroids of clusters. (The selection can be made by random choice of k patterns as initial centroids or by random choice of k points in the feature space.)
2. Assign each pattern of the sample set to a cluster on the basis of the smallest distance between the pattern and the cluster centroid.
3. For clusters created in Step 2. compute new centroids.
4. Repeat Steps 2. and 3. until clusters are stabilized. (We say that clusters are stabilized, if pattern assignments do not change in a successive step (or changes

[24] Various metrics are introduced in Appendix G.2.

[25] Hugo Steinhaus—a Polish mathematician, a professor of Jan Kazimierz University in Lwów (now Lviv, Ukraine) and Wrocław University, a Ph.D. student of David Hilbert, a co-founder of the Lwów School of Mathematics (together with, among others, Stefan Banach and Stanisław Ulam). His work concerns functional analysis (Banach-Steinhaus theorem), geometry, and mathematical logic.

[26] James B. MacQueen—a psychologist, a professor of statistics at the University of California, Los Angeles. His work concerns statistics, cluster analysis, and Markov processes.

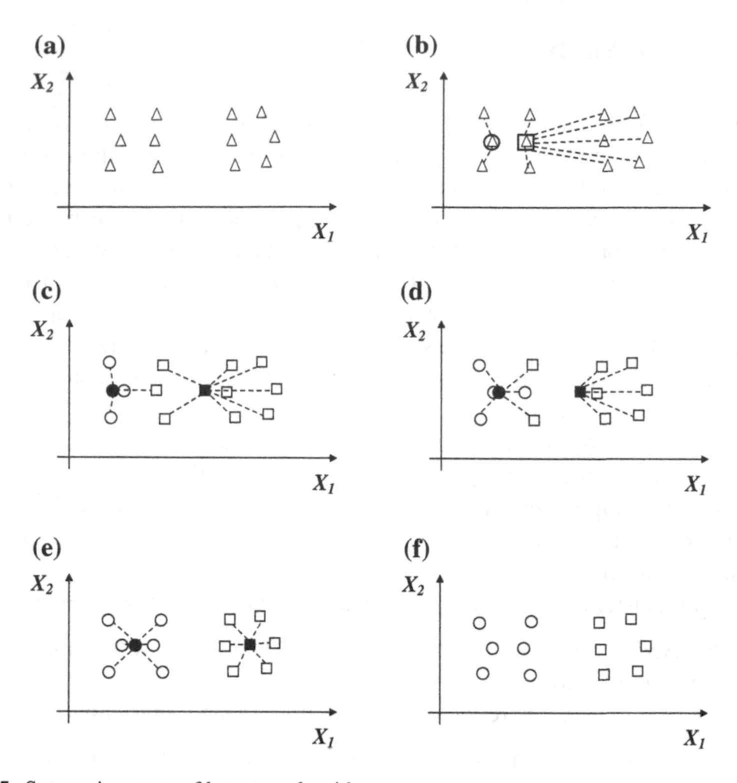

Fig. 10.5 Successive steps of k-means algorithm

are below a certain threshold) or centroids do not change (or changes are below a certain threshold).)

Let us consider an example of the k-means algorithm, which is shown in Fig. 10.5. A placement of patterns in a feature space is shown in Fig. 10.5a. Let us assume that we would like to group the patterns into two clusters, i.e., $k = 2$. (Elements of these clusters are marked either with circles or rectangles.) We assume that we have selected cluster centroids randomly as shown in Fig. 10.5b. The distance between the two leftmost patterns and the centroid of the "circle" cluster is smaller than the distance between these patterns and the centroid of the "rectangle" cluster. Therefore, they are assigned to this cluster (cf. Fig. 10.5b). The remaining patterns are assigned to the "rectangle" cluster, because they are closer to its centroid than to the centroid of the "circle" cluster. (Assignments of patterns to centroids are marked with dashed lines.) After that we compute new centroids for both clusters. We have marked them with a black circle (the first cluster) and a black rectangle (the second cluster). As we can see in Fig. 10.5c, the centroid of the "rectangle" cluster has moved to the right significantly and the centroid of the "circle" cluster has moved to the left a little bit. After setting the new centroids we assign patterns to clusters anew, according to the closest centroid. This time the "circle" cluster absorbs one pattern which was a "rectangle" previously. (It was the initial centroid of "rectangles".) In Fig. 10.5d we can see the movement of both centroids to the right and the absorption of two "rectangle" patterns by the "circle" cluster. The final placement of the centroids is shown in Fig. 10.5e. After this both clusters are stabilized. The effect of the grouping is shown in Fig. 10.5f.

The idea of *hierarchical cluster analysis* was defined by Stephen C. Johnson[27] in 1967 [150]. In such an approach we do not predefine the number of clusters. Instead of this, we show how clusters defined till now can be merged into bigger clusters (*an agglomerative approach*[28]) or how they can be decomposed into smaller clusters (*a divisive approach*[29]).

Let us assume that a sample set consists of M patterns. The scheme of an agglomerative method can be defined in the following way.

1. Determine M initial clusters, which consist of a single pattern. Compute the distances between pairs of such clusters as the distances between their patterns.
2. Find the nearest pair of clusters. Merge them into one cluster.
3. Compute distances between this newly created cluster and the remaining clusters.
4. Repeat Steps 2. and 3. until one big cluster containing all M patterns is created.

An agglomerative scheme is shown in Fig. 10.6. (Successive steps are shown from left to right.) The feature space is one-dimensional (a feature X_1). Firstly, we merge the first two one-element clusters (counting from the top), because they are the nearest to each other. In the second step we merge the next two one-element clusters. In the third step we merge the second two-element cluster with the last one-element cluster, etc. Let us notice that if clusters contain more than one element, we should define a method to compute a distance between them. The most popular methods include:

- the *single linkage method*—the distance between two clusters A and B is computed as the distance between the two nearest elements E_A and E_B belonging to A and B, respectively,
- the *complete linkage method*—the distance between two clusters A and B is computed as the distance between the two farthest elements E_A and E_B belonging to A and B, respectively,
- the *centroid method*—the distance between two clusters A and B is computed as the distance between their centroids.

Successive steps of a divisive scheme are shown in Fig. 10.6 from right to left.

---

[27] Stephen Curtis Johnson—a researcher at Bell Labs and AT&T, then the president of USENIX. A mathematician and a computer scientist. He has developed cpp—a C language compiler, YACC—a UNIX generator of parsers, int—a C code analyzer, and a MATLAB compiler.

[28] In this case we begin with clusters containing single patterns and we can end up with one big cluster containing all patterns of a sample set.

[29] In this case we begin with one big cluster containing all patterns of a sample set and we can end up with clusters containing single patterns.

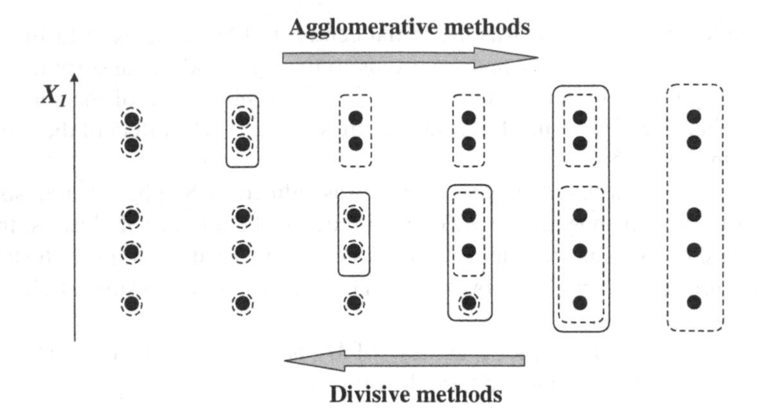

Fig. 10.6 Agglomerative methods and divisive methods in hierarchical cluster analysis

Bibliographical Note

Monographs [28, 78, 79, 106, 171, 309] are good introductions to pattern recognition. Cluster analysis methods are presented in [4, 85, 127].

Chapter 11
Neural Networks

As we have mentioned in the previous chapter, the neural network model (NN) is sometimes treated as one of the three approaches to pattern recognition (along with the approach introduced in the previous chapter and syntactic-structural pattern recognition). In fact, as we will see in this chapter, various models of (artificial) neural networks are analogous to standard pattern recognition methods, in the sense of their mathematical formalization.[1]

Nevertheless, in spite of these analogies, neural network theory is distinguished from standard pattern recognition because of the former original methodological foundations (connectionism), the possibility of implementing standard algorithms with the help of network architectures and a variety of learning techniques.

A lot of different models of neural networks have been developed till now. A taxonomy of these models is usually troublesome for beginners in the area of neural networks. Therefore, in this chapter we introduce notions in a hierarchical ("bottom-up") step-by-step way. In the first section we introduce a generic model of a neuron and we consider the criteria used for defining a typology of artificial neurons. Basic types of neural networks are discussed in Sect. 2. A short survey of the most popular specific models of neural networks is presented in the last section.

[1]Anil K. Jain—an eminent researcher in both these areas of Artificial Intelligence pointed out these analogies in a paper [148] published in 2000. Thus, there are the following analogies: linear discriminant functions—one-layer perceptron, Principal Component Analysis—auto-associative NN, non-linear discriminant functions—multilayer perceptron, etc.

M. Flasiński, *Introduction to Artificial Intelligence*,
DOI 10.1007/978-3-319-40022-8_11

11.1 Artificial Neuron

At the end of the nineteenth century Santiago Ramón y Cajal[2] discovered that a brain consists of neural cells, then called *neurons*. The structure of a neuron is shown in Fig. 11.1a. A neuron consists of a *cell body* (called the *perikaryon* or *soma*) and cellular extensions of two types. Extensions called *dendrites* are thin branching structures. They are used for the transmission of signals from other neurons to the cell body. An extension called an *axon* transmits a signal from the cell body to other neurons.

Communication among neurons is done by transmitting electrical or chemical signals with the help of *synapses*. Research led by John Carew-Eccles[3] discovered the mechanism of communication among neurons. Transmission properties of synapses are *controlled* by chemicals called *neurotransmitters* and synaptic signals can be excitatory or inhibitory.

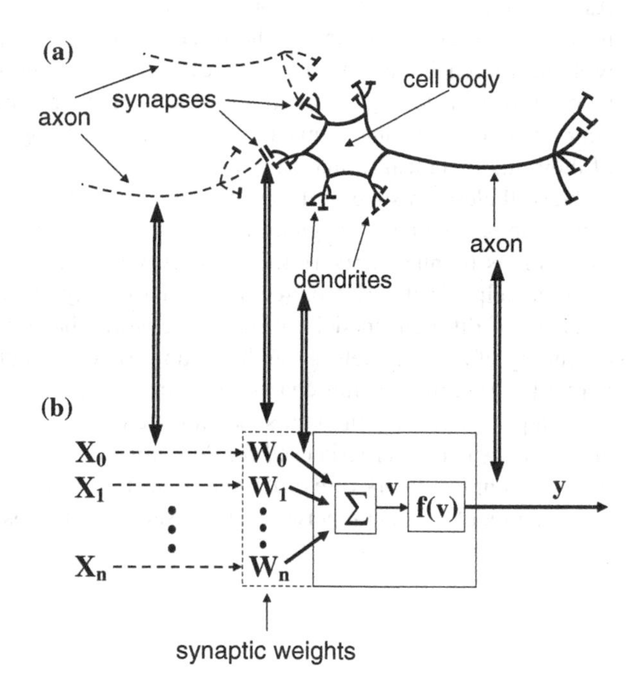

Fig. 11.1 **a** The structure of a neuron, **b** the scheme of an artificial neuron

[2]Santiago Ramón y Cajal—an eminent histologist and neuroscientist, a professor of universities in Valenzia, Barcelona, and Madrid. In 1906 he received the Nobel Prize (together with Camillo Golgi) for research into neural structures.

[3]John Carew Eccles—a professor of neurophysiology at the University of Otago (New Zealand), Australian National University, and the University at Buffalo. In 1963 he was awarded the Nobel Prize for research into synaptic transmission.

At the same time, Alan Lloyd Hodgkin[4] and Andrew Fielding Huxley[5] led research into the process of initiating an action potential, which plays a key role in communication among neurons. They performed experiments on the huge axon of an Atlantic squid. The experiments allowed them to discover the mechanism of this process. At the moment when the total sum[6] of excitatory postsynaptic potential reaches a certain threshold, an *action potential* occurs in the neuron. The action potential is then emitted by the neuron (we say that the neuron *fires*) and it is propagated via its axon to other neurons. As was later shown by Bernard Katz[7] the action potential is generated according to the *all or none* principle, i.e., either it occurs fully or it does not occur at all. In Fig. 11.1a a neuron as described above is marked with a solid line, whereas axons belonging to two other neurons which send signals to it are marked with dashed lines.

As we have mentioned already in Sect. 3.1, an (artificial) *neural network, NN*, is a (simplified) model of a brain treated as a structure consisting of neurons. The model of an *artificial neuron* was developed by Warren S. McCulloch and Walter Pitts in 1943 [198]. It is shown in Fig. 11.1b. We describe its structure and behavior on the basis of the notions which have been introduced for a biological neuron above. Input signals $X_0, X_1, \ldots, X_n$ correspond to neural signals sent from other neurons. We assume (for technical reasons) that $X_0 = 1$. These signals are represented by an *input vector* $\mathbf{X} = (X_0, X_1, \ldots, X_n)$.

In order to compute the total sum of affecting input signals on the neuron, we introduce a *postsynaptic potential function* g. In our considerations we assume that the function g is in the form of a sum. This means that input signals are multiplied by *synaptic weights* $W_0, W_1, \ldots, W_n$, which define a *weight vector* $\mathbf{W} = (W_0, W_1, \ldots, W_n)$. Synaptic weights play the role of the *controller* of transmission properties of synapses by analogy to a biological neuron. The weights set some inputs to be *excitatory synapses* and some to be *inhibitory synapses*. The multiplication of input signals by weights corresponds to the enhancement or weakening of signals sent to the neuron from other neurons. After the multiplication of input signals by weights, we sum the products, which gives a signal v:

$$v = g(\mathbf{W}, \mathbf{X}) = \sum_{i=0}^{n} W_i X_i. \tag{11.1}$$

---

[4] Alan Lloyd Hodgkin—a professor of physiology and biophysics at the University of Cambridge. In 1963 he was awarded the Nobel Prize for research into nerve action potential. He was the President of the Royal Society.

[5] Andrew Fielding Huxley—a professor of physiology and biophysics at the University of Cambridge. In 1963 he was awarded the Nobel Prize (together with Alan Lloyd Hodgkin). He was a grandson of the biologist Thomas H. Huxley, who was called "Darwin's Bulldog" because he vigorously supported the theory of evolution during a famous debate with the Bishop of Oxford Samuel Wilberforce in 1860.

[6] In the sense that multiple excitatory synapses act on the neuron.

[7] Bernard Katz—a professor of biophysics at University College London. He was awarded the Nobel Prize in physiology and medicine in 1970.

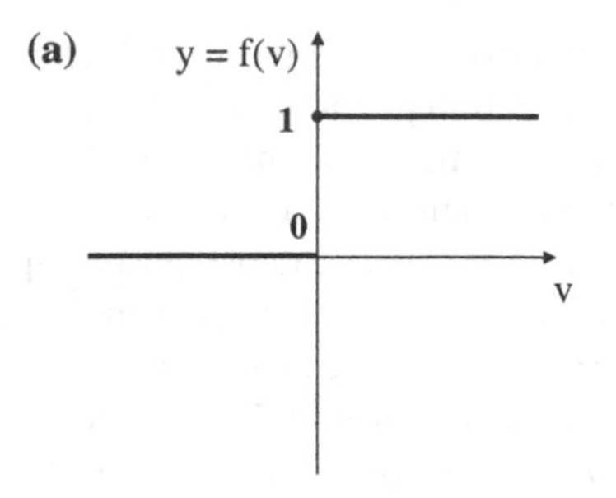

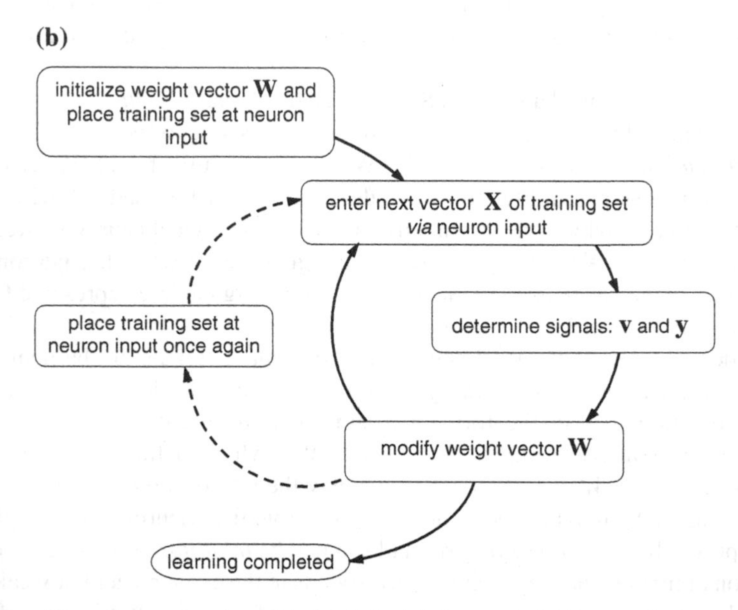

Fig. 11.2 **a** The activation function of the McCulloch-Pitts neuron, **b** the general scheme of neuron learning

This signal corresponds to the total sum of excitatory postsynaptic potential. Thus, we have to check whether it has reached the proper threshold which is required for activating the neuron. We do this with the help of the *activation (transfer) function* f, which generates an *output signal* y for a signal v, i.e.,

$$y = f(v). \tag{11.2}$$

McCulloch and Pitts used the *Heaviside step function*, usually denoted by $\mathbf{1}(v)$ (cf. Fig. 11.2a), as the activation function. It is defined in the following way:

$$\mathbf{1}(v) = \begin{cases} 1, & \text{if } v \geq 0, \\ 0, & \text{if } v < 0. \end{cases} \tag{11.3}$$

As we can see the Heaviside step function gives 0 for values of a signal v which are less than zero, otherwise it gives 1. Thus, the threshold is set to 0. In fact, we can set any threshold with the help of the synaptic weight W_0, since we have assumed that the signal $X_0 = 1$.

In the generic neuron model the *output function out* is the third function (in a sequence of signal processing). It is used in advanced models. In the monograph we assume that it is the identity function, i.e., $out(y) = y$. Therefore, we omit it in further considerations.

The brain is an organ which can learn, so (artificial) neural networks also should have this property. The general scheme of *neuron learning* is shown in Fig. 11.2b. A neuron should learn to react in a proper way to *patterns* (i.e., feature vectors of patterns) that are shown to it.[8] We begin with a random initialization of the weight vector $\mathbf{W}$ of the neuron. We place the training set at the neuron input. Then, we start the main cycle of the learning process.

We enter the next vector $\mathbf{X}$ of the training set[9] via the neuron input. The neuron computes a value v for this vector according to a formula (11.1) and then it determines the value y according to the given activation function.

The output signal y is the reaction of the neuron to a pattern which has been shown. The main idea of the learning process consists of modifying the weights of the neuron, depending on its reaction to the pattern shown. This is done according to the chosen learning method.[10] Thus, in the last step of the main cycle we modify the weight vector $\mathbf{W}$ of the neuron. Then, we enter the next pattern of the training set, etc.

After showing all feature vectors of the training set to the neuron, we can decide whether it has learned to recognize patterns. We claim it has learned to recognize patterns if its weights are set in such a way that it reacts to patterns in a correct way. If not, we have to repeat the whole cycle of the learning process, i.e., we have to place the training set at the neuron input once again and we have to begin showing vectors once again. In Fig. 11.2b this is marked with dashed arrows.

Methods of neuron learning can be divided into the following two groups:

- supervised learning,
- unsupervised learning.

In *supervised learning* the training set is of the form:

$$U = (\,(\mathbf{X}(1), u(1)), (\mathbf{X}(2), u(2)), \ldots, (\mathbf{X}(M), u(M))\,), \qquad (11.4)$$

where $\mathbf{X}(j) = (X_0(j), X_1(j), \ldots, X_n(j)),\ j = 1, \ldots, M$, is the jth input vector and $u(j)$ is the signal which should be generated by the neuron after input of this vector (according to the opinion of a *teacher*). We say that the neuron reacts properly

[8] Showing patterns means entering their feature vectors *via* the neuron input.

[9] The first one is the first vector of the training set.

[10] Basic learning methods are introduced later.

to the vectors shown, if for each pattern $\mathbf{X}(j)$ it generates an output signal $y(j)$ which is equal to the signal $u(j)$ required by the teacher (accurate within a small error).

In *unsupervised learning* the training set is of the form:

$$U = (\mathbf{X}(1), \mathbf{X}(2), \ldots, \mathbf{X}(M)). \tag{11.5}$$

In this case the neuron should modify the weights itself in such a way that it generates the same output signal for *similar* patterns and it generates various output signals for patterns which are *different* from one another.[11]

Let us consider supervised learning with the example of the *perceptron* introduced by Frank Rosenblatt in 1957 [246]. The *bipolar step function* is used as the activation function for the perceptron. It is defined in the following way (cf. Fig. 11.3):

$$f(v) = \begin{cases} 1, & \text{if } v > 0, \\ -1, & \text{if } v \leq 0. \end{cases} \tag{11.6}$$

where v is computed according to formula (11.1). Learning, i.e., modifying the perceptron weights, is performed according to the following principle.
If at the jth step of learning $y(j) \neq u(j)$, then new weights (for the $(j+1)$th step) are computed according to the following formula:

$$W_i(j+1) = W_i(j) + u(j)X_i(j), \tag{11.7}$$

where $X_i(j)$ is the ith coordinate of the vector shown in the jth step and $u(j)$ is the output signal required for this vector. Otherwise, i.e., if $y(j) = u(j)$, the weights do not change, i.e., $W_i(j+1) = W_i(j)$.[12]

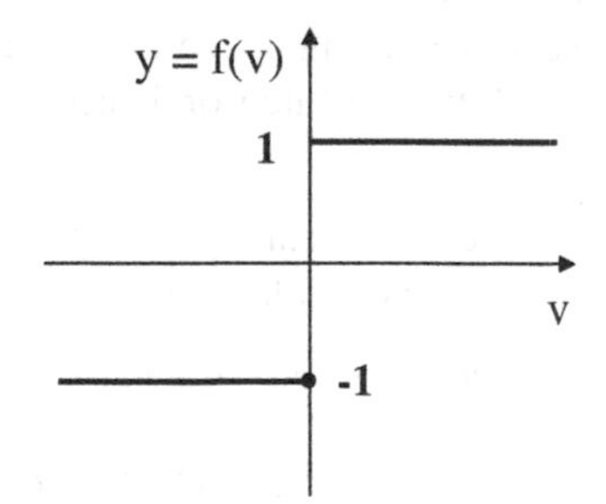

Fig. 11.3 The activation function of the perceptron

[11]The reader will easily notice analogies to *pattern recognition* and *cluster analysis* which have been introduced in the previous chapter.

[12]In order to avoid confusing the indices of the training set elements we assume that after starting a new cycle of learning we re-index these elements, i.e., they take the subsequent indices. Of course, the first weight vector of the new cycle is computed on the basis of the last weight vector of the previous one.

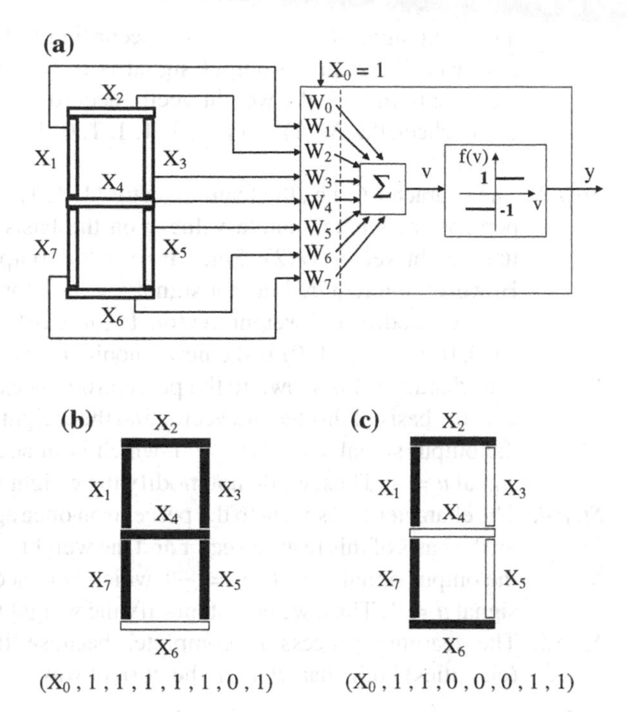

Fig. 11.4 **a** The connection of LED display segments to the perceptron input, **b** the display of the character *A* and the corresponding feature vector, **c** the display of the character *C* and the corresponding feature vector

Let us assume that we would like to use a perceptron for recognizing characters which are shown by an LED display as depicted in Fig. 11.4a. The display consists of seven segments. If a segment is switched on, then it sends a signal to the perceptron according to the scheme of connections which is shown in Fig. 11.4a. Thus, we can denote the input vector by $\mathbf{X} = (X_0, X_1, X_2, X_3, X_4, X_5, X_6, X_7)$.

Let us assume that we would like to teach the perceptron to recognize two characters, *A* and *C*.[13] These characters are represented by the following input signals: $\mathbf{X}_A = (X_0, 1, 1, 1, 1, 1, 0, 1)$ and $\mathbf{X}_C = (X_0, 1, 1, 0, 0, 0, 1, 1)$[14] (cf. Fig. 11.4b, c). In the case of the character *A* the perceptron should generate an output signal $u = 1$ and in the case of the character *C* the perceptron should generate an output signal $u = -1$. Let us assume that $\mathbf{W}(1) = (W_0(1), W_1(1), W_2(1), W_3(1), W_4(1), W_5(1), W_6(1), W_7(1)) = (0, 0, 0, 0, 0, 0, 0, 0)$ is the initial weight vector.[15]

Let us track the subsequent steps of the learning process.

Step 1. The character *A*, i.e., the feature vector (1, 1, 1, 1, 1, 1, 0, 1) is shown to the perceptron. We compute a value v on the basis of this feature vector and the initial weight vector according to formula (11.1). Since $v = 0$, we get

[13] A single perceptron with n inputs can be used for dividing the n-dimensional *feature space* into two areas corresponding to two classes.

[14] Let us remember that $X_0 = 1$ according to our earlier assumption.

[15] In order to show the *idea* of perceptron learning in a few steps, we make *convenient assumptions*, e.g., that the randomly selected initial weight vector is of such a form.

an output signal $y = f(v) = -1$ according to formulas (11.2) and (11.6). However, the required output signal is $u = 1$ for the character A. Thus, we have to modify the weight vector according to formula (11.7). One can easily check that $\mathbf{W}(2) = (1, 1, 1, 1, 1, 1, 0, 1)$ is the new (modified) weight vector.[16]

Step 2. The character C, i.e., the feature vector $(1, 1, 1, 0, 0, 0, 1, 1)$ is shown to the perceptron. We compute a value v on the basis of this feature vector and the weight vector $\mathbf{W}(2)$. Since $v = 4$, the output signal $y = f(v) = 1$. However, the required output signal $u = -1$ for the character C. Thus, we have to modify the weight vector. It can easily be checked that $\mathbf{W}(3) = (0, 0, 0, 1, 1, 1, -1, 0)$ is the new (modified) weight vector.[17]

Step 3. The character A is shown to the perceptron once again. We compute a value v on the basis of this feature vector and the weight vector $\mathbf{W}(3)$. Since $v = 3$, the output signal $y = f(v) = 1$ which is in accordance with the required signal $u = 1$. Thus, we do not modify the weight vector, i.e., $\mathbf{W}(4) = \mathbf{W}(3)$.

Step 4. The character C is shown to the perceptron once again. We compute a value v on the basis of this feature vector and the weight vector $\mathbf{W}(4)$. Since $v = -1$, the output signal $y = f(v) = -1$ which is in accordance with the required signal $u = 1$. Thus, we do not modify the weight vector, i.e., $\mathbf{W}(5) = \mathbf{W}(4)$.

Step 5. The learning process is complete, because the perceptron recognizes (classifies) both characters in the correct way.

Let us notice that the weight vector obtained as a result of the learning process,

$$\mathbf{W} = (W_0, W_1 = 0, W_2 = 0, W_3 = 1, W_4 = 1, W_5 = 1, W_6 = -1, W_7 = 0),$$

has an interesting interpretation. The *neutral* weights $W_1 = W_2 = W_7 = 0$ mean that features X_1, X_2, and X_7 are the same in both patterns. The *positive* weights $W_3 = W_4 = W_5 = 1$ enhance features X_3, X_4, and X_5, which occur (are switched on) in the pattern A and not in the pattern C. On the other hand, the *negative* weight $W_6 = -1$ weakens the feature X_6, which occurs (is switched on) in the pattern C and not in the pattern A.

Although the perceptron is one of the earliest neural network models, it is still an object of advanced research because of its interesting learning properties [25].

After introducing the basic notions concerning construction, behavior, and learning an artificial neuron, we discuss differences among various types of artificial neurons. A typology of artificial neuron models can be defined according to the following four criteria:

- the structured functional scheme,
- the rule used for learning,
- the kind of the activation function,
- the kind of postsynaptic potential function.

A scheme of a neuron of a certain type which presents its functional components (e.g., an adder computing the value of the postsynaptic potential function,

---

[16] We add the vector $\mathbf{X}_A$ to the vector $\mathbf{W}(1)$, because $u = 1$.

[17] We subtract the vector $\mathbf{X}_C$ from the vector $\mathbf{W}(2)$, because $u = -1$.

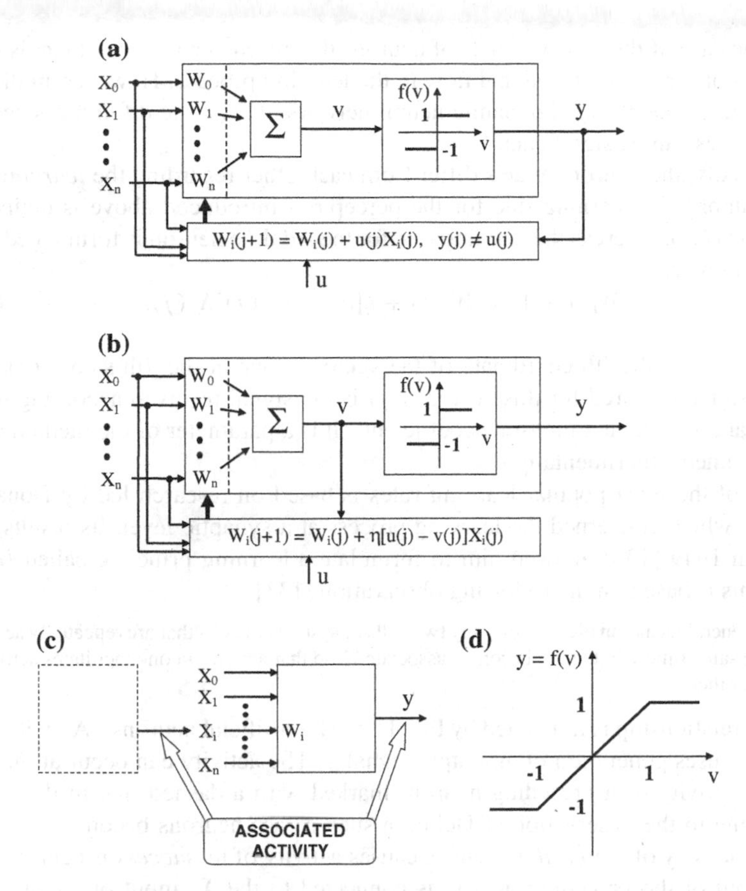

Fig. 11.5 Structured functional schemes of **a** a perceptron and **b** an Adaline neuron; **c** the scheme of Hebb's rule, **d** a piecewise linear activation function

a component generating the value of the activation function, etc.) and data/signal flows among these components is called a *structured functional scheme*.[18] Such a scheme for a perceptron is shown in Fig. 11.5a. A structured functional scheme for an *Adaline* (Adaptive Linear Neuron) introduced by Bernard Widrow[19] and Marcian E. "Ted" Hoff[20] in 1960 [313] is shown in Fig. 11.5b. One can easily notice that in the Adaline scheme the signal v is an input signal of the learning component. (In the perceptron model the signal y is used for learning.)

[18] There is no standard notation for structured functional schemes. Various drawing conventions are used in monographs concerning neural networks.

[19] Bernard Widrow—a professor of electrical engineering at Stanford University. He invented, together with T. Hoff, the least mean square filter algorithm (LMS). His work concerns pattern recognition, adaptive signal processing, and neural networks.

[20] In 1971 Ted Hoff, together with Stanley Mazor, Masatoshi Shima, and Federico Faggin, designed the first microprocessor—Intel 4004.

In the case of these two models of neurons the difference between them is not so big and concerns only the signal flow in the learning process. However, in the case of advanced models, e.g., dynamic neural networks [117], the differences between the schemes can be significant.

Secondly, the neuron models differ from each other regarding the *learning rule*. For example, the learning rule for the perceptron introduced above is defined by formula (11.7), whereas the learning rule for the Adaline neuron is formulated in the following way:

$$W_i(j+1) = W_i(j) + \eta[u(j) - v(j)]X_i(j), \tag{11.8}$$

where $X_i(j)$ is the ith coordinate of the vector shown at the jth step, $u(j)$ is the output signal required for this vector, $v(j)$ is the signal received according to rule (11.1), and η is the learning-rate coefficient. (It is a parameter of the method which is determined experimentally.)

One of the most popular learning rules is based on research led by Donald O. Hebb,[21] which concerned the learning process at a synaptic level. Its results, published in 1949 [133] allowed him to formulate a learning principle called *Hebb's rule*. This is based on the following observation [133]:

> The general idea is an old one, that any two cells or systems of cells that are repeatedly active at the same time will tend to become "associated", so that activity in one facilitates activity in the other.

This relationship is illustrated by Fig. 11.5c (for artificial neurons). Activity of the neuron causes generation of an output signal y. The activity can occur at the same time as activity of a preceding neuron (marked with a dashed line in the figure). According to the observation of Hebb, in such a case neurons become *associated*, i.e., the activity of a *preceding* neuron causes activity of its *successor* neuron. Since the output of the preceding neuron is connected to the X_i input of its successor, the association of the two neurons is obtained by increasing the weight W_i. This relationship can be formulated in the form of *Hebb's rule* of learning:

$$W_i(j+1) = W_i(j) + \eta y(j)X_i(j), \tag{11.9}$$

where η is the learning-rate coefficient. Let us notice that the input signal $X_i(j)$ is equated here with activity of the preceding neuron which causes the generation of its output signal. Of course, both neurons are self-learning, i.e., without the help of a *teacher*.

The *kind of activation function* is the third criterion for defining a taxonomy of neurons. We have already introduced activation functions for the McCulloch-Pitts neuron (the Heaviside step function) and the perceptron (the bipolar step function). Both functions operate in a very *radical* way, i.e., by a step. If we want a less radical operation, we define a *smooth* activation function. For example, the piecewise linear activation function shown in Fig. 11.5d is of the following smooth form:

[21] Donald Olding Hebb—a professor of psychology and neuropsychology at McGill University. His work concerns the influence of neuron functioning on psychological processes such as learning.

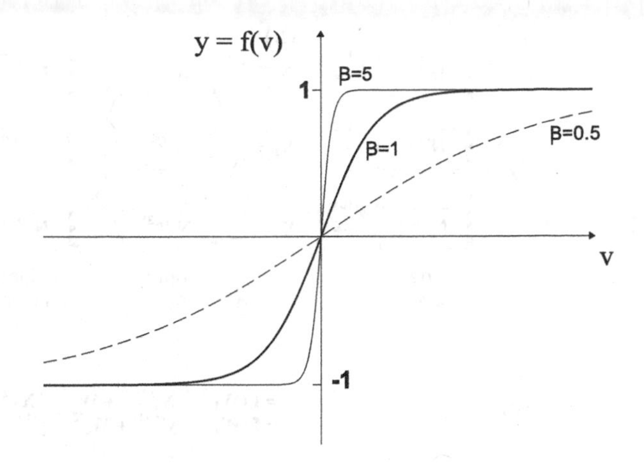

Fig. 11.6 The sigmoidal activation function

$$f(v) = \begin{cases} 1; & \text{if } v > 1, \\ v, & \text{if } -1 \leq v \leq 1, \\ -1, & \text{if } v < -1. \end{cases} \quad (11.10)$$

In the case of a *sigmoidal neuron* we use the *sigmoidal activation function*, which is even smoother (cf. Fig. 11.6). It is defined by the following formula (the bipolar case):

$$f(v) = tanh(\beta v) = \frac{1 - e^{\beta v}}{1 + e^{-\beta v}}. \quad (11.11)$$

As we can see in Fig. 11.6, the greater the value of the parameter β is, the more rapidly the output value changes and the function is more and more similar to the piecewise linear function. In the case of advanced neuron models we use more complex functions such as e.g., radial basis functions or functions based on the Gaussian distribution (we discuss them in Sect. 11.3).

The *kind of postsynaptic potential function* is the last criterion introduced in this section. It is used for computing the total sum of the affecting input signals of the neuron. In this section we have assumed that it is of the form of a sum, i.e., it is defined by formula (11.1). In fact, such a form is used in many models of neurons. However, other forms of postsynaptic potential function can be found in the literature in the case of advanced models such as Radial Basis Function neural networks, fuzzy neural networks, etc.

11.2 Basic Structures of Neural Networks

As we have already mentioned, we use (artificial) neurons for building structures called (artificial) neural networks. A *one-layer neural network* is the simplest structure. Its scheme is shown in Fig. 11.7a. Input signals are usually sent to all neurons of such a structure. In the case of a *multi-layer neural network*, neurons which belong

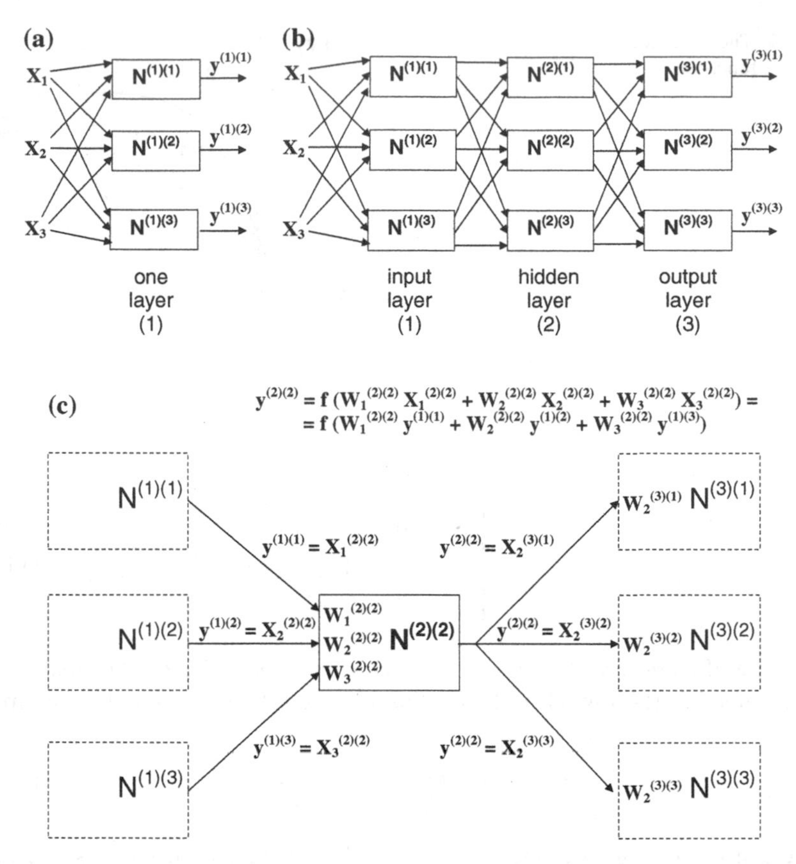

Fig. 11.7 Examples of feedforward neural networks: **a** one-layer, **b** multi-layer (three-layer); **c** the scheme for computing an output signal in a multi-layer network

to the rth layer send signals to neurons which belong to the $(r + 1)$th layer. Neurons which belong to the same layer cannot communicate (cf. Fig. 11.7b). The first layer is called the *input layer*, the last layer is called the *output layer*, and intermediate layers are called *hidden layers*. Networks (one-layer, multi-layer) in which signals flow in only one direction are called *feedforward neural networks*. Let us assume the following notations (cf. Fig. 11.7b, c). $N^{(r)(k)}$ denotes the kth neuron of the rth layer, $y^{(r)(k)}$ denotes its output signal. Input signals to the neuron $N^{(r)(k)}$ are denoted by $X_i^{(r)(k)}$, where $i = 1, \ldots, n$, n is the number of inputs to this neuron,[22] and its weights are denoted by $W_i^{(r)(k)}$. Let us notice also (cf. Fig. 11.7b, c) that the following holds:

$$y^{(r-1)(k)} = X_k^{(r)(p)} \tag{11.12}$$

[22]In our considerations we omit the input $X_0^{(r)(k)}$, because it equals 1 and we omit the weight $W_0^{(r)(k)}$, because it can be (as a constant) taken into account by modifying an activation threshold.

for any pth neuron of the rth layer. In other words, for any neuron which belongs to the rth layer its kth input is connected with the output of the kth neuron of the preceding layer, i.e., of the $(r-1)$th layer.[23]

An example of computing the output signal for a neuron of a feedforward network is shown in Fig. 11.7c. As we can see, for a neuron $N^{(2)(2)}$ the output signal is computed according to formulas (11.1) and (11.2), i.e., $y^{(2)(2)} = f(W_1^{(2)(2)}X_1^{(2)(2)} + W_2^{(2)(2)}X_2^{(2)(2)} + W_3^{(2)(2)}X_3^{(2)(2)})$. In the general case, we use the following formula for computing the output signal for a neuron $N^{(r)(k)}$ taking into account the relationship (11.12):

$$y^{(r)(k)} = f(\sum_i W_i^{(r)(k)}X_i^{(r)(k)}) = f(\sum_i W_i^{(r)(k)}y^{(r-1)(i)}). \tag{11.13}$$

The *backpropagation* method was published by David E. Rumelhart, Geoffrey E. Hinton and co-workers in 1986 [252]. It is the basic learning technique of feedforward networks. At the first step output signals for neurons of the output layer L are computed by subsequent applications of formula (11.3) for neurons which belong to successive layers of the network. At the second step we compute errors $\delta^{(L)(k)}$ for every kth neuron of the Lth (output) layer according to the following formula (cf. Fig. 11.8a):

$$\delta^{(L)(k)} = (u^{(k)} - y^{(L)(k)})\frac{df(v^{(L)(k)})}{dv^{(L)(k)}}, \tag{11.14}$$

where $u^{(k)}$ is the correct (required) output signal for the kth neuron of the Lth layer and f is the activation function.

Then, errors of neurons of the given layer are *propagated backwards* to neurons of a preceding layer according to the following formula (cf. Fig. 11.8b):

$$\delta^{(r)(k)} = \sum_m (\delta^{(r+1)(m)}W_k^{(r+1)(m)})\frac{df(v^{(r)(k)})}{dv^{(r)(k)}}, \tag{11.15}$$

where m goes through the set of neurons of the $(r+1)$th layer. In the last step we compute new weights $W_i^{'(r)(k)}$ for each neuron $N^{(r)(k)}$ on the basis of the computed errors in the following way (cf. Fig. 11.8c):

$$W_i^{'(r)(k)} = W_i^{(r)(k)} + \eta\delta^{(r)(i)}X_i^{(r)(k)} = W_i^{(r)(k)} + \eta\delta^{(r)(i)}y^{(r-1)(i)}, \tag{11.16}$$

where η is the learning-rate coefficient. A mathematical model of learning with the help of the backpropagation method and the derivation of formulas (11.14)–(11.16) are contained in Appendix H.

Fundamental problems of neural network learning include determining a stopping condition for a learning process, how to compute the error of learning, determining

[23] Thus, we could omit an index of the neuron in the case of input signals, retaining only the index of the layer, i.e., we could write $X_i^{(r)}$ instead of $X_i^{(r)(k)}$.

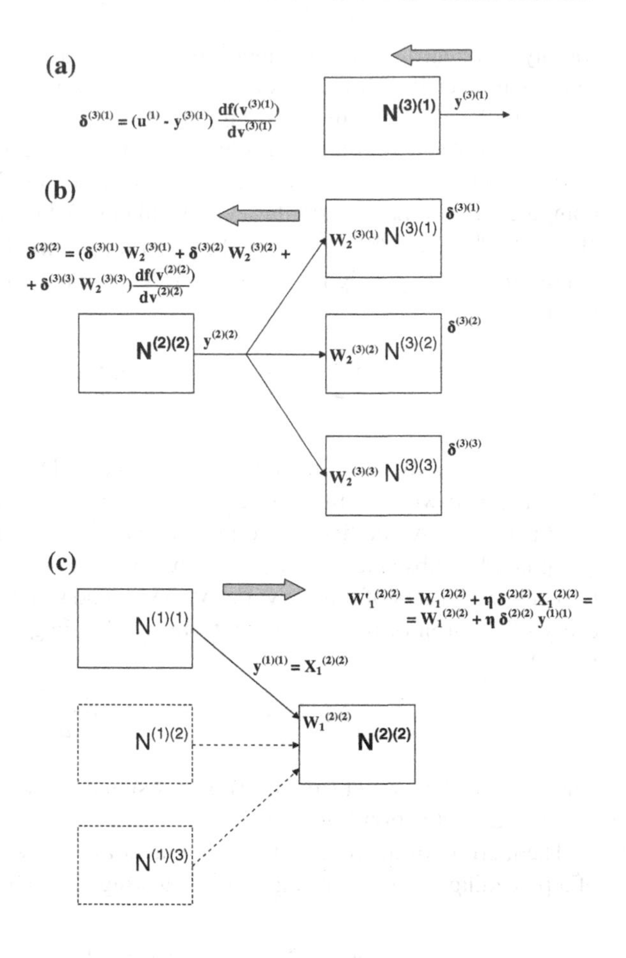

Fig. 11.8 Backpropagation learning: **a** computing the error for a neuron of the output layer, **b** error propagation, **c** computing a weight

initial weights, and speeding up the learning process. We do not discuss these issues in the monograph, because of its introductory nature. The reader is referred to monographs included in the bibliographical note at the end of this chapter.

A neural network can be designed in such a way that it contains connections from some layers to preceding layers. In such a case input signals can be propagated from later processing phases to earlier phases. Such networks are called *recurrent neural networks*, from Latin *recurrere–running back*. They have great computing power that is equal to the computing power of a Turing machine [156]. The first recurrent neural network was introduced by John Hopfield in 1982 [140]. Neurons of a *Hopfield network* are connected as shown in Fig. 11.9 (for the case of three neurons).[24] Input signals are directed multiply to the input of the network and signals

[24]We usually assume, following the first paper of Hopfield [140], that neuron connections are symmetrical, i.e., $W_i^{(1)(k)} = W_k^{(1)(i)}$. However, in generalized models of Hopfield networks this property is not assumed.

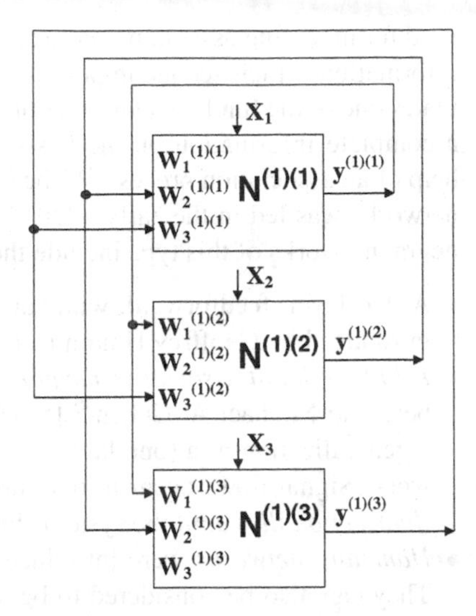

Fig. 11.9 A recurrent Hopfield network

recur for some time until the system stabilizes. Hopfield networks are also used as associative memory networks, which are introduced in the next section.

In general, recurrent networks can be multi-layer networks. In 1986 Michael I. Jordan[25] proposed a model, then called the *Jordan network* [151]. In this network, apart from an input layer, a hidden layer and an output layer, an additional *state layer* occurs. Inputs of neurons of the state layer are connected to outputs of neurons of the output layer and outputs of neurons of the state layer are connected to inputs of neurons of the hidden layer. Jordan networks are used for modeling human motor control.

A similar functional structure occurs in the *Elman network* defined in the 1990s by Jeffrey L. Elman[26] [84]. The main difference of this model with respect to the Jordan network consists of connecting inputs of neurons of an additional layer, called here the *context layer*, with outputs of neurons of the hidden layer (not the output layer). However, outputs of neurons of the context layer are connected to inputs of neurons of the hidden layer, as in the Jordan model. Elman networks are used in Natural Language Processing (NLP), psychology, and physics.

11.3 Concise Survey of Neural Network Models

This section includes a concise survey of the most popular models of neural networks.

Autoassociative memory networks are used for storing pattern vectors in order to recognize similar patterns with the help of an association process. They can be

[25] Michael Irwin Jordan—a professor of computer science and statistics at the University of California, Berkeley and the Massachusetts Institute of Technology. His achievements concern self-learning systems, Bayesian networks, and statistical models in AI.

[26] Jeffrey Locke Elman—an eminent psycholinguist, a professor of cognitive science at the University of California, San Diego.

used for modeling associative storage[27] in computer science. Processing incomplete information is their second important application field. In this case the network simulates one of the fundamental functionalities of the brain, that is the ability to restore a complete information on the basis of incomplete or distorted patterns with the help of an association process.[28] The original research into autoassociative memory networks was led in the early 1970s by Teuveo Kohonen [164]. The most popular neural networks of this type include the following models[29]:

- A two-layer, feedforward, with supervised learning *Hinton network*, which was introduced by Geoffrey Hinton in 1981 [136].
- A *bidirectional associative memory, BAM*, network proposed by Stephen Grossberg and Michael A. Cohen[30] [54].[31] The BAM model can be considered to be a generalization of a (one-layer) Hopfield network for a two-layer recurrent network. Signal flows occur in one direction and then the other in alternate cycles (*bidirectionally*) until the system stabilizes.
- *Hamming networks* were introduced by Richard P. Lippmann[32] in 1987 [183]. They can also be considered to be a generalization of Hopfield networks with a three-layer recurrent structure. Input and output layers are feedforward and the hidden layer is recurrent. Their processing is based on minimizing the Hamming distance[33] between an input vector and model vectors stored in the network.[34]

Self-Organizing Maps, SOMs, were introduced by Teuvo Kohonen [165] in 1982. They are used for cluster analysis, discussed in Sect. 10.7. Kohonen networks generate a discrete representation called a *map* of low dimensionality (maps are usually two- or three-dimensional) on the basis of elements of a learning set.[35] The map shows *clusters* of vectors belonging to the learning set. In the case of Self-Organizing Maps we use a specific type of unsupervised learning, which is called *competitive learning*.

---

[27] Associative storage allows a processor to perform high-speed data search.

[28] For example, if somebody mumbles, we can *guess* correct words. If we see a building that is partially obscured by a tree, we can *restore* a view of the whole building.

[29] Apart from the Hopfield networks introduced in the previous section.

[30] Michael A. Cohen—a professor of computer science at Boston University, Ph.D. in psychology. His work concerns Natural Language Processing, neural networks, and dynamical systems.

[31] The BAM model was developed significantly by Bart Kosko [170], who has been mentioned in Chap. 1.

[32] Richard P. Lippmann—an eminent researcher at the Massachusetts Institute of Technology. His work concerns speech recognition, signal processing, neural networks, and statistical pattern recognition.

[33] The Hamming metric is introduced in Appendix G.

[34] Lippmann called the model the Hamming network in honor of Richard Wesley Hamming, an eminent mathematician whose works influenced the development of computer science. Professor Hamming programmed the earliest computers in the Manhattan Project (the production of the first atomic bomb) in 1945. Then, he collaborated with Claude E. Shannon at the Bell Telephone Laboratories. Professor Hamming has been a founder and a president of the Association for Computing Machinery.

[35] This set can be defined formally by (11.15).

Output neurons compete in order to be activated during the process of showing patterns. Only the best neuron, called the *winner*, is activated. The *winner* is the neuron for which the distance between its weight vector and the shown vector is minimal. Then, in the case of competitive learning based on a *WTA* (*Winner Takes All*) strategy only the weights of the winner are modified. In the case of a *WTM* (*Winner Takes Most*) strategy weights are modified not only for the winner, but also for its neighbors ("the winner takes most [of a prize], but not all").

ART (Adaptive Resonance Theory) neural networks are used for solving problems of pattern recognition. They were introduced by Stephen Grossberg and Gail Carpenter [42] for solving one of the most crucial problems of neural network learning, namely to increase the number of elements of a learning set. In case we increase the number of elements of the learning set,[36] the learning process has to start from the very beginning, i.e., including patterns which have already been shown to the network. Otherwise, the network could *forget* about them. Learning, which corresponds to cluster analysis introduced in Sect. 10.7, is performed in ART networks in the following way. If a new pattern is *similar* to patterns belonging to a certain class, then it is added to this class. However, if it is not *similar* to any class, then it is not added to the *nearest* class, but a new class is created for this pattern. Such a strategy allows the network to preserve characteristics of the classes defined so far. For the learning process a *vigilance parameter* is defined, which allows us to control the creation of new classes. With the help of this parameter, we can divide a learning set into a variety of classes which do not differ from each other significantly or we can divide it into a few generalized classes.

Probabilistic neural networks defined by Donald F. Specht in 1990 [283] recognize patterns on the basis of probability density functions of classes in an analogous way to statistical pattern recognition introduced in Sect. 10.5.

Boltzmann machines[37] can be viewed as a precursor of probabilistic neural networks. They were defined by Geoffrey E. Hinton and Terrence J. Sejnowski[38] in 1986 [137].

Radial Basis Function, RBF, neural networks are a very interesting model. In this model activation functions in the form of *radial basis functions*[39] are defined for each neuron separately, instead of using one global activation function. If we use a *standard* neural network with one activation function (the step function, the sigmoidal function, etc.), then we divide the feature space into subspaces (corresponding to classes) in a *global* way with the help of all the neurons which take part in the process. This is consistent with the idea of a *distributed connectionist network model* introduced in

---

[36]Of course, this is recommended, since the network becomes "more experienced".

[37]Named after a way of defining a probability according to the Boltzmann distribution, similarly to the *simulated annealing* method introduced in Chap. 4.

[38]Terrence "Terry" Joseph Sejnowski—a professor of biology and computer science and director of the Institute of Neural Computation at the University of California, San Diego (earlier at California Institute of Technology and John Hopkins University). John Hopfield was an adviser of his Ph.D. in physics. His work concerns computational neuroscience.

[39]The value of a radial basis function depends only on the distance from a certain point called the *center*. For example, the Gaussian function can be used as a radial basis function.

Sect. 3.1. Meanwhile, the behavior of the basis function changes around the center radially. It allows a neuron to "isolate" (determine) a subarea of the feature space in a *local* way. Therefore, RBF networks are sometimes called *local networks*, which is a reference to the *local connectionist network model* introduced in Sect. 3.1.

Bibliographical Note

There are a lot of monographs on neural networks. Books [27, 50, 87, 129, 257, 324] are good introductions to this area of Artificial Intelligence.

Chapter 12
Reasoning with Imperfect Knowledge

If we reason about propositions in AI systems which are based on classic logic, we use only two possible logic values, i.e., *true* and *false*. However, in the case of reasoning about the real (physical) world such a two-valued evaluation is inadequate, because of the aspect of uncertainty. There are two sources of this problem: *imperfection of knowledge* about the real world which is gained by the system and *vagueness of notions* used for describing objects/phenomena of the real world.

We discuss models which are applied for solving the problem of *imperfect knowledge* in this chapter.[1] There are three aspects of the imperfection of knowledge: *uncertainty of knowledge* (information can be uncertain), *imprecision of knowledge* (measurements of signals received by the AI system can be imprecise) and *incompleteness of knowledge* (the system does not know all required facts).

In the first section the model of *Bayesian inference* based on a probability measure is introduced. This measure is used to express our uncertainty concerning knowledge, not for assessing the degree of truthfulness of propositions. *Dempster-Shafer theory*, which allows us to express a lack of complete knowledge, i.e., our ignorance, with specific measures is considered in the second section. Various models of non-monotonic reasoning can also be applied for solving the problem of incompleteness of knowledge. Three such models, namely *default logic*, *autoepistemic logic*, and *circumscription reasoning* are discussed in the third section.

12.1 Bayesian Inference and Bayes Networks

In Sect. 10.2 we have discussed the use of the Bayesian *probability a posteriori* model[2] for constructing a classifier in statistical pattern recognition. In this section we interpret notions of the Bayesian model in a different way, in another application context.

[1] Models applied for solving a problem of vague notions are introduced in the next chapter.

[2] Mathematical foundations of probabilistic reasoning are introduced in Appendix I.

M. Flasiński, *Introduction to Artificial Intelligence*,
DOI 10.1007/978-3-319-40022-8_12

Let e be an observation of some event (situation, behavior, symptom, etc.).[3] Let $h_1, h_2, \ldots, h_n$ be various (distinct) hypotheses which can explain the occurrence of the observation e. Let us consider a hypothesis h_k with an a priori *probability* (i.e., without knowledge concerning the observation e) of $P(h_k)$. Let us assume that the probability of an occurrence of the observation e assuming the truthfulness of the hypothesis h_k, i.e., the *conditional probability* $P(e|h_k)$, is known. Then, the *a posteriori probability*, i.e., the probability of the hypothesis h_k assuming an occurrence of the observation e, is defined by the following formula:

$$P(h_k|e) = \frac{P(e|h_k) \cdot P(h_k)}{P(e)}, \tag{12.1}$$

where $P(e)$ is the probability of an occurrence of the observation e given hypotheses $h_1, h_2, \ldots, h_n$. The probability $P(e)$ is computed according to the following formula:

$$P(e) = \sum_{i=1}^{n} P(e|h_i) \cdot P(h_i). \tag{12.2}$$

Let us analyze this model with the help of the following example.[4] Let us assume that we would like to diagnose a patient *John Smith*. Then $h_1, h_2, \ldots, h_n$ denote possible disease entities.[5] We assume that the bird flu, denoted with h_p, is spreading throughout our country. The a priori probability of going down with the bird flu can be evaluated as the percentage of our countrymen who have the bird flu.[6] Then, let an observation e mean the patient has a temperature which is more than 39.5 °C. The probability that a patient having the bird flu has a temperature above 39.5 °C, i.e., $P(e|h_p)$, can be evaluated as the percentage of our countrymen having the bird flu who also have a temperature which is more than 39.5 °C.

Now, we can diagnose *John Smith*. If he has a temperature higher than 39.5 °C, i.e., we observe an occurrence of the symptom e, then the probability that he has gone down with the bird flu, $P(h_p|e)$, can be computed with the help of formula (12.1).[7]

Of course, making a hypothesis on the basis of one symptom (one observation) is not sound. Therefore, we can extend our formulas to the case of m observations $e_1, e_2, \ldots, e_m$. If we assume that the observations are conditionally independent

---

[3] Such an observation is represented as a *fact* in a knowledge base.

[4] Of course, all examples are simplified.

[5] Strictly speaking, h_i means making a diagnosis (hypothesis) that the patient has the disease entity denoted by an index i.

[6] Let us notice that it is really an a priori probability in the case of diagnosing *John Smith*, because for such an evaluation we do not take into account any symptoms/factors concerning him.

[7] Let us notice that in order to use $P(e)$ in formula (12.1) we have to compute this probability with formula (12.2). Thus, a priori probabilities $h_1, h_2, \ldots, h_n$ should be known for all disease entities. We should also know the probabilities that a patient having an illness denoted by an index $i = 1, 2, \ldots, n$ has a temperature which is more than 39.5 °C, i.e., $P(e|h_i)$. We are able to evaluate these probabilities if we have corresponding statistical data.

given each hypothesis $h_i, i = 1, \ldots, n$,[8] then we obtain the following formula for the probability of a hypothesis h_k given observations $e_1, e_2, \ldots, e_m$:

$$P(h_k|e_1, e_2, \ldots, e_m) = \frac{P(e_1|h_k) \cdot P(e_2|h_k) \cdot \cdots \cdot P(e_m|h_k) \cdot P(h_k)}{P(e_1, e_2, \ldots, e_m)}, \tag{12.3}$$

where $P(e_1, e_2, \ldots, e_m)$ is the probability of observations $e_1, e_2, \ldots, e_m$ occurring given hypotheses $h_1, h_2, \ldots, h_n$. This probability is computed according to the following formula:

$$P(e_1, e_2, \ldots, e_m) = \sum_{i=1}^{n} P(e_1|h_i) \cdot P(e_2|h_i) \cdot \cdots \cdot P(e_m|h_i) \cdot P(h_i). \tag{12.4}$$

Summing up, the Bayesian model allows us to compute the probability of the truthfulness of a given hypothesis on the basis of observations/facts which are stored in the knowledge base. We will come back to this model at the end of this section, when we present Bayes networks. First, however, we introduce basic notions of probabilistic reasoning.

In probabilistic reasoning a problem domain is represented by a set of *random variables*.[9] For example, in medical diagnosis random variables can represent symptoms (e.g., *a body temperature*, *a runny nose*), disease entities (e.g., *hepatitis*, *lung cancer*), risk factors (*smoking*, *excess alcohol*), etc.

For each random variable its domain (i.e., a set of events for which it is defined) is determined. For example, in the case of car diagnosis for a variable *Engine failure cause* we can determine its domain in the following way:

$$Engine\,failure\,cause\colon \langle piston\,seizing\,up,\ timing\,gear\,failure,\ starter\,failure,\ exhaust\,train\,failure,\ broken\,inlet\,valve\rangle.$$

Random variables are often *logic (Boolean) variables*, i.e., they take either the value *1* (*true* (T)) or *0* (*false (F)*). Sometimes we write *smoking* in case this variable takes the value 1 and we write ¬ *smoking* otherwise.

For a random variable which describes a problem domain we define its *distribution*. The distribution determines the probabilities that the variable takes specific values. For example, assuming that a random variable *Engine failure cause* takes values: *1*, *2*, …, *5* for the events listed above, we can represent its distribution with the help of the one-dimensional table shown in Fig. 12.1a.[10] Let us notice that the probabilities should add up to 1.0.

---

[8]This means that $P(e_1, e_2, \ldots, e_m|h_i) = P(e_1|h_i) \cdot P(e_2|h_i) \cdot \cdots \cdot P(e_m|h_i)$.

[9]In this chapter we consider discrete random variables. Formal definitions of a random variable, a random vector, and distributions are contained in Appendix B.1.

[10]In the first column of the table elementary events are placed. For each elementary event the value which is taken by the random variable is also defined.

(a)

Engine failure cause	P(Engine failure cause)
1 (piston seizing up)	**0.05**
2 (timing gear failure)	**0.1**
3 (starter failure)	**0.4**
4 (exhaust train failure)	**0.3**
5 (broken inlet valve)	**0.15**

(b)

	Unstable gas flow	¬ Unstable gas flow
Gas supply subsystem failure	**0.008**	**0.01**
¬ Gas supply subsystem failure	**0.002**	**0.98**

X_2

X_1

(c)

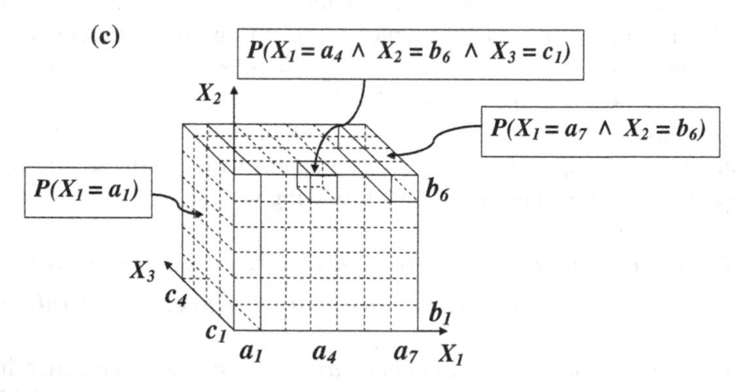

Fig. 12.1 **a** An example of the distribution of a random variable, **b** an example of the distribution of a two-dimensional random vector, **c** the scheme of the table of the joint probability distribution

In the general case, if there are n random variables $X_1, X_2, \ldots X_n$, which describe a problem domain, they create a *random vector* $(X_1, X_2, \ldots X_n)$. In such a case we define the *distribution of a random vector*. This determines all the possible combinations of values that can be assigned to all variables. The distribution of a random vector $(X_1, X_2, \ldots X_n)$ is called the *joint probability distribution, JPD*, of random variables $X_1, X_2, \ldots X_n$.

In the case of two discrete random variables X_1 and X_2 taking values from domains which have m_1 and m_2 elements respectively, their joint probability distribution can be represented by a two-dimensional $m_1 \times m_2$ table $P = [p_{ij}], i = 1, \ldots, m_1$, $j = 1, \ldots, m_2$. An element p_{ij} of the table determines the probability that the variable X_1 takes the value i and the variable X_2 takes the value j. For example, if there are two logical variables, *Unstable gas flow* and *Gas supply subsystem failure*, then their joint probability distribution can be represented as shown in Fig. 12.1b. Then, for example

$$P(Unstable\ gas\ flow \wedge \neg Gas\ supply\ subsystem\ failure) = 0.002.$$

Similarly to the one-dimensional case, the probabilities of all cells of the table should add up to 1.0. Let us notice that we can determine probabilities not only for *complete* propositions which concern a problem domain, i.e., for propositions which include all variables of a random vector with values assigned. We can also determine probabilities for propositions containing only some of the variables. For example, we can compute the probability of the proposition ¬ *Unstable gas flow* by adding the probabilities of the second column of the table, i.e., we can sum the probabilities for all the values which are taken by the other variable *Gas supply subsystem failure*[11]:

$$P(\neg Unstable\ gas\ flow) = 0.01 + 0.98 = 0.99.$$

Returning to an n-dimensional random vector $(X_1, X_2, \ldots X_n)$, the joint probability distribution of its variables is represented by an n-dimensional table. For example, the scheme of such a table is shown in Fig. 12.1c. As we can see, the variables take values $X_1 = a_1, a_2, \ldots, a_7$, $X_2 = b_1, b_2, \ldots, b_6$, $X_3 = c_1, c_2, c_3, c_4$. Each elementary cell of the table contains the probability for the proposition including all the variables. Thus, for example for the proposition $X_1 = a_4 \wedge X_2 = b_6 \wedge X_3 = c_1$ the probability included in the elementary cell defined by the given coordinates of the table is determined. The probability of the proposition $X_1 = a_7 \wedge X_2 = b_6$ is computed by adding the probabilities included in the cells which belong to the rightmost upper "beam". (It is defined according to the marginal distribution for variables X_1 and X_2, whereas X_3 takes any values.) For example, the probability of the proposition $X_1 = a_1$ is computed by adding the probabilities included in the cells which belong to the leftmost "wall". (It is defined according to the marginal distribution for the variable X_1, whereas X_2 and X_3 take any values.)

There are two disadvantages of using the table of the joint probability distribution. Firstly, we should be able to evaluate all values of a random vector distribution. This is very difficult and sometimes impossible in practice. Secondly, it is inefficient, since in practical applications we have hundreds of variables and each variable can take thousands of values. Thus, the number of cells of the table of the joint probability

[11] Let us assume that a random vector (X_1, X_2) is given, where X_1 takes values $a_1, \ldots, a_{m1}$ and X_2 takes values $b_1, \ldots, b_{m2}$. If we are interested only in the distribution of one variable and the other variable can take any values, then we talk about the *marginal distribution* of the first variable. Then, for example the marginal distribution of the variable X_1 is determined in the following way: $P(X_1 = a_i) = P(X_1 = a_i, X_2 = b_1) + \cdots + P(X_1 = a_i, X_2 = b_{m2})$, $i = 1, \ldots, m_1$.
The marginal distribution for the second variable X_2 is determined in an analogous way. For an n-dimensional random vector we can determine the marginal distribution for any subset of variables, assuming that the remaining variables take any values.

distribution can be huge. For example, if there are n variables and each variable can take k values on average, then the number of cells is k^n. Now, we can come back to the Bayesian model, which inspired Pearl [222] to define a method which allows probabilistic reasoning without using the joint probability distribution.

The Pearl method is based on a graph representation called a *Bayes network*. A Bayes network is a directed acyclic graph.[12] Nodes of the graph correspond to random variables, which describe a problem domain. Edges of the graph represent a direct dependency between variables. If an edge goes from a node labeled by X_1 to a node labeled by X_2, then a direct cause-effect relation holds for the variable X_1 (a direct cause) and the variable X_2 (an effect). We say the node labeled by X_1 is the *predecessor* of the node labeled by X_2. Further on, the node labeled by X is equated with the random variable X.

In a Bayes network, for each node which has predecessors we define a table showing an influence of the predecessors on this node. Let a node X_i have p predecessors, $X_{i1}, \ldots, X_{ip}$. Then, the conditional probabilities of all the possible values taken by the variable X_i depending on all possible values of the variables $X_{i1}, \ldots, X_{ip}$ are determined by the table. For example, let a node X_3 have two predecessors X_1 and X_2. Let these variables take values as follows: $X_1 = a_1, \ldots, a_{m1}$, $X_2 = b_1, \ldots, b_{m2}$, $X_3 = c_1, \ldots, c_{m3}$. Then, the table defined for the node X_1 is of the following form ($p_{(i)(j)}$ denotes the corresponding probability):

X_1	X_2	$P(X_3 \mid X_1, X_2)$		
		c_1	$\ldots$	c_{m3}
a_1	b_1	$p_{(1)(1)}$	$\ldots$	$p_{(1)(m3)}$
$\vdots$	$\vdots$	$\vdots$	$\vdots$	$\vdots$
a_1	b_{m2}	$\ldots$	$\ldots$	$\ldots$
$\vdots$	$\vdots$	$\vdots$	$\vdots$	$\vdots$
a_{m1}	b_1	$\ldots$	$\ldots$	$\ldots$
$\vdots$	$\vdots$	$\vdots$	$\vdots$	$\vdots$
a_{m1}	b_{m2}	$p_{(m1 \cdot m2)(1)}$	$\ldots$	$p_{(m1 \cdot m2)(m3)}$

Let us notice that the values of all probabilities in any row of the table should add up to 1.0. If a node X has no predecessors, then we define the table of the distribution of the random variable X as we have done for the example table shown in Fig. 12.1a.

We consider an example of a Bayes network for logical variables.[13] Let us notice that if variables X_1, X_2 in the table above are logical, then there are only four combinations of (logic) values, i.e., *1-1* (i.e. *True-True*), *1-0*, *0-1*, *0-0*.[14] If the variable X_3 is also a logical variable, we can write its value only if it is *True*, because we can

[12] That is, there are no directed cycles in the graph.

[13] In order to simplify our considerations, without loss of generality of principles.

[14] In our examples they are denoted *T-T*, *T-F*, *F-T*, *F-F*, respectively.

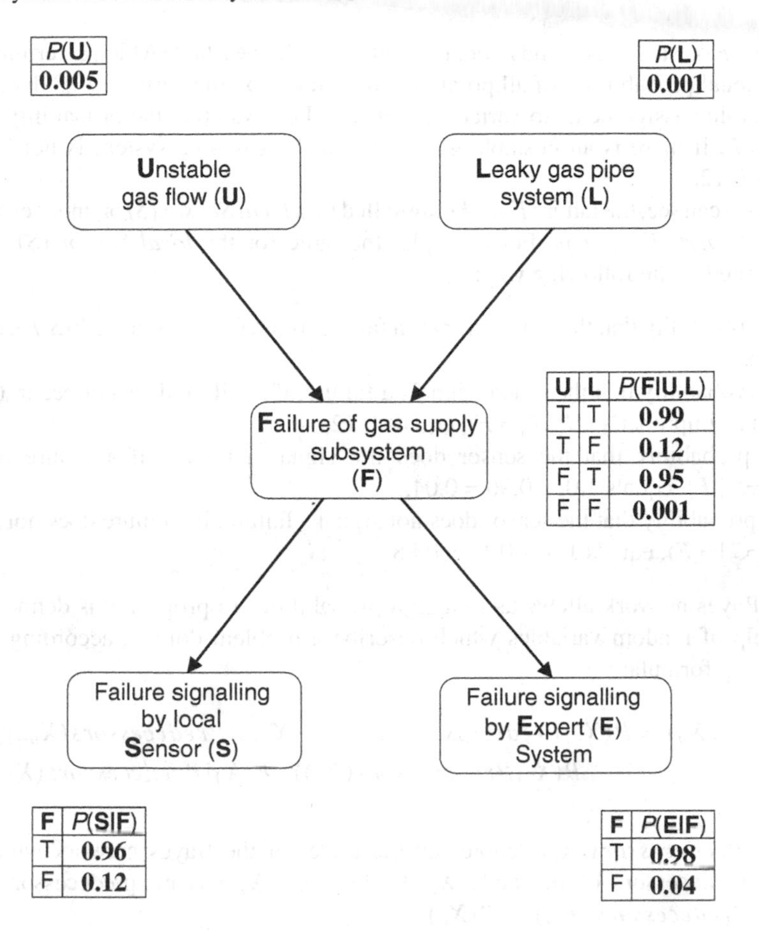

Fig. 12.2 An example of a Bayes network

compute the value corresponding to *False*, taking into account the fact that the two values should add up to 1.0.

An example of a Bayes network defined for diagnosing a gas supply subsystem is shown in Fig. 12.2. The two upper nodes of the network represent the logic random variables *Unstable gas flow* and *Leaky gas pipe system* and they correspond to possible causes of a failure. Tables of distributions of these variables contain probabilities of the causes occurring. For example, the probability of a leaky gas pipe system $P(L)$ equals 0.001. The table for the variable L determines the whole distribution, because the probability that leaking does not occur $P(\neg L)$ is defined in an *implicit* way as the complement of $P(L)$.[15]

Each of the causes U and L can result in *Failure of gas supply subsystem* (F), which is represented by edges of the network. The edges denote the *direct* dependency of the

[15]We can compute it as follows: $P(\neg L) = 1.0 - P(L) = 1.0 - 0.001 = 0.999$.

variable F on variables U and L. For the node F we define a table which determines the conditional probabilities of all possible values taken by the variable F,[16] depending on all value assignments to variables U and L. For example, the probability of the failure F, if there is an unstable gas flow and the gas pipe system is not leaking equals 0.12.

As we can see, the failure F can be signalled by a *local Sensor (S)* or, independently, by an *Expert (E) System*. For example, the table for the *local Sensor (S)* can be interpreted in the following way:

- the probability that the sensor signals a failure, if the failure occurs, $P(S|F)$, equals 0.96,
- the probability that the sensor signals a failure, if a failure does not occur, (i.e., it signals improperly), $P(S \mid\neg F)$, equals 0.12,
- the probability that the sensor does not signal a failure, if a failure occurs, $P(\neg S \mid F)$, equals $1.0 - 0.96 = 0.04$,
- the probability that the sensor does not signal a failure, if a failure does not occur, $P(\neg S \mid\neg F)$, equals $1.0 - 0.12 = 0.88$.

A Bayes network allows us to assign probabilities to propositions defined with the help of random variables which describe a problem domain according to the following formula:

$$P(X_1, \dots X_n) = P(X_n|Predecessors(X_n)) \cdot P(X_{n-1}|Predecessors(X_{n-1})) \cdot \dots \\ \dots \cdot P(X_2|Predecessors(X_2)) \cdot P(X_1|Predecessors(X_1)), \tag{12.5}$$

where $Predecessors(X_i)$ denotes all the nodes of the Bayes network which are direct predecessors of the node X_i. If the node X_i has no predecessors, then $P(X_k|Predecessors(X_k)) = P(X_k)$.

Formula (12.5) says that if we want to compute the probability of a proposition defined with variables $X_1, \dots, X_n$, then we should multiply conditional probabilities representing dependency of $X_i, i = 1, \dots, n$, only for those variables which influence X_i *directly*.

For example, if we want to compute the probability that neither the local sensor nor the expert system signals the failure in case there is an unstable gas flow and the gas pipe system is no leaking, i.e.,

$$U\ , \neg L\ , F\ , \neg S\ , \neg E\ ,$$

then we compute it according to formula (12.5) and the network shown in Fig. 12.2 as follows:

---

[16]The variable F is a logical variable. Therefore, it is sufficient to determine the probabilities when F equals *True*. Probabilities for the value *False* are complements of these probabilities.

$$P(U, \neg L, F, \neg S, \neg E) = P(U) \cdot P(\neg L) \cdot P(F|U, \neg L) \cdot P(\neg S|F) \cdot P(\neg E|F) = 0.005 \cdot 0.999 \cdot 0.12 \cdot 0.04 \cdot 0.02 = 0.00000047952.$$

Finally, let us consider the main idea of constructing a Bayes network that allows us to use formula (12.5) for simplified probabilistic reasoning without using the joint probability distribution. Let network nodes be labeled by variables $X_1, \ldots, X_n$ in such a way that for a given node its predecessors have a lower index. In fact, using formula (12.5), we assume that the event represented by the variable X_i is conditionally independent[17] from earlier events[18] which are *not* its *direct* predecessors, assuming the events represented by its *direct* predecessors[19] have occurred. This means that we should define the structure of the network in accordance with this assumption if we want to make use of formula (12.5). In other words, if we add a node X_i to the Bayes network, we should connect it with all the nodes among $X_1, \ldots, X_{i-1}$ which influence it *directly* (and only with such nodes). Therefore, Bayes networks should be defined in strict cooperation with domain (human) experts.

12.2 Dempster-Shafer Theory

As we have shown in the previous section, Bayes networks allow AI systems to reason in a more efficient way than the standard models of probability theory. Apart from the issue of the efficiency of inference based on imperfect knowledge, the problem of incompleteness of knowledge makes the construction of a reasoning system difficult. In such a situation, we do not know all required facts and we suspect that the lack of complete information influences the quality of the reasoning process. Then, the problem of *expressing lack of knowledge* arises, since we should be able to differentiate between uncertainty concerning knowledge possessed and *our ignorance* (i.e., our awareness of the lack of some knowledge). This problem was noticed by Arthur P. Dempster.[20] To solve it he proposed a model based on the concept of *lower* and *upper probability* in the late 1960s [67]. This model was then developed by Glenn Shafer[21] in 1976 [271]. Today the model is known as *Dempster-Shafer Theory*, *belief function theory*, or the *mathematical theory of evidence*.[22]

---

[17] Conditional independence of variables is defined formally by Definition I.10 in Appendix I.

[18] *Earlier* in the sense of indexing nodes of the network.

[19] We have denoted such predecessors by $Predecessors(X_i)$.

[20] Arthur Pentland Dempster—a professor of statistics at Harvard University. John W. Tukey (the Cooley-Tukey algorithm for Fast Fourier Transforms) was an adviser of his Ph.D. thesis. His work concerns the theory introduced in this section, cluster analysis, and image processing (the *EM* algorithm).

[21] Glenn Shafer—a professor of statistics at Rutgers University. Apart from the development of DST, he proposed a new approach to probability theory based on game theory (instead of measure theory).

[22] In the context of reasoning with incomplete knowledge, *evidence* means information gained by an AI system at some moment which is used as a *premise of inference*.

A complete specification of the probability model is required in the Bayesian approach. On the contrary, in Dempster-Shafer Theory a model can be specified in an incomplete way. The second difference concerns the interpretation of the notion of *probability*[23] and, in consequence, a different way of computing it in a reasoning model. In the Bayesian approach we try to compute the probability that a given proposition (hypothesis) is true. On the contrary, in DST we try to compute the probability saying how available information, which creates the premises of our reasoning, supports our *belief* about the truthfulness of a given proposition (hypothesis). A "probability" interpreted in such a way is measured with the help of a *belief function*, usually denoted Bel.[24]

For example, let us assume that I have found *The Assayer* by Galileo in an unknown antique shop in Rome. I would like to buy it, because I like old books. On the other hand, I am not an expert. So I do not know whether it is genuine. In other words, I do not possess any information concerning the book. In such a case we should define a belief function Bel in the following way according to Dempster-Shafer Theory[25]:

$$Bel(genuine) = 0 \text{ and } Bel(\neg genuine) = 0.$$

Fortunately, I have recalled that my friend Mario, who lives in Rome, is an expert in old books. Moreover, he has a special device which allows him to perform tests. So I have phoned him and I have asked him to help me. Mario has arrived. He has brought two devices. The first one has been made to confirm the authenticity of old books according to certain criteria. The second one has been made to question the authenticity of old books according to other criteria. After taking measurements of the book, he has told me that he believes with a 0.9 degree of certainty that the book is genuine as indicated by the first device. On the other hand, he believes with 0.01 degree of certainty that the book is fake as indicated by the second device. This time the belief function Bel should be computed in the following way:

$$Bel(genuine) = 0.9 \text{ and } Bel(\neg genuine) = 0.01.$$

So I have bought the book.

According to the Dempster-Shafer approach, the belief function Bel is a *lower probability*. The *upper probability* is called a *plausibility function* Pl, which for a proposition S is defined as follows:

$$Pl(S) = 1 - Bel(\neg S).$$

[23] We mean an intuitive interpretation of this notion, not in the sense of probability theory.

[24] Basic definitions of Dempster-Shafer Theory are included in Appendix I.3.

[25] Let us notice that a probability measure P has the following property: $P(\neg genuine) = 1 - P(genuine)$. This property does not hold for a belief function Bel.

Thus, the plausibility function says how strong the evidence is against the proposition S.[26] Coming back to our example, we can compute the plausibility function for *genuine* in the following way:

$$Pl(genuine) = 1 - Bel(\neg genuine) = 1 - 0.01 = 0.99.$$

Summing up, in Dempster-Shafer Theory we define two probability measures Bel and Pl for a proposition. In other words, an interval $[Bel\ ,\ Pl]$ is determined for the proposition. In our example this interval is [0 , 1] for *genuine* before getting the advice from Mario and [0.9 , 0.99] after that. The width of the interval $[Bel\ ,\ Pl]$ for the proposition represents the degree of completeness/incompleteness of our information, which can be used in a reasoning process. If we receive more and more information (evidence) the interval becomes narrow. Rules which allow us to take into account new evidence for constructing a belief function are defined in Dempster-Shafer Theory as well [67, 271].

12.3 Non-monotonic Reasoning

Reasoning models based on classical logic are *monotonic*. This means that after adding new formulas to a model the set of its consequences is not reduced. Extending the set of formulas can cause the possibility of inferring additional consequences, however all consequences that have been inferred previously are sound. In the case of AI systems which are to be used for reasoning about the real (physical) world, such a reasoning scheme is not valid, because our beliefs (assumptions) are often based on uncertain and incomplete knowledge.

For example, I claim "my car has good acceleration".[27] I can use this proposition in a reasoning process, since I have no information which contradicts it. However, I have just got a new message that my car has been crushed by a bulldozer. This means that the claim "my car has good acceleration" should be removed from the set of my beliefs, as well as all propositions which have been previously inferred on the basis of this claim. Thus, the new proposition has not extended my set of beliefs. It has reduced this set. As we can see, common-sense logic which is used for reasoning about the real (physical) world is *non-monotonic*. Now, we introduce three non-monotonic models, namely default logic, autoepistemic logic, and circumscription reasoning.

[26] The stronger evidences are the less a value of $Pl(S)$ is.

[27] Let us assume that I have only one car.

Default logic was defined by Raymond Reiter[28] in 1980 [239]. It is a formalism which is more adequate for reasoning in AI systems than classical logic. Let us notice that even such seemingly simple and obvious propositions as "Mammals do not fly", if expressed in First Order Logic, i.e.,

$$\forall x[is\_mammal(x) \Rightarrow does\_not\_fly(x)],$$

is false, because there are some mammals (bats), which fly. Of course, sometimes defining a list of *all* the exceptions is impossible in practice. Therefore, in default logic, apart form standard rules of inference[29] *default inference rules* are defined. In such rules a *consistency requirement* is introduced. This is of the form *"it is consistent to assume that $P(x)$ holds"*, which is denoted by $\mathbf{M}\, P(x)$. For our example such a rule can be formulated in the following way:

$$\frac{is\_mammal(x)\ :\ \mathbf{M}\, does\_not\_fly(x)}{does\_not\_fly(x)}$$

which can be interpreted as follows: "If x is a mammal and it is consistent to assume that x does not fly, then x does not fly". In other words: "If x is a mammal, then x does not fly in the absence of information to the contrary".

Reiter introduced a very convenient rule of inference for knowledge bases, called the *Closed-World Assumption, CWA*, in 1978 [240]. It says that the information included in a knowledge base is a complete description of the world, i.e., if something is not known to be true, then it is false.

Autoepistemic logic was formulated by Robert C. Moore[30] in 1985 [206] as a result of research which was a continuation of studies into modal non-monotonic systems led by Drew McDermott[31] and Jon Doyle in 1980 [199]. The main idea of this logic can be expressed as follows. Reasoning about the world can be based on our introspective knowledge/beliefs. For example, from the fact that I am convinced that I am not the husband of Wilma Flinstone, I can infer that I am not the husband of Wilma Flinstone, because I would certainly know that I am the husband of Wilma Flinstone, if I was the husband of Wilma Flinstone. Autoepistemic logic can be viewed as a modal logic containing an operator *"I am convinced that"*. In such logic sets of beliefs are used instead of sets of facts.

A non-monotonic logic called *circumscription* was constructed by John McCarthy in 1980 [196]. We introduce its main idea with the help of our example proposition

---

[28]Raymond Reiter—a professor of computer science and logic at the University of Toronto. His work concerns non-monotonic reasoning, knowledge representation models, logic programming, and image analysis.

[29]*Standard rules* means such rules as the ones introduced in Chap. 6.

[30]Robert C. Moore—a researcher at Microsoft Research and NASA Ames Research Center, Ph.D. in computer science (MIT). His work concerns NLP, artificial intelligence, automatic theorem proving, and speech recognition.

[31]Drew McDermott—a professor of computer science at Yale University. His work concerns AI, robotics, and pattern recognition.

concerning mammals. This time, however, in order to handle the problem defined above we introduce the predicate *is_peculiar_mammal*(x). Now, we can express our proposition in First Order Logic in the following way:

$$\forall x[\mathit{is\_mammal}(x) \land \neg \mathit{is\_peculiar\_mammal}(x) \Rightarrow \mathit{does\_not\_fly}(x)].$$

Of course, we may not know whether a specific mammal is peculiar. Therefore, we minimize the extension of such a predicate as *is_peculiar_mammal*(x), i.e., we minimize its extension only to the set of objects which are known to be peculiar mammals. For example, if *Zazu* is not in this set, then the following holds: $\neg$*is_peculiar_mammal*(*Zazu*), which means that *does_not_fly*(*Zazu*). Let us notice an analogy to the concept of Closed-World Assumption introduced above.

As we have mentioned at the beginning of this section, sometimes a non-monotonic-reasoning-based system should remove a certain proposition as well as propositions inferred on the basis of this proposition after gaining new information. One question is: "Should all the propositions inferred on the basis of such a proposition be removed?" If these propositions can be inferred only from a removed proposition, then of course they should also be removed. However, the system should not remove those propositions which can be inferred without using a removed proposition. In order to solve this problem practically Jon Doyle[32] introduced *Truth Maintenance Systems, TMS*, in 1979 [73]. Such systems can work according to various scenarios. The simplest scenario consists of removing all the conclusions inferred from a removed proposition-premise and repeating the whole inference process for all conclusions. However, this simple scenario is time-consuming. An improved version consists of remembering the chronology of entering new information and inferring propositions in the system. Then, after removing some proposition-premise P, only those conclusions are removed which have been inferred after storing the proposition-premise P in the knowledge base.

Remembering sequences of *justifications* for conclusions is an even more efficient method. If any proposition is removed, then all justifications, which can be inferred *only* on the basis of this proposition are also removed. If, after such an operation, a certain proposition cannot be justified, then it is invalidated.[33] This scenario is a basis for *Justification-based Truth Maintenance Systems, JTMSs*. They were defined by Doyle in 1979 [73].

[32] Jon Doyle—a professor of computer science at the Massachusetts Institute of Technology, Stanford University, and Carnegie-Mellon University. His work concerns reasoning methods, philosophical foundations of Artificial Intelligence, and AI applications in economy and psychology.

[33] Such a proposition does not need to be removed (physically) from the knowledge base. It is enough to mark that the proposition is invalid (currently). If, for example, the removed justification is restored, then the system needs only to change its status to valid.

In 1986 Johan de Kleer[34] introduced a new class of truth maintenance systems called *Assumption-based Truth Maintenance Systems, ATMSs* [66]. Whereas in systems based on justifications a consistent *image* of the world is stored (consisting of premises and *justified* propositions), all justifications that have been assumed in the knowledge base are maintained in an ATMS (maybe some of them currently as *invalid*). Thus, the system maintains all *assumptions* that can be used for inferring a given proposition. The system can justify a given proposition given a certain set of assumptions, called a *world*. Such an approach is especially useful if we want the system to change its view depending on its set of assumptions.

The issue of maintaining a knowledge base when new data frequently come into an AI system is closely related to the *frame problem* formulated by McCarthy and Patrick J. Hayes[35] in 1969 [195]. The issue concerns defining efficient formalisms for representing elements of a world description which *do not* change during an inference process.

Bibliographical Note

The monograph [223] is a good introduction to Bayesian inference and networks. A description of Dempster-Shafer Theory can be found in the classic book [271]. A concise introduction to non-monotonic reasoning in AI can be found in [39].

[34] Johan de Kleer—a director of Systems and Practices Laboratory, Palo Alto Research Center (PARC). His work concerns knowledge engineering, model-based reasoning, and AI applications in qualitative physics.

[35] Patrick John Hayes—a British computer scientist and mathematician, a professor of prestigious universities (Rochester, Stanford, Essex, Geneva). His work concerns knowledge representation, automated inference, philosophical foundations of AI, and semantic networks.

Chapter 13
Defining Vague Notions in Knowledge-Based Systems

The second reason for the unreliability of inference in AI systems, apart from imperfection of knowledge, is *imperfection of the system of notions* which is used for a description of the real (physical) world. Standard methods of computer science are based on models developed in mathematics, technical sciences (mechanics, automatic control, etc.), or natural sciences (physics, astronomy, etc.). In these sciences precise (crisp) notions are used for describing aspects of the world. A problem arises if we want to apply computer science methods for solving problems in branches which use notions that are less precise (e.g., psychology, sociology)[1] or for solving problems of everyday life, in which we use popular terms. In such a case there is a need to construct formalized models, in which such *vague notions* are represented in a precise way.

There are two basic aspects of the problem of vagueness of notions in knowledge-based systems. The first aspect concerns the *unambiguity* of a notion which results from its subjective nature. Notions relating to the age of a human being, e.g., young, old, are good examples here. In this case, taking into account the subjective nature of such a vague notion by introducing a measure which grades "being young (old)" seems to be the best solution. The *theory of fuzzy sets* presented in the first section is based on this idea of a measurement which grades objects belonging to vague notions.

The second aspect concerns the *degree of precision (detail, accuracy)* used during a process of notion formulation. This degree should be appropriate (adequate) to the nature of the considered phenomena/objects. The term *appropriate* means that the system should distinguish between phenomena/objects which are considered to belong to different categories and it should not distinguish between phenomena/objects which are treated as belonging to the same category (of course, with

[1]This results from the more complex subject of research in these branches.

M. Flasiński, *Introduction to Artificial Intelligence*,
DOI 10.1007/978-3-319-40022-8_13

respect to the set of features assumed). *Rough set theory* is often used for solving this problem in AI. This theory is presented in the second section.[2]

13.1 Model Based on Fuzzy Set Theory

First of all, we introduce a *universe of discourse* U, which is the set of all the considered elements. If we want to represent a notion $\mathcal{A}$ in a knowledge base, then we can define it as the set:

$$A = \{x : x \text{ has features which are consistent with the notion } \mathcal{A}, \;\; x \in U\}. \quad (13.1)$$

Now, in order to decide whether an element $y \in U$ fulfills the conditions of the definition of a notion $\mathcal{A}$ we should have a *membership function* for the set A, which is denoted μ_A. In the case of a *crisp notion*, which is a notion for which we are able to define features allowing the system to distinguish objects/phenomena which have the properties of this notion from those which do not have them, the membership function is equal to the characteristic function χ_A of the set A. This function is defined in the following way.

$$\chi_A(x) = \begin{cases} 0, & x \notin A, \\ 1, & x \in A. \end{cases} \quad (13.2)$$

However, in the case of a *vague notion*, which is not defined in an precise and unambiguous way, a characteristic function is not an appropriate formalism for solving such a problem. This observation allowed Lotfi A. Zadeh to formulate *fuzzy set theory* in 1965 [321]. In this theory the membership function for a set A is defined in the following way:

$$\mu_A : U \longrightarrow [0, 1]. \quad (13.3)$$

This function assigns values 0 and 1 to an element $x \in U$ according to formula (13.2) for the function χ_A. However, in case of partial membership of an element in the set A it assigns a value s, which belongs to the interval $(0, 1)$. This value is called the *grade of membership* of the set A. If the membership function is defined in such a way, then the set is called a *fuzzy set*.

Let us consider the following example of how to represent vague notions with the help of fuzzy sets. Let us assume that we want to characterize a *human being's age* with vague notions *young* ($\mathcal{Y}$), *adult* ($\mathcal{A}$), and *old* ($\mathcal{O}$). Let us restrict our considerations to the age interval $[0, \;101]$. Then, these vague notions can be represented by fuzzy sets Y, A, O, respectively, and, in fact, by the corresponding membership functions[3] μ_Y, μ_A, μ_O. For example, these functions can be defined as shown in

[2]Formal definitions of fuzzy set and rough set theories are contained in Appendix J.

[3]A membership function for a set determines this set.

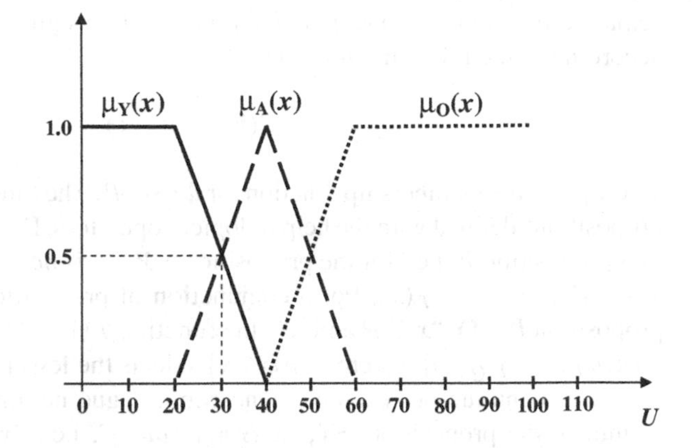

Fig. 13.1 Membership functions for fuzzy sets for the example of a human being's age: μ_Y—for the notion *young*, μ_A—for the notion *adult*, μ_O—for the notion *old*

Fig. 13.1. (The function μ_Y is marked with a solid line, the function μ_A—with a dashed line, the function μ_O—with a dotted line.) People of age 0–20 are definitely considered to be young ($\mu_Y(x) = 1.0, x \in [0, 20]$) and people of age more than 60 are definitely considered to be old ($\mu_O(x) = 1.0, x \in [60, 101]$). A typical adult is 40 years old ($\mu_A(40) = 1.0$). If a person has completed 20 years and progresses in age, then we consider her/him less and less young. If a person completes 30 years, we consider her/him adult rather than young ($\mu_Y(x) < \mu_A(x), x \in [30, 40]$). If a person completes 40 years we definitely do not consider her/him to be young ($\mu_Y(x) = 0.0, x \in [40, 101]$), etc. As we can see fuzzy sets allow us to represent vague notions in a convenient way.

The formalism used in the example is called a *linguistic variable* in fuzzy set theory [40, 322]. The term *human being age* is called a *name of a linguistic variable*. The vague notions *young* ($\mathcal{Y}$), *adult* ($\mathcal{A}$), and *old* ($\mathcal{O}$) are called *linguistic values*. A mapping which assigns a fuzzy set to a vague notion, i.e., to a linguistic value, is called a *semantic function*.

Nowadays, fuzzy set theory is well developed and formalized. Apart from complete characteristics of operations with fuzzy sets, e.g., the union of fuzzy sets, the intersection of fuzzy sets, etc., such models as fuzzy numbers and operations with them, fuzzy relations, and fuzzy grammars and automata have been developed. Since this book is an introduction to the field, we do not discuss these issues here. Later, we introduce two formalisms which are very important in AI, namely fuzzy logic and fuzzy rule-based systems.

The possibility of the occurrence of vague notions is a basic difference between the propositions considered in *fuzzy logic* [105, 192, 249, 322] and propositions of a classical logic. In consequence, instead of assigning two logical values *1* (*True*) and *0* (*False*) to propositions, we can assign values which belong to the interval [0, 1].[4] Similarly as for sets, we define the truth degree function T. Let **P** be a proposition: "*x is* $\mathcal{P}$", where $\mathcal{P}$ is a vague notion for which a fuzzy set P is determined by a

[4]Fuzzy logic is an example of *multi-valued logic* with an infinite number of values, which has been introduced by Polish logician and mathematician Jan Łukasiewicz.

semantic function. The *truth degree function* T assigns a value to a proposition $\mathbf{P}$ according to the following formula.

$$T(\mathbf{P}) = \mu_P(x), \tag{13.4}$$

where μ_P is the membership function for the set P. The function T can be extended to propositions defined with the help of logical operators. For example, for the negation of a proposition $\mathbf{P}$, i.e., for the proposition $\neg\mathbf{P}$, "x *is not* $\mathcal{P}$", the function $T(\neg\mathbf{P}) = 1 - T(\mathbf{P}) = 1 - \mu_P(x)$. For a conjunction of propositions $\mathbf{P}$ and $\mathbf{Q}$, i.e., for the proposition $\mathbf{P} \wedge \mathbf{Q}$, "$x$ *is* $\mathcal{P}$ and $\mathcal{Q}$", the function $T(\mathbf{P} \wedge \mathbf{Q}) = min\{T(\mathbf{P}), T(\mathbf{Q})\} = min\{\mu_P(x), \mu_Q(x)\}$, where $min\{x, y\}$ selects the lesser value of x and y.[5]

Let us continue our example concerning vague notions of human age. We can formulate the proposition $\neg\mathbf{Y}$, "x *is not young*", i.e., "x *is not* $\mathcal{Y}$". The function $T(\neg\mathbf{Y}) = 1 - \mu_Y(x)$ is shown in Fig. 13.2a. As we can see, the function T is defined in an intuitive way. Now, let us define this function for the more complex case of the conjunction $\mathbf{Y} \wedge \mathbf{A}$, "$x$ *is young and adult*", i.e., "x *is* $\mathcal{Y}$ and $\mathcal{A}$", which is shown in Fig. 13.2b. As we can see, the function $T(\mathbf{Y} \wedge \mathbf{A})$ equals 0 in the interval [0, 20]. (In this interval the function $\mu_A(x)$ gives a lower value than $\mu_Y(x)$ and it equals 0.) This is consistent with our intuition, since we do not consider somebody who belongs to this interval a "mature/adult person". Analogously, the function $T(\mathbf{Y} \wedge \mathbf{A})$ equals 0 in the interval [40, 101]. (In this interval the function $\mu_Y(x)$ gives a lower value than $\mu_A(x)$ and it equals 0.) This is also consistent with our intuition, since we do not consider somebody who belongs to this interval a young person. For the interval [20, 40] we have two cases. In the subinterval [20, 30] we use the lower value as the value of $T(\mathbf{Y} \wedge \mathbf{A})$, i.e., we use the value of the function $\mu_A(x)$. It increases in this subinterval and it reaches its maximum at the age of 30 years. At this point the charts of membership functions $\mu_Y(x)$ and $\mu_A(x)$ meet and they determine the best representative of *being young and adult*. In the second subinterval, i.e., [30, 40] the value of the function $T(\mathbf{Y} \wedge \mathbf{A})$, which measures the truth degree of *being adult and (at the same time) young*, decreases. Although, in this subinterval, as time goes by, we are more and more experienced, we are older and older (unfortunately).

In Chap. 9, we have presented rule-based reasoning.[6] In the Zadeh theory the concept of a *fuzzy rule-based system* [191, 322] has been developed. The rules of such a system are of the form:

$$R^k : \ \mathsf{IF} x_1^k \ \mathsf{is} \ \mathcal{A}_1^k \wedge \ldots \wedge x_n^k \ \mathsf{is} \ \mathcal{A}_n^k \ \mathsf{THEN} \ y^k \ \mathsf{is} \ \mathcal{B}^k,$$

where $x_1^k, \ldots, x_n^k, y^k$ are linguistic variables and $\mathcal{A}_1^k, \ldots, \mathcal{A}_n^k, \mathcal{B}^k$ are linguistic values corresponding to fuzzy sets $A_1^k, \ldots, A_n^k, B^k$.

Now, we present an example of reasoning in a fuzzy rule-based system. We do this using one of the first models of fuzzy reasoning. It was introduced by Ebrahim

---

[5]A formal definition of the function T for logical operators is contained in Appendix J.

[6]If the reader has omitted Chap. 9, the rest of this section may be difficult to understand.

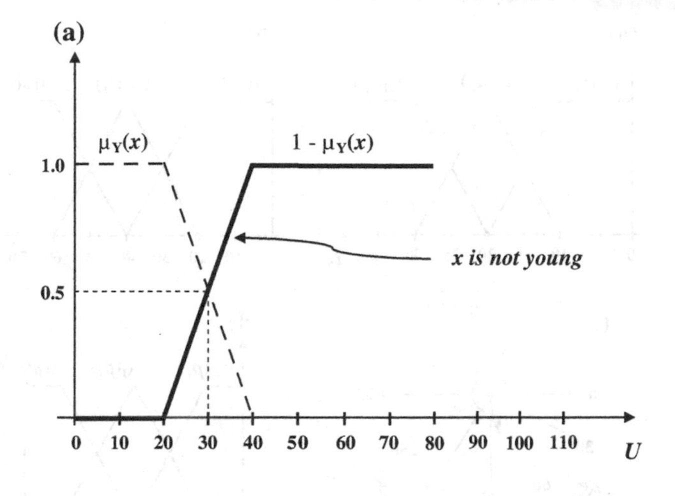

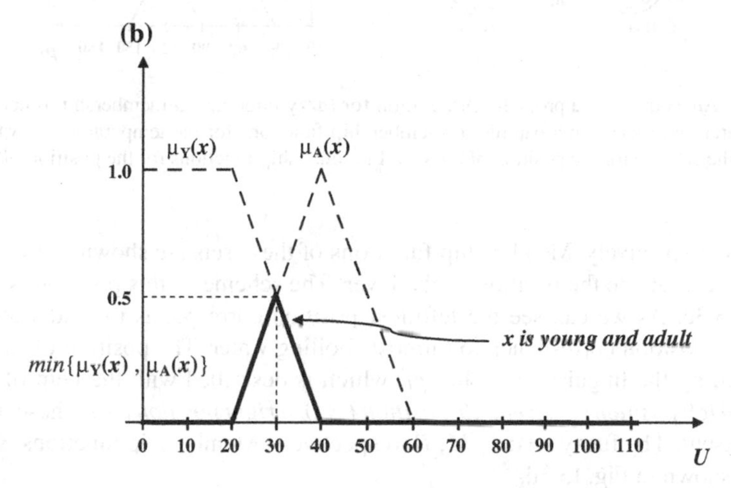

Fig. 13.2 Examples of membership functions for fuzzy sets defined with logical operators: **a** negation, **b** conjunction

Mamdani[7] in 1975 [191]. Let us assume that we want to design an expert system which controls the position of a lever regulating the temperature of water in a bathtub faucet. We assume that the linguistic variable t_b denoting the water temperature in the bathtub is described with the help of vague notions (linguistic values) *(too) low* ($\mathcal{L}$), *proper* ($\mathcal{P}$), *(too) high* ($\mathcal{H}$). These notions are represented by fuzzy sets L, P, H, respectively. Membership functions of these sets are shown in Fig. 13.3a. The water temperature in the faucet t_f is described with the help of vague notions *cold* ($\mathcal{C}$), *warm* ($\mathcal{W}$), *(nearly) boiling* ($\mathcal{B}$). These notions are represented by fuzzy sets

[7] Ebrahim Mamdani—a professor of electrical engineering and computer science at Imperial College, London. A designer of the first fuzzy controller. His work mainly concerns fuzzy logic.

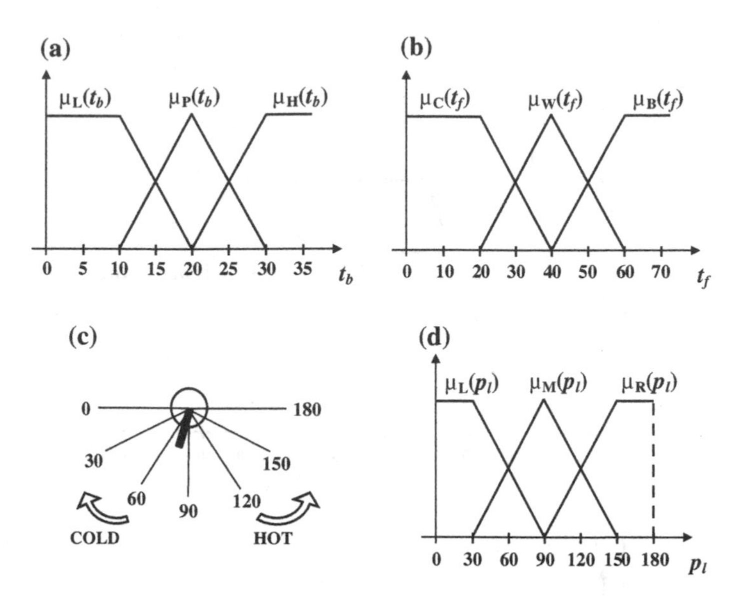

Fig. 13.3 An example of a problem formulation for fuzzy inference: **a** membership functions for the temperature of water in a bathtub, **b** membership functions for the temperature of water in a faucet, **c** the scheme for the position of a lever, **d** membership functions for the position of a lever

C, W, B, respectively. Membership functions of these sets are shown in Fig. 13.3b. Now, we can define the position of the lever. The scheme of this position is shown in Fig. 13.3c. As we can see the leftmost position corresponds to cold water, the rightmost position corresponds to (almost) boiling water. The position of the lever is defined by the linguistic variable p_l, which is described with the help of vague notions *a left position ($\mathcal{L}$), a middle position ($\mathcal{M}$), a right position ($\mathcal{R}$).* These notions are represented by fuzzy sets: L, M, R, respectively. Membership functions of these sets are shown in Fig. 13.3d.

We present reasoning in the system with the help of the following two rules.

$$R^1 : \ \text{IF} t_b \ \text{is} \ proper \wedge t_f \ \text{is} \ warm \ \text{THEN} \ p_l \ \text{should be in} \ middle\_position,$$

$$R^2 : \ \text{IF} t_b \ \text{is} \ high \wedge t_f \ \text{is} \ boiling \ \text{THEN} \ p_l \ \text{should be in} \ left\_position.$$

Now, we can start reasoning. Although a fuzzy rule-based system reasons on the basis of fuzzy sets, it receives data and generates results in the form of numerical values. Therefore, it should convert these numerical values into fuzzy sets and vice versa. Let us assume that the system makes the first step of reasoning on the basis of the first rule. There are two variables, t_b and t_f, and two linguistic values, $\mathcal{P}$ and $\mathcal{W}$, which are represented by membership functions μ_P and μ_W, respectively, in the condition of the first rule. Let us consider the first pair: $t_b - \mu_P$. Let us assume that the temperature of water in the bathtub equals $t_b = T_b = 28\,°\text{C}$. We

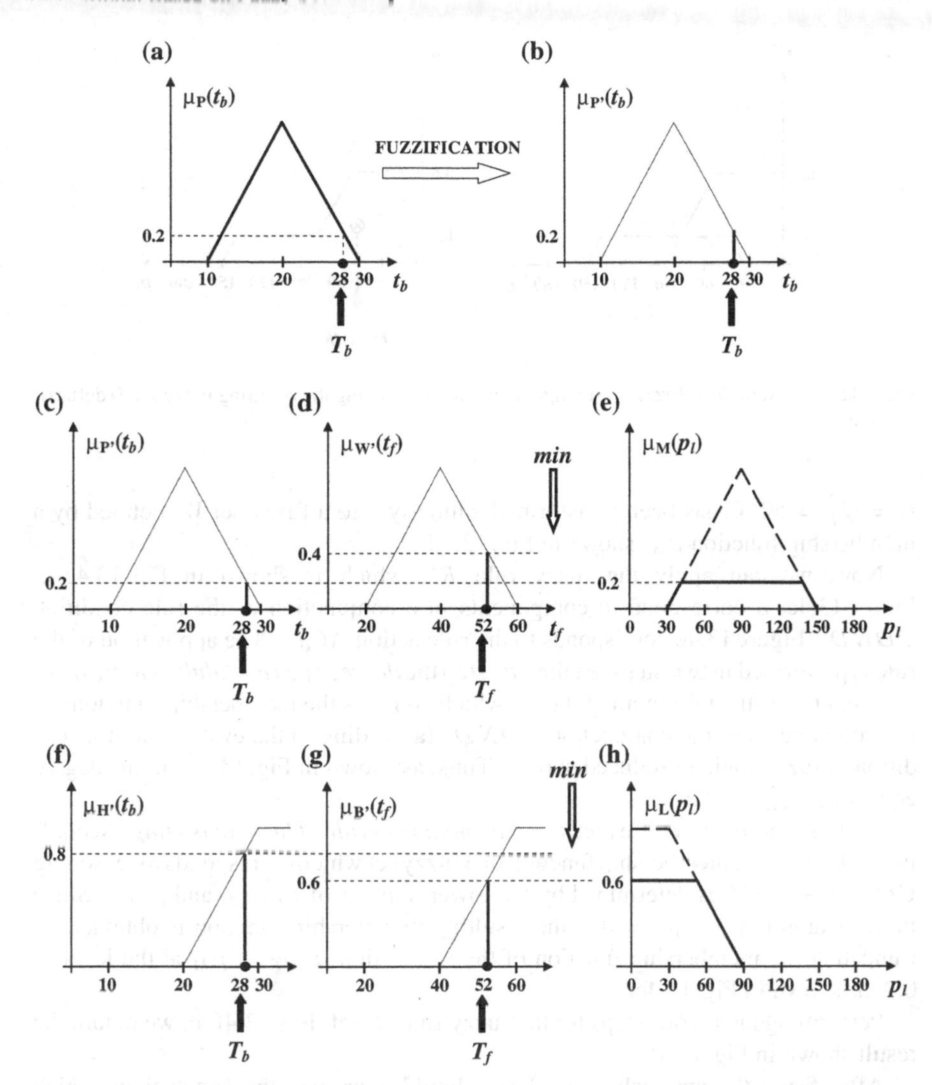

Fig. 13.4 An example of fuzzy reasoning: **a–b** fuzzification, **c–e** application of the first rule, **f–h** application of the second rule

have to convert the number T_b into a fuzzy set on the basis of the membership function μ_P. We do this with the help of the simplest *fuzzification operation*, which is called a singleton fuzzification. It converts the number T_b into a fuzzy set P' defined by the membership function which equals $\mu_{P'}(T_b)$ for T_b and equals 0 for other arguments. This operation is shown in Fig. 13.4a, in which a value $\mu_P(28) = 0.2$ is determined, and in Fig. 13.4b, in which the new membership function $\mu_{P'}$ is defined. This function determines the new fuzzy set P', which is the result of the fuzzification of the number $T_b = 28$. Then, let us assume that the temperature of water in the faucet

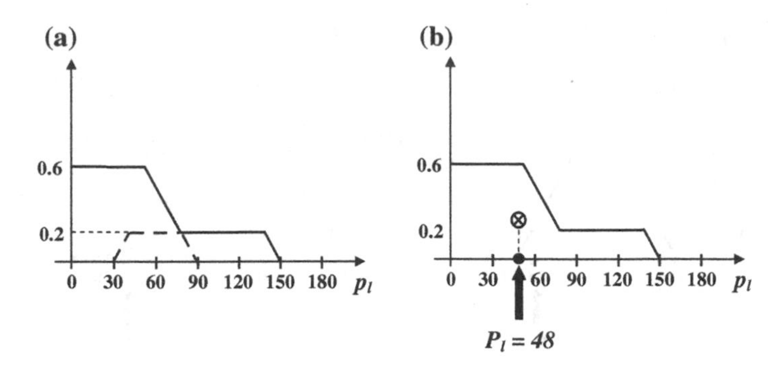

Fig. 13.5 An example of fuzzy reasoning—cont.: **a** constructing the resulting fuzzy set, **b** defuzzification

$t_f = T_f = 52\,°\text{C}$ has been transformed similarly into a fuzzy set W' defined by a membership function $\mu_{W'}$ shown in Fig. 13.4d.

Now, we can apply the fuzzy rule R^1, which is shown in Fig. 13.4c–e. Figure 13.4c, d correspond to components in a conjunction of the rule condition $COND^1$. Figure 13.4e corresponds to the rule action ACT^1. The application of the rule is performed in two steps. At the first step the *degree of a rule fulfillment* μ_{COND^1} is computed as the minimum of the set which contains the membership functions of the components of the conjunction $COND^1$ (according to the evaluation of a condition in fuzzy logic introduced above). Thus, as shown in Fig. 13.4c, d, this degree equals $\mu_{COND^1} = 0.2$.

At the second step we use the *Mamdani minimum rule of fuzzy reasoning*. According to this rule the membership function of a fuzzy set which corresponds to reasoning $COND \rightarrow ACT$ is determined by the lower number of μ_{COND} and μ_{ACT}. Since in our example $\mu_{COND^1} = 0.2$, the resulting membership function is obtained by truncating the membership function of the rule action $\mu_{ACT} = \mu_M$ at the level of 0.2, as shown in Fig. 13.4e.

Performing analogous steps for the fuzzy rule R^2, cf. Fig. 13.4f–h, we obtain the result shown in Fig. 13.4h.

After firing the applicable rules[8] we should aggregate the conclusions which have been obtained. Let us notice that the conclusion is of the form of truncated membership functions. In our example, membership functions for the conclusions are marked with solid lines in Fig. 13.4e, h. The operation of aggregating these functions is shown in Fig. 13.5a. As we can see, this time we use the maximum operation over the set of conclusions, i.e., we define the final membership function taking the greatest value of the aggregated functions.

This final membership function determines a fuzzy set, which is the result of fuzzy reasoning. However, we need a numeric value representing the angle of the position

[8] A fuzzy rule is not applicable at a reasoning cycle if its degree of fulfillment equals 0.

of the lever, cf. Fig. 13.3c (not a fuzzy set). Therefore, at the end we should perform a *defuzzification operation*, which determines a numerical value on the basis of a fuzzy set.

For example, a defuzzification operation with a centroid involves computing the center of gravity of the area under a curve determined by the final membership function, which is shown in Fig. 13.5b. The coordinate of this center on the axis of the linguistic variable (the variable p_l in the case of our example) determines the resulting numerical value. In Fig. 13.5b the center of gravity is marked by a circle with a cross inside. As we can see the resulting numerical value, which represents the correct angle of the lever is $p_l = P_l = 48\,°C$. Thus, our expert system sets the lever in this position after fuzzy reasoning.

13.2 Model Based on Rough Set Theory

In 1982 Zdzisław Pawlak proposed an alternative, or rather complementary, approach [216] for representing vague notions. This approach, called rough set theory, has been developed considerably since then [218, 219, 220]. As we have mentioned at the beginning of this chapter, fuzzy set theory is used for solving the problem of ambiguity of notions, whereas in rough set theory vagueness of notions is considered in the aspect of their degree of precision (detail, accuracy).[9] The degree of precision of characterization of a notion should be *adequate* for the problem considered. In rough set theory this *adequacy* is described as a feature of a system which allows it to distinguish between phenomena/objects which are considered to belong to different categories and not to distinguish between phenomena/objects which belong to the same category. Let us consider this feature with the help of the following example.

Let the domain containing objects to be considered, called the *universe of discourse* U, consist of passenger cars. Let us assume that we analyze the problem of placing new car service stations for various car marques in various regions of Poland, such as the Warsaw region, the Cracow region, etc. For such an analysis we can define a *set of attributes describing objects*[10] $A = \{marque\_of\_car, region\_of\_registration\}$. For every attribute we should define a *set of possible values*, i.e., the *domain of the attribute*. Then, for an object x belonging to the universe U an expression *marque_of_car* (x) =*BMW* denotes the fact that x *is a BMW model*. A partition of the universe U into subareas determined by the given set of attributes[11] is shown in Fig. 13.6a. We say that objects which belong to the same subarea (e.g., two BMW cars registered in the Warsaw region and marked with white dots in Fig. 13.6a) are *indiscernible objects*

[9] Rough set theory is discussed in this book only from the AI point of view. However, this theory is applied much more broadly and it is interpreted in a more general way in computer science.

[10] Passenger cars are objects.

[11] A set of attributes means *a set of attributes together with their domains*.

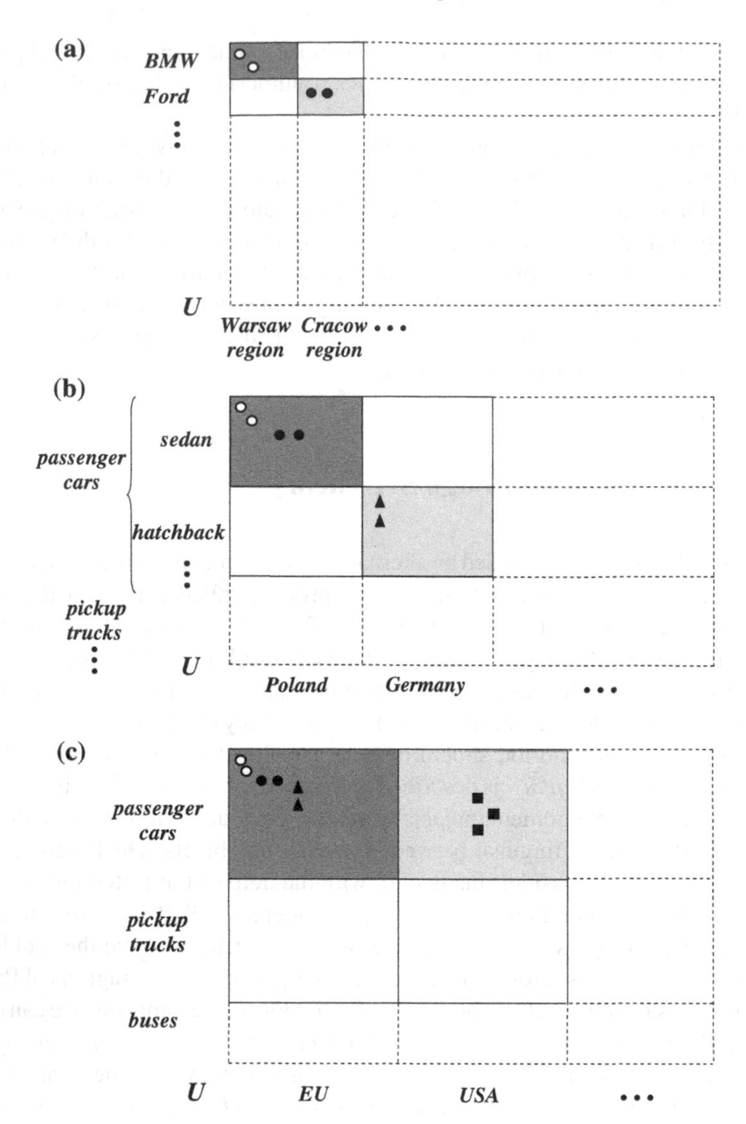

Fig. 13.6 An example of controlling the level of generality of considerations with the help of a set of attributes

with respect to the given set of attributes.[12] A subarea of the universe determined in this way is called an *elementary set*. Elementary sets are called also *knowledge*

[12]In fact, in rough set theory these two cars are not distinguishable. In order to distinguish between them one should introduce an additional attribute, e.g., the registration number. However, in the context of our problem this is not necessary.

granules. They correspond to elementary notions, which can be used for defining more complex notions, including vague notions.

The set of attributes is used for controlling the degree of precision of the definition of elementary notions. In other words, it is used for determining the degree of granularity of the universe of discourse. For example, let us consider a universe of cars and a manager responsible for sales. He/she wants to know the preferences of customers from the EU countries for different types of car (sedan, hatchback, etc.). Then, an adequate set of attributes should be defined as $A = \{type\_of\_car, country\_of\_registration\}$. In this case, a partition of the universe into elementary sets is defined as shown in Fig. 13.6b. As we can see, the degree of granularity of the universe is less than in the previous case, i.e., knowledge granules are bigger. This results from using more general attributes. Now, two BMW sedans from the Warsaw region are *indiscernible* from two Ford sedans from the Cracow region. Of course, we can go to a higher level of abstraction and consider types of cars with respect to bigger markets, like the EU market or the US market, as shown in Fig. 13.6c. Now, the granules are even bigger. This means that some cars which have been distinguishable previously are now indistinguishable.

In our example we have controlled the degree of precision of definitions of elementary notions by determining more/less general attributes. In fact, we often control it by increasing/decreasing the number of attributes. The fewer attributes we use, the less the granularity. For example, if we remove the attribute *marque_of_car* in Fig. 13.6a, then the granularity of the universe is reduced. (Then, it is partitioned only into vertical granules, which contain all passenger cars in one region of Poland.)

Now, we can consider the issue of defining vague notions in rough set theory. A vague notion $\mathcal{X}$ is defined by two crisp notions: a *lower approximation* $\underline{B}\mathcal{X}$ and an *upper approximation* $\overline{B}\mathcal{X}$. A lower approximation $\underline{B}\mathcal{X}$ is defined as a set $\underline{B}X$ which contains knowledge granules that *necessarily* are within the scope of a vague notion $\mathcal{X}$. An upper approximation $\overline{B}\mathcal{X}$ is defined as a set $\overline{B}X$ which contains knowledge granules that *possibly* are within the scope of a vague notion $\mathcal{X}$.

Let us consider this way of representing a vague notion with the following example. Let us assume that we construct a system which selects ripe plums on the basis of two attributes: hardness and color. The vague notion *ripe plum* ($\mathcal{R}$) is represented by the set R shown in Fig. 13.7a, which is determined with the help of ripe plums denoted by black dots. Unripe plums are denoted by white dots. The border of the set R is also marked in this figure. The area of the whole rectangle represents the universe of discourse U. After defining attributes together with their domains, the universe U is divided into the knowledge granules shown in Fig. 13.7b. The lower approximation $\underline{B}\mathcal{R}$ is defined as the set $\underline{B}R$, which contains knowledge granules marked with a grey color in Fig. 13.7b. These granules contain only ripe plums. In Fig. 13.7c the set $\overline{B}R$, which corresponds to an upper approximation $\overline{B}\mathcal{R}$, is marked with a grey color. The set $\overline{B}R$ contains all ripe plums and also some unripe plums. However, using such a degree of granularity of the universe we are not able to distinguish between ripe and unripe plums for three granules. These three granules determine a *boundary region*, which is marked with a grey color in Fig. 13.7d. It contains objects which cannot be assigned to R in an unambiguous way.

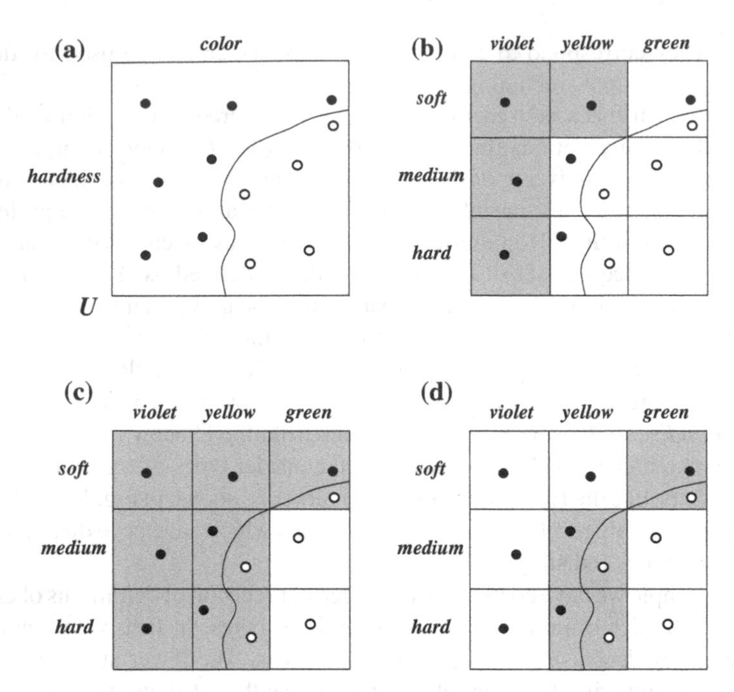

Fig. 13.7 An example of basic notions of rough set theory: **a** a universe of discourse with a set R marked, **b** a lower approximation of the set R, **c** an upper approximation of the set R, **d** the boundary region of the set R

Finally we can define a *rough set*. For a rough set a lower approximation is different from an upper approximation assuming a fixed granularity of the universe. In other words, the boundary region is not empty for a rough set. On the other hand, for a *crisp (exact) set* the lower approximation is equal to the upper approximation, i.e., its boundary region is the empty set.

Usually after defining a vague notion with the help of a lower and an upper approximation we would like to know how *good* our approximation is. There are several measurements of approximation quality in rough set theory. The *coefficient of accuracy of approximation* is one of them. It is calculated by dividing the number of objects belonging to the lower approximation by the number of objects belonging to the upper approximation. In our example it is equal to $4/10 = 0.4$. Let us notice that this coefficient belongs to the interval [0, 1]. For a crisp set it equals 1.

The right choice of attributes is a very important issue in the application of rough set theory in AI. The number of attributes should be as small as possible (for reasons of computation efficiency), yet sufficient for a proper granulation of the universe. Methods for determining an optimum number of attributes have been developed in rough set theory.

Rough sets can be used, like fuzzy sets, for implementing rule-based systems. Methods for automatic generation of rules on the basis of lower and upper approximations have been defined as well [277].

Bibliographical Note

Foundations of fuzzy set theory are presented in [63, 75, 257, 323]. Fuzzy logic and fuzzy rule-based systems are discussed in [249].

A good introduction to rough set theory can be found in [71, 128, 217, 229, 257, 278].

Chapter 14
Cognitive Architectures

In this part of the monograph we present various *methods* used for problem solving by artificial intelligence systems. This chapter, however, does not include a description of any method, but it contains a discussion on the possible *structure of an artificial intelligence system*. In computer science such a structure, called a *system architecture*, results from a model defined according to the principles of software engineering. In AI, however, a different approach to designing a system architecture has been used since the 1980s. In this approach we try to define an AI system architecture on the basis of the model of cognitive abilities assumed. Thus, so-called *cognitive architectures* have been introduced to the methodology of system design. Since they are very important for the further development of the AI field, we present them in this chapter.

The cognitive architecture of an AI system is defined in such a way that its structure is based on a given cognitive model and its computational processes correspond to mental/cognitive processes defined in this model. In other words, a cognitive architecture is a computer science model that corresponds to a cognitive model. It plays two basic roles. Firstly, it is a generic pattern for the practice of constructing AI systems. Secondly, it can be used for verifying hypotheses of some models of cognitive science.[1]

In the first section we introduce a generic architecture which is based on the concept of an *agent*, whereas in the second section we present *multi-agent systems*, which are constructed with the help of this concept.

---

[1] This verification is made via simulation experiments performed with the help of a corresponding AI system.

M. Flasiński, *Introduction to Artificial Intelligence*,
DOI 10.1007/978-3-319-40022-8_14

14.1 Concept of Agent

It is difficult to point to the psychological theory or theory of mind which has influenced the development of the concept of agent in Artificial Intelligence the most. Surely, we can see in this concept the behavioral idea of explaining the behavior of organisms in terms of *stimulus-response*. We can also find the influence of the Piaget[2] constructivist theory of cognitive development, especially mechanisms of *assimilation* and *accommodation*[3] of concepts. Some authorities in AI claim that functionalism and the Dennett approach to intentionality[4] have had an impact on this model as well [256].

Neglecting some differences among notions of *agent* in the literature, we define it as a system which influences its environment in an *autonomous* way on the basis of perceived information in order to achieve required goals. The expression *autonomous* plays a key role here and means that an intentional influence on the environment should be based on the *experience* an agent *gained itself*. In other words, an agent should have an ability of *self-learning* and it should make use of this ability to cognize an unknown environment. Let us explain this concept with the help of the following example.

Let us return to our example of searching for a route to the exit in a labyrinth, which has been introduced in Chap. 4. Let us consider a certain *world of labyrinth paths*, which is shown in Fig. 14.1a. As one can see this world consists of paths which run from north (N) to south (S) and sometimes turn east (E). Then, we reach a corridor which runs to the west (W) to the exit. (On the east side of the map there are a lot of similar paths and there is no exit on this side.) In this world, at the beginning of each path we place an agent. We expect it to find a route to the exit. Of course, since the agent is autonomous, we do not deliver any specific knowledge about this world to it. Instead, we assume it has the ability to perceive, basic *meta-knowledge*, and the ability to move.

Let us begin with defining the ability to perceive. Let us assume that the agent perceives the world one step ahead and it can recognize an empty space or a wall. Additionally, the agent knows the direction (N-E-S-W) it is moving (see Fig. 14.1b), i.e., it has an *internal compass*. The ability to move is defined with the following operations: go one step ahead, go one step back, turn right, turn left.

The meta-knowledge of the agent is modeled with the help of rules of the form introduced in Chap. 9. The rules define basic principles which allow the agent to exit the labyrinth. For example, we can define the following rules:

IF *I see* an empty space THEN *I go one step ahead*,
IF *I see* a wall THEN *I turn right*.

---

[2]Jean Piaget—a professor of psychology at the Rousseau Institute in Geneva. He is known primarily for his theory of cognitive development and epistemological research with children.

[3]Information perceived by humans is *assimilated* according to pre-existing cognitive schemas. However, if this is not possible, because the information perceived does not fit these schemas, they are altered to take into account the new experience, which is called *accommodation*.

[4]These theories are introduced in the next part of the book.

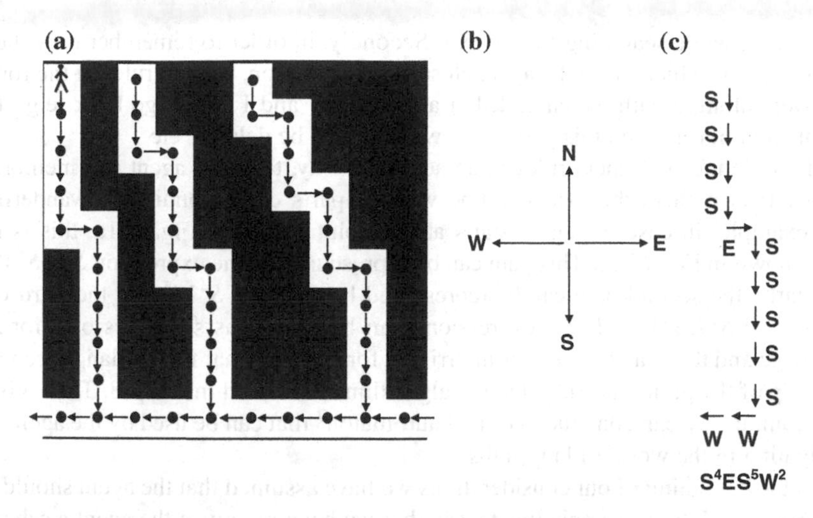

Fig. 14.1 **a** An exemplary world of labyrinths, **b** possible steps of an agent, **c** an example of a route to the exit

We have assumed above that the agent is autonomous. Therefore, we should give it cognitive abilities that can be used for constructing knowledge schemas. (However, we should not give these schemas to it.) For example, we can require that the agent should gain knowledge defined with the rules presented above on its own, i.e., via experiments made in its environment. In such a case, we might define the agent as a *child* at the age of developing sensory-motor abilities.[5] Additionally, we should give a body to the agent, so it experiences sensations. Then, we should define meta-rules, such as, for example, the following ones:

IF *I make a step ahead* $\wedge$ *my body hurts* THEN *I go back*,

IF *as a result of the activity* A *my body hurts* THEN *I do not repeat* A.

We should also define mechanisms which allow the agent to generate specialized rules on the basis of meta-rules. Such mechanisms are modeled in the area of cognitive architectures.[6]

If the agent generates the rules, such as the ones defined at the beginning of our considerations, then they will be too general from a practical point of view. They do not include any information that is specific to the world of the labyrinth the agent *lives* in. For example, the agent may walk to the north (although it should never do it), it may try to go east more than one step (it should not do it, because in paths running from north to south, after turning east it should go south), etc.

Thus, let us give new cognitive abilities to the agent which enable it to create *schemas* of a specific world. Firstly, we allow the agent to gather experience, i.e., to

[5]We discuss kinesthetic intelligence in the next chapter.

[6]For example, such mechanisms are implemented in cognitive architecture systems, such as Soar and ACT-R mentioned in the first chapter.

store paths while searching for the exit. Secondly, in order to remember only those parts of paths which bring the agent closer to the solution, we add rules of the form: if I went along a path which ended at a cul-de-sac and I had to go back (e.g., the agent went north), then this part of my walk should be deleted, etc.

If we define such meta-rules in an adequate way, then the agent can memorize the shortest path to the exit, i.e., one without parts corresponding to wandering. For example, in case the agent starts at the point shown in Fig. 14.1a, this is the path shown in Fig. 14.1c. This path can be represented by the expression $S^4ES^5W^2$. Similarly, the second path can be represented by $S^2ES^3ES^4W^5$ and the third one by $SES^2ES^2ES^4W^9$. These expressions can be treated as sentences of a formal language and they can be used for inferring a formal grammar for this language with the help of the grammatical induction algorithm introduced in Chap. 8. Then, given the grammar we can construct a formal automaton[7] that can be used by the agent for navigating in the world of labyrinths.

At the beginning of our considerations we have assumed that the agent should be *autonomous*. In our example this means that we have not given the agent a scheme representing the world with the help of the formal automaton, but we have given the agent the ability to construct such an automaton. Thanks to this approach, if the agent goes to another world of labyrinths, such as, for example, the *rotated* world of labyrinths shown in Fig. 14.2a,[8] it should be able to solve the problem as well. Having cognitive meta-rules, the agent is able to construct a corresponding formal automaton after some time. As a result, such an automaton would help the agent to find the shortest path to the exit, like the one shown in Fig. 14.2b.

We can give cognitive abilities to the agent which are based on various methods of Artificial Intelligence. The exemplary agent[9] shown in Fig. 14.2c is implemented with the help of three different models. After perceiving information concerning its environment (in this case it is an environment consisting of various devices and equipment), the agent updates its representation of this environment, which is defined as a *semantic network*. Then, it reasons about its situation in this environment with the help of a *graph automaton* (*parser*) in order to detect events which require its response. After recognizing the type of the event, the agent infers an adequate action with the help of a *rule base*. Proposed actions are sent to a *steering command generator*, which is used to influence the environment. The agent has the ability to learn by inferring a formal grammar,[10] which is the knowledge base for a graph automaton (parser).

---

[7] As is discussed in Chap. 8, as well.

[8] Our agent should be able to go to any world of labyrinths, assuming paths in such a world are characterized by some regularities (i.e., they can be described by some principles).

[9] The example is discussed in the paper Flasiński M.: Automata-based multi-agent model as a tool for constructing real-time intelligent control systems. *Lecture Notes in Artificial Intelligence 2296* (2002), 103–110.

[10] This is an ETPL(k) graph grammar introduced in Chap. 8 and formally defined in Appendix E.

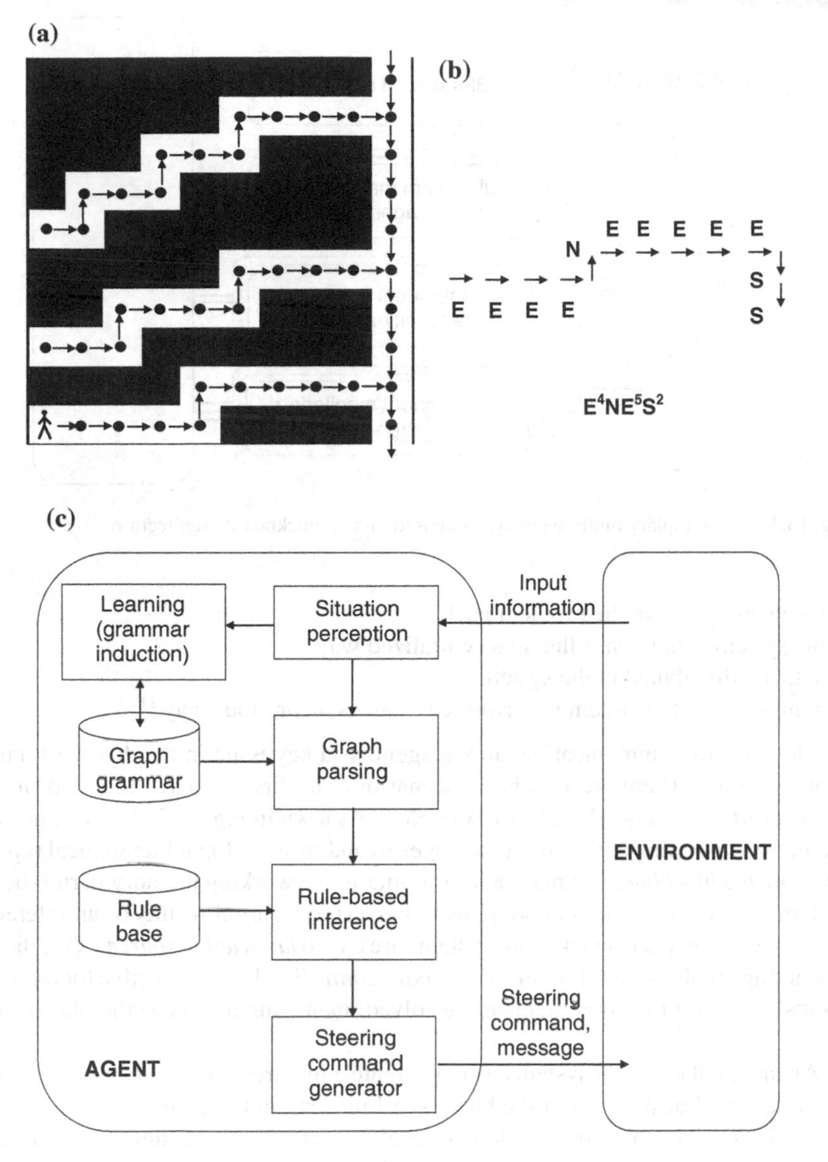

Fig. 14.2 **a** A rotated world of labyrinths, **b** the route to the exit in the rotated world, **c** an example of an agent structure

14.2 Multi-agent Systems

For solving very difficult problems we can implement teams of interacting agents. Such teams are called *multi-agent systems* (*MAS*). There are four basic features of a multi-agent system [295]:

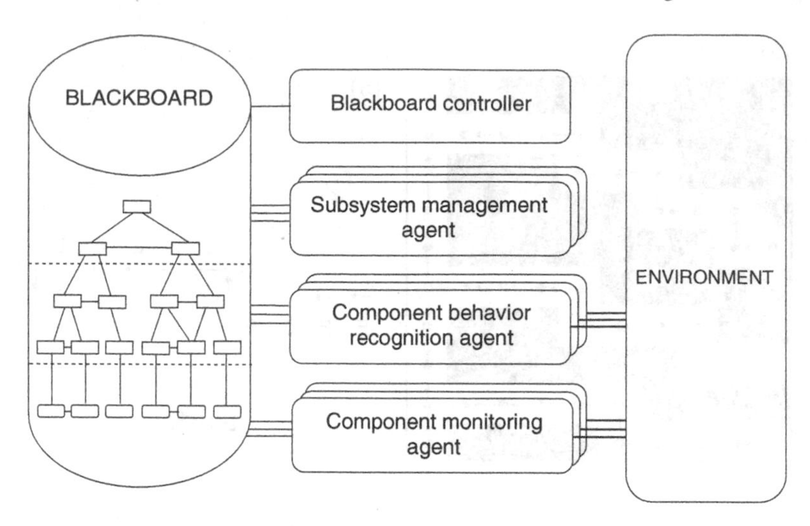

Fig. 14.3 An exemplary multi-agent system based on the blackboard architecture

- no agent can solve the problem itself,
- the system is not controlled in a centralized way,
- data are distributed in the system,
- computing in the system is performed in an asynchronous way.[11]

The form of communication among agents is a key issue in the theory of multi-agent systems. There are two basic scenarios. The first scenario is based on the *blackboard architecture* [132]. An example[12] is shown in Fig. 14.3. The agents communicate via writing and reading messages to and from a global hierarchical repository called a *blackboard*. It plays a role similar to the working memory in rule-based systems and it can contain hypotheses to be verified, partial results of an inference process, etc. Agents in blackboard systems are called *knowledge sources*. One distinguished agent plays the role of the *blackboard controller*. It is responsible for focusing agents' attention to subproblems to be solved, managing access to the blackboard, etc.

Agents in blackboard systems usually create a hierarchical structure corresponding to levels of abstraction of the blackboard information structure. As one can see in Fig. 14.3, agents placed at the lowest level perform simple monitoring of components of complex equipment, which is the environment of the multi-agent system.

[11] A multi-agent system is a distributed system, i.e., it is a system consisting of independent computing components. An asynchronous way of computing means here that we do not assume any time restrictions for performing computations by these components. Of course, if we were able to impose time restrictions (the synchronous model), then we would be able to control the whole computing process. Unfortunately, in multi-agent systems we cannot assume what length of time an agent requires to solve its subproblem.

[12] The example is discussed in the paper: Behrens U., Flasiński M., et al.: Recent developments of the ZEUS expert system. *IEEE Trans. Nuclear Science 43* (1996), 65–68.

Each such agent reads information describing the current situation in the component supervised. Then, it identifies the current state of this component and performs the action of sending a message about this state to an agent on a higher level. Thus, the lowest-level agent performs according to the principle: *perception—action*. An agent of this kind is called a *reflex agent*.

At a higher level are placed agents which recognize the behavior of components in time series. Such agents are not only able to perceive a single event, but they can also monitor processes, i.e., they can remember sequences of events. As a result of monitoring a process the *internal state* of such an agent can change.[13] In some states the agent can perform certain actions. An agent of this type is called a *reflex agent with internal states* (or *model-based agent*).[14]

Since components of the equipment are grouped into subsystems which should be supervised by the multi-agent system, managing agents are placed at the highest level of the hierarchy. These agents make use of information delivered by agents of lower levels in order take make optimum decisions. An agent of this level makes decisions after communicating (via the blackboard) with other agents of this level, because the subsystem it supervises does not work independently from other subsystems of the equipment. A managing agent should be able to determine a goal which should be achieved in any specific situation in the environment. Therefore, an agent of this kind is called a *goal-based agent*.

In the example above a *utility-based agent* has not been defined. An agent of this type determines goals in order to reach the maximum *satisfaction* (treated as a *positive emotion*) after their achievement. The main idea of this agent is based on the *appraisal theory of emotions* introduced in the twentieth century by Magda Arnold.[15] According to this theory an evaluation (appraisal) of a perceived situation results in an emotional response, which is based on this evaluation. In the case of an utility-based agent, however, the main problem concerns ascribing numerical values to its internal (*emotional*) states.

In the second scenario of communication among agents, the *message passing* model is used. This model is based on *speech-act theory*, inspired by Wittgenstein's *Sprachspielen* concept.[16] The model was introduced by John L. Austin[17] in the second half of the twentieth century [13]. According to this theory, uttering certain sentences,

---

[13]Changes of internal states of an agent can be simulated with the help of a finite automaton introduced in Chap. 8.

[14]The difference between a *reflex agent* and a *reflex agent with internal states* is analogous to the difference between a human being who reacts immediately, in a non-reflexive manner, after perceiving a stimulus and a human being who does not react at once, but observes a situation and then takes an action if his/her internal state changes (e.g., from being calm to being annoyed).

[15]Magda B. Arnold—a professor at Loyola University in Chicago and Harvard University. Her work mainly concerns psychology of emotions. She was known as a indefatigable woman. In her nineties she was climbing hills. She died when she was nearly 99 years old.

[16]The *Sprachspielen* concept is introduced in Chap. 15.

[17]John Langshaw Austin—a professor of philosophy at Oxford University, one of the best known scientists of the British analytic school. His work concerns philosophy of language, philosophy of mind, and philosophy of perception.

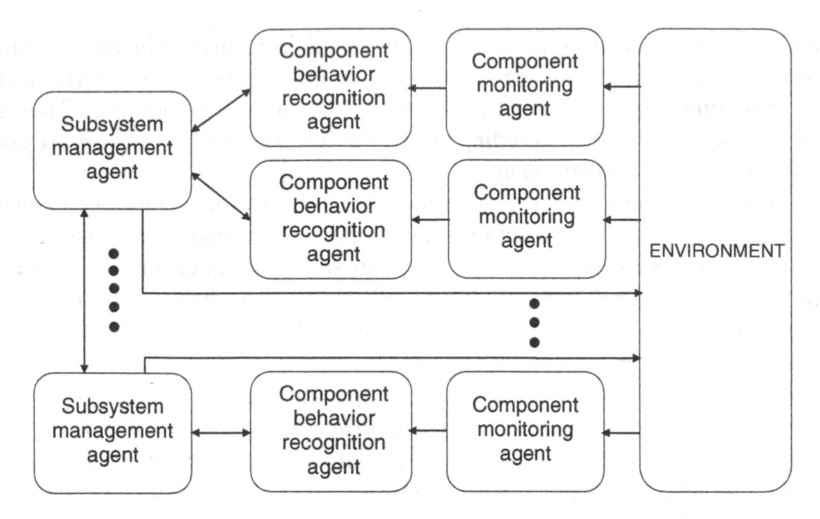

Fig. 14.4 An exemplary multi-agent system based on message passing

called by Austin *performatives*, in certain circumstances is not just *saying* something, but performing an action of a certain kind. For example, uttering a marriage formula by a priest results in a marriage act. In the case of multi-agent systems, performatives correspond to *communication acts*, which are performed by an agent in relation to other agents. For example, they can be of the form of a query concerning something, a demand to perform some action, a promise of performing some action, etc. In this model agents communicate according to predefined rules, which should ensure the performative result of the messages sent.

An example of the scheme of a multi-agent system based on the message passing model is shown in Fig. 14.4.[18] As one can see, agents interact directly with each other. These interactions are performed by sending performative messages, which result in influencing the environment in the required way.

Bibliographical Note

A good introduction to multi-agent systems can be found in [80, 88, 274, 312, 319]. The prospects for future development of cognitive architectures are discussed in [77].

[18]The example is discussed in the paper Flasiński M.: Automata-based multi-agent model as a tool for constructing real-time intelligent control systems. *Lecture Notes in Artificial Intelligence 2296* (2002), 103–110.

Part III
Selected Issues in Artificial Intelligence

Chapter 15
Theories of Intelligence in Philosophy and Psychology

Basic approaches to the simulation of sensual/intellectual cognitive abilities, such as problem solving, pattern recognition, constructing knowledge representations, learning, etc. have been presented in previous parts of the book. IT systems constructed on the basis of these approaches are called *Artificial Intelligence systems*. If one is asked "What is Artificial Intelligence?", we could, therefore, answer that from a computer science point of view Artificial Intelligence is a feature of IT systems constructed on the basis of such approaches. Of course, nobody would be satisfied with such an answer, because such a definition does not explain the heart of the matter. We discuss this issue of Artificial Intelligence in this part of the monograph.

As we will see the notion of *intelligence* is defined in various ways in philosophical and psychological theories, as are related concepts such as *mind*, *cognition*, *knowledge*, etc. It seems that this is a reason for the disputes about the term *artificial intelligence*. Therefore, we present philosophical and psychological interpretations of these basic notions in this chapter.

In the first section we introduce the main philosophical approaches to issues of cognition and mind. This presentation is necessary for discussing various ideas on artificial intelligence in Chap. 17. Various definitions and models of intelligence in psychology are presented in the second section. They will be used primarily for determining a list of cognitive/mental abilities. This list will be used for presenting application areas of AI systems in Chap. 16.

15.1 Mind and Cognition in Epistemology

Issues of mind, cognition, and intelligence are studied in an area of philosophy called *epistemology*. In this section we limit our presentation of views of philosophers only to those which can form a basis for a discussion about Artificial Intelligence.

The distinction between *sense perception* and *mental perception* was introduced by Parmenides of Elea in the fifth century B.C. Plato (427 B.C.–347 B.C.) claimed that before the soul was embodied the intellect perceived perfect ideas in a direct

M. Flasiński, *Introduction to Artificial Intelligence*,
DOI 10.1007/978-3-319-40022-8_15

way. Such an intuitive direct cognition *via* an insight into the heart of the matter is called *noesis* (νόησις [nóesis]). The embodied soul can recall this knowledge (during a perception of the world), which is called *anamnesis* (ἀνάμνησις [ánámnesis]).[1] Mathematical knowledge (διάνοια [diánoia]) is placed in the knowledge hierarchy at a lower level than noesis. These two types of knowledge are considered to be *justified true belief*, which together is called *episteme* (ἐπιστήμη [épistéme]). Contrary to episteme, the physical world is cognized via senses in an uncertain, *doxa*-type way (δόξα [doksa]) [228], as a *shadow* of reality (the allegory of the cave).

Contrary to Plato, Aristotle (384 B.C.–322 B.C.) denied that humans have innate ideas.[2] The *intellect (mind)* (νοῦς [noús], *intellectus*) is "the part of the soul by which it knows and understands". During the understanding process the intellect operates in the following way. Things are perceived through senses. On the basis of sense experience the intellect forms mental images called *phantasms*. The intellect transforms phantasms into concepts (notions) by *abstracting*. Abstraction is the act of isolating the *universal concepts* (*intelligible species, intelligibles*), which are intellectual abstracts, from the phantasms. The *active intellect* (νοῦς ποιητικός [noús poietikós], *intellectus agens*) *illuminates* the phantasm, and in this way the universal concept created from the phantasm is grasped by the *potential intellect* (νοῦς παθητικός [noús pathetikós], *intellectus in potentia*) [9].

Concepts create a hierarchy. A concept of a species is defined by giving its nearest genus (*genus proximus*) and its specific difference (*differentia specifica*), which distinguishes members of this species from other members of this nearest genus. In this way concepts create a hierarchy, which can be represented by a concept tree.[3] Linguistic representations of concepts, i.e., *terms* can be joined to form *propositions*. A proposition is a statement which is either true or false. Propositions can also create a kind of hierarchy, which is based on the relation *premise–conclusion*. An inference from a more general premise to a less general premise is of a *deductive* nature and is consistent with the logical order. Aristotle developed the theory of *syllogism* that involves deductive reasoning in which the conclusion is inferred from two premises. However, in the psychological order of cognizing reality a general proposition is derived from specific observations. Such reasoning is called *induction* [10].

St. Thomas Aquinas (1225–1274) reinterpreted and developed Aristotelian epistemology [285]. There are two great powers of the *mind* (*mens*): the *intellect* (*intellectus*), which is a cognitive power, and the *will* (*voluntas*), which is an appetitive[4] power. There are *three generic operations (acts) of the intellect* (*tres operationes rationis*). The *simple apprehension* of what something is (*indivisibilium intelligentia*), which consists of comprehending a concept by abstracting, is the first act. *Pronouncing a judgment* (*componere et dividere*), which consists of stating a proposition that is affirmed or denied, is the second operation. The third act, called *reasoning*

---

[1] Let us recall that the view that certain abilities are inborn is called *nativism*.

[2] Let us recall that such a view is called *genetic empiricism*.

[3] We have introduced such concept trees in the form of *semantic networks* representing *ontologies* in Sect. 7.1.

[4] Aquinas defines *appetite* as "all forms of internal inclination".

(*ratiocinare*), consists of proceeding from one proposition to another according to logical rules (principles). Thus, *reason* (*ratio*), in the sense of the abstract noun, can be defined as the reasoning function of an intellect or cognizing by discursive reasoning.

St. Thomas Aquinas distinguished, after Aristotle, the *practical mind* (*intellectus practicus*), which is used by the human being for planning, defining strategies for activities, decision making, etc. from the *theoretical mind* (*intellectus speculativus*), which allows the human being to understand and contemplate. A distinction can also be made between *dispositional intellect* (*intellectus in habitu*), which has basic universal concepts and is prepared for intellection and *actualized (achieved) intellect* (*intellectus adeptus*), which has already achieved knowledge.[5] St. Thomas Aquinas defined *intelligence* (*intelligentia*) as "*the act itself of intellect, which is understanding*"[6] [285]. In other words, intelligence means a cognitive act, which is performed by achieved intellect.

Mental perception characterized in such a way is preceded by sense perception. St. Thomas Aquinas distinguished *external senses* from *internal senses*. The external sensorium includes sight, hearing, taste, smell, and touch. There are four internal senses. *Common sense* (*sensus communis*) perceives objects of the external senses and synthesizes them into a coherent representation. *Imagination* (*imaginatio, phantasia*) produces a mental image of something in its absence. *Memory* (*vis memorativa*) stores perceptions which have been cognized and evaluated with respect to the interests of the perceiver. Such perceptions are called *intentions* by St. Thomas Aquinas. They can be called to mind at will. An evaluation of a perception with respect to the interests of the perceiver, i.e., whether it is beneficial (useful) or harmful, is performed by the *cognitive power*, called also *particular reason* (*vis cogitativa, ratio particularis*). Cogitative power thus enables human beings to react to stimuli in an adequate way, which is a necessary condition of proper behavior in the environment.[7]

William of Ockham, Occam (1288–1348) is known for the *principle of parsimony* (*lex parsimoniae*), also called *Ockham's Razor*. It states that "entities should not be multiplied beyond necessity" (*Entia non sunt multiplicanda sine necessitate*), which means that if two theories can be used to draw the same conclusions, then the simpler theory is better. According to Ockham incorrect reasoning often results from an incorrect, from a logical point of view, use of the language. Thus, logical analysis is more important than speculative discourse [213]. Therefore, he sometimes is considered a forerunner of *analytic philosophy*.

René Descartes (1596–1650) addressed the problem of the relationship between mind and matter, called the *mind-body problem*, which is one of the key issues discussed in the area of Artificial Intelligence. He claimed there is a rigid distinction

[5]This distinction was adopted by European scholastic philosophy from Islamic philosophy. Ibn Sina, Avicenna (980–1037) already used both concepts: *intellectus in habitu* (*al-'aql bi-l-malakah*) and *intellectus adeptus* (*al-'aql al-mustafâd*). Then, these concepts were adopted and reinterpreted by scholastics (St. Albertus Magnus, St. Thomas Aquinas).

[6]Adding also that sometimes "*the separate substances that we call angels are called intelligences*".

[7]An evaluation of a perception w.r.t. the interests of a perceiver is performed by animals by natural instinct. In the case of animals we talk about *(natural) estimative power* (*vis aestimativa*).

between *mind* (*res cogitans*), which is a nonmaterial substance, and *matter* (*res extensa*). Knowledge is certain, if it is clear and distinct (*clair et distinct*) [69]. Mathematical knowledge is clear and distinct. Therefore, knowledge systems should be constructed in the same way as in mathematics. Genuine knowledge is acquired by reason with the help of *innate ideas* (*ideae innatae*) and is independent of sensory experience.[8]

Thomas Hobbes (1588–1679) denied the existence of any nonmaterial substance (*materialistic monism*) [138]. According to him, a human being is a complex machine. Hobbes resolved the Cartesian mind-body problem by claiming that consciousness can be reduced to a bodily activity. Cognitive operations are of a mechanical nature, i.e., sensory experience consists of a mechanical effect on a body. *Reasoning* is a form of *computation*, i.e., it is a manipulation of symbols.

Baruch Spinoza (1632–1677) proposed another solution to the mind-body problem [284]. He defined body and mind as two aspects, attributes (*attributa*) of one universal substance (*neutral monism*). Although there is no causal interaction between mental and material phenomena, they occur in parallel, i.e., they are *programmed* in such a way that if some mental event takes place, then a *parallel* material event occurs (*psychophysical parallelism*).

As a consequence of his theory of monads, Gottfried Wilhelm Leibniz (1646–1716) claimed that all knowledge is acquired independently of any experience (*radical apriorism*) [176]. According to him, every valid proposition is an analytical proposition. So, every valid proposition can be proved. Therefore, Leibniz carried out research into defining a *universal symbolic language* (*ars characteristica, scientia generalis*), in which all valid sentences would be decidable mechanically.[9]

Contrary to the views of Descartes and Leibniz, John Locke (1632–1704) maintained that knowledge is obtained by *sensory experience* only[10] [185]. An experience can concern the external world influencing our (external) senses, and then he talks about *sensation*. It can also concern our mind, i.e., it can relate to awareness of our own intellectual activity, which is called *reflection* (inner sense) by Locke. Sensation is less certain than reflection, however the former precedes the latter. In fact, Locke pointed out an important knowledge source, which is *introspection*.

David Hume (1711–1776) defined two kinds of knowledge: *relations among ideas* and *matters of fact*. The first kind, which includes mathematics and other abstract models, is certain, however uninformative. The second kind, which concerns propositions about the nature of existing objects, so interesting for us, is not certain, unfortunately [144]. Hume was sceptical about our ability to acquire knowledge by reason. He claimed that interesting knowledge about the external world is acquired by inductive inference, which is based on our natural instinct.

---

[8]Let us recall that such a view is called *methodological rationalism (apriorism)*.

[9]If Leibniz had succeeded, then philosophers, instead of disputing, would formulate their opinions with the help of *ars characteristica* and then *compute* a solution. It seems to the author that it is better that Leibniz never completed his research.

[10]Let us recall that such a view is called *methodological empiricism*.

Immanuel Kant (1724–1804) tried to bring together both empiricist and rationalist views concerning cognition. He claimed that cognition is based on both experience and a priori ideas, since *"Thoughts without content are empty, intuitions without concepts are blind."* [154]. He distinguished the following three kinds of propositions. An *analytic proposition* a priori expresses only what is contained in the concept of its subject. Thus, such propositions are used to explain knowledge already existing in our mind, since either it is contained in the definition of the subject or it can be derived from this definition. Such propositions are independent from experience and they are present in our minds (a priori).

The two remaining kinds of propositions, called synthetic propositions, expand our knowledge, because they add new attributes to their subjects. A *synthetic proposition a posteriori* is formulated on the basis of experience gained.[11] A *synthetic proposition* a priori expands our knowledge and is certain. Empiricists denied that there are such propositions. However, Kant claimed that such propositions exist.[12] In order to explain how synthetic propositions a priori are created in our mind, he developed *transcendental philosophy*.[13] According to this philosophy, *sensations* (*Empfindung*) perceived by the senses are combined with the help of a priori *pure forms of sensuous intuition* (*Anschauung*), i.e., space and time, into their mental representations (*Vorstellung*). Then, *intellect* (understanding faculty) (*Verstand*), using a priori categories of the understanding,[14] combines these representations into *concepts*. Finally, *judgment faculty* formulates propositions with the help of a priori rules.

John Stuart Mill (1806–1873) considered inductive reasoning to be more important for knowledge acquisition than deductive reasoning. General principles should be known in order to use deduction. As an empiricist, Mill claimed that experience is a source of knowledge. However, experience concerns a single event. Thus, one should use inductive reasoning in order to formulate general principles. Mill developed schemes of inductive reasoning which are helpful for defining causal relationships, called *Mill canons of induction* [202].

Franz Brentano (1838–1917) claimed that *experience* is the only source of knowledge, however, experience in the form of an *internal perception (introspection)*. He reintroduced, after scholastic philosophy, the concept of *intentionality*, which is a property of a mental act meaning directing at a certain (intentional) object (from Latin *intendere*—to direct toward something) [38]. Due to intentionality, which is a property of human minds, we can distinguish mental phenomena from physical phenomena.

To Edmund Husserl (1859–1938), a founder of *phenomenology*, *intuition* was the basic source of knowledge [145]. Intuition is a condition of both deduction and

---

[11] Empirical propositions in physics, chemistry, biology, etc. are synthetic propositions *a posteriori*. However, from the point of view of logic these propositions are uncertain.

[12] Kant considered mathematical propositions (e.g., theorems) to be synthetic propositions a priori.

[13] *Transcendental* means here transcending experience.

[14] Kant defined twelve such categories.

induction, since it provides these two types of reasoning with premises. Deduction and induction can be used only for indirectly deriving the truth.

Within *analytic philosophy* Ludwig Wittgenstein (1889–1951) and his colleagues forming the *Vienna Circle*[15] tried to define a formalized language that could be precise and unambiguous enough to become a language for a *unified science* (*Einheitswissenschaft*).[16] The possibility of empirically determining the *true* meaning of linguistic expressions was the basic assumption of this research. In his later works Wittgenstein claimed that the separation of the meaning of expressions from their *usage* in *live* language is impossible and notions can be meaningful, even if they are not defined precisely [316]. This results from the fact that the (natural) language used by us is a set of *language-games* (*Sprachspielen*),[17] which proceed according to specific rules and logic.

Although Kurt Gödel (1906–1978) was a logician and mathematician rather than a philosopher, his work considerably influenced epistemology. In 1931 he published his first *limitation (incompleteness) theorem* [110]. It says that if a formal system including the arithmetic of natural numbers is consistent,[18] then it is incomplete.[19] Intuitively speaking, Gödel showed that in non-trivial formal theories there exist true sentences that cannot be proved in these theories.

15.2 Models of Intelligence in Psychology

As we have seen in a previous section, there are various views concerning the nature of cognition and mind in philosophy. Where the concept of intelligence is concerned, we meet a similar situation in psychology.[20] For the purpose of our considerations about artificial intelligence, we propose a definition which is a synthesis of the scopes of definitions known from the literature. Thus, (human) *intelligence* can be defined as a set of abilities that allow one:

[15] The *Vienna Circle* was an influential group of philosophers, mathematicians, and physicists founded in 1920s. Its best-known members were, among others Kurt Gödel, Rudolph Carnap, Moritz Schlick, and Otto Neurath.

[16] The issue of such a language had already been of great importance for William of Ockham and Gottfried Wilhelm Leibniz mentioned above.

[17] A *language-game* consists of a sequence of expressions and actions in a certain context. For example, the dialogue of a man and a woman in case he wants to pick her up or the conversation of a professor and a student during an exam at a university. Thus, our life can be treated as a sequence of language-games. Each language-game is characterized by its specific grammar, just as each game is characterized by its specific rules.

[18] A formal system is consistent if it does not contain a contradiction.

[19] A formal system is complete if for any sentence belonging to the system either the sentence or its negation is provable in this system.

[20] In this section we describe selected psychological models of intelligence. A selection is made with respect to further discussion of the concept of *artificial intelligence*. Studies of intelligence in psychology can be found e.g., in [108, 290, 291].

- to adapt oneself to a changing environment, and
- to perform cognitive activity, which consists of creating and operating abstract structures.

In the definition we have deliberately not listed such abilities, because, as we see further on, there are differences among various psychological models with respect to this aspect. We will try to identify such abilities within each succeeding model discussed below.

The second constituent of our definition has been formulated, since here we do not want to ascribe intelligence to all biological organisms which possess adaptation mechanisms. The expression *abstract structure* corresponds, in principle, to the psychological terms *abstract concept*, *mental representation*, and *cognitive structure* (in cognitive psychology). However, we assume that abstract structures do not necessarily need to be created in order to represent knowledge about an environment, since we do not want to exclude abstract constructs defined e.g., in logic and mathematics from our considerations.

There are two basic types of models of intelligence in psychology. In *hierarchical models* mental abilities are layered and create a hierarchical structure. By contrast, in *multiple aptitude models* mental abilities are treated as a set of equivalent and independent factors. Firstly, we discuss models which belong to the first approach.

The *two-factor theory of intelligence* introduced by Charles E. Spearman[21] in 1923 [282] is a generic theory for hierarchical models. In the theory one general factor g, which corresponds to *general intelligence*, and several specific factors $s_1, s_2, \ldots, s_n$, representing mental abilities that are used for solving various types of problems, are distinguished.

General intelligence (*g factor*) was characterized by Spearman with the help of three *noegenetic principles*[22] [142]. The principle of *apprehension of experience* says that the apprehension of the meaning of a received perception is the first operation of general intelligence. Basic elements which describe a problem, called *fundaments*, are created due to this operation. During the second operation, according to the principle of *eduction of relations*,[23] relations between fundaments are discovered. According to the principle of *eduction of correlates*, the third operation consists of discovering, via reasoning, further relations on the basis of the relations identified during the second operation. These further relations are not discerned directly *via* the eduction of correlates [142].

Let us notice that the operations defined by noegenetic principles can be interpreted as comprehending concepts, pronouncing judgments,[24] and reasoning (in the sense of proceeding from one proposition to another). Thus, they are analogous to

---

[21] Charles Edward Spearman—a professor of psychology at University College London. He is also known for his contribution to statistics (Spearman's rank correlation).

[22] Spearman used the term *noegenetic* with reference to the Platonian notion *noesis*, which was introduced in the previous section. Let us recall that *noesis* means intuitive direct cognition via an insight into the heart of the matter [142].

[23] Here, *eduction* means discovering something, from Latin: *educere*.

[24] In this case, pronouncing judgments by defining relations.

the three generic acts of the intellect (*tres operationes rationis*) defined by Aquinas, which have been presented in a previous section. According to Alfred Binet[25] pronouncing judgments is the fundamental operation [26]. We will come back to this opinion in the last chapter, in which the *issue of artificial intelligence* is discussed.

In principle, designers of AI systems do not try to simulate general intelligence.[26] Instead, methods which simulate specific mental abilities of a human being are developed. Later, we make a short survey of psychological models of intelligence which identify specific mental abilities. A discussion of a simulation of these abilities in AI systems is contained in the next chapter.

In 1997 Linda S. Gottfredson[27] performed an analysis of the mental abilities which are tested during a personnel selection process [114]. These abilities correspond to research subareas in Artificial Intelligence. The following mental skills were identified as important: *problem solving*, *reasoning decision making*, *planning*, *(verbal) linguistic abilities*, and *learning ability*.

Half a century before the research of Gottfredson, Louis L. Thurstone[28] identified a model of *primary mental abilities* [303]. This model belongs to the multiple aptitude approach. For our further considerations we choose *perception* (treated as effective pattern recognition) from the set of these abilities. Such factors as number facility, spatial visualization, and memorizing (associative memory) are implemented in *standard* IT systems. Reasoning and linguistic abilities have been identified already on the basis of the Gottfredson model. It is worth pointing out that Thurstone influenced the area of pattern recognition remarkably, mainly due to his statistical model in signal detection theory [302]. In the 1930s he identified such crucial issues of modern pattern recognition as recognition of animated patterns and 3D pattern recognition performed on the basis of various views perceived by a moving observer.

In 1947 Jean Piaget[29] identified *kinesthetic intelligence* during research into stages of a child's mental development [226]. It is related to both locomotion and manipulation abilities and their development takes place from birth to two years (the sensorimotor stage). Piaget distinguished two kinds of manipulation: *unspecific manipulation*, which is not adequate to the specificity of an object, and *specific manipulation*,[30] which involves *tuning* of movements to an object's specificity.

---

[25] Alfred Binet—a professor of psychology at the Sorbonne, a director of the Laboratory of Experimental Psychology at the Sorbonne (1891–1894). He is known as the inventor of the first intelligence test (the IQ test).

[26] Apart from designers of *generic* cognitive architectures.

[27] Linda Susanne Gottfredson—a professor of educational psychology at the University of Delaware and a sociologist. She is known for the public statement *Mainstream Science of Intelligence* signed by 51 researchers, in which she claims that mental skills are different for various races (statistically).

[28] Louis Leon Thurstone—a professor of psychology at the University of Chicago and a mechanical engineer (master's in mechanical engineering from Cornell University). One of the fathers of psychometrics and psychophysics. He was the President of the American Psychological Association.

[29] A note on Jean Piaget has been included in the previous chapter, in which we have discussed notions of *assimilation* and *accommodation* and their influence on the concept of *agent*.

[30] Specific manipulation develops from 8 months.

In psychology *specific* types of intelligence are also considered. In 1920 Edward L. Thorndike[31] distinguished *social intelligence* [300]. This kind of intelligence can be defined as a set of abilities which allow one to establish interpersonal relations, to achieve social adjustment, and to influence people. These abilities can be divided into *social awareness* (e.g., the ability to interpret people's behavior) and *social facility* (e.g., the ability to generate behavior, which informs others about our internal state) [112].

Emotional intelligence is related to social intelligence. It is defined as a set of abilities which allow one to perceive others' emotions, to control ones own emotions and to use emotions in mental processes and during problem solving.

Recently research into implementing AI systems which simulate *human creativity*, such as, e.g., musical creativity or visual creativity, has been carried out as well. Such systems try to simulate *creative abilities* in order to develop original ideas, visual art, or solutions (e.g., in architecture). *Structure of Intellect theory* introduced by Joy P. Guilford[32] in 1967 [116] is one of the most important psychological models relating to creativity. The following intellectual processes are defined in this model[33]: cognition, memory operations, convergent production, divergent production, and evaluation of information. Whereas a *convergent production* consists of deriving a single valid solution for a given standard problem (i.e., *problem solving* in the sense used in AI), during a *divergent production* one draws original ideas via *creative* generation of multiple possible solutions. The quality of a divergent production can be assessed on the basis of such criteria as the number of ideas generated, the variety of approaches used for solving a problem, and the originality of the ideas.

In cognitive psychology *mental (cognitive) processes* are studied to model intelligence. This approach is interesting with respect to research into AI systems, since a *process*, interpreted as a transformation of certain information structures into other ones, is a fundamental concept of computer science. The interpretation of a mental process in psychology is analogous to the one used in computer science. A mental process is used for creating and transforming a cognitive structure, which is a mental representation of a certain aspect of an external reality. In particular, cognitive structures can represent our *knowledge*.[34] Thus, in Artificial Intelligence they correspond to *models of knowledge representation*, which will be discussed in the next chapter.

Mental processes can be divided into several basic categories, e.g., perception processes, attention processes, memory processes, and thinking processes. Apart

---

[31] A note on Edward L. Thorndike, a pioneer of the connectionist approach, has been included in Sect. 3.1.

[32] Joy Paul Guilford—a professor of psychology at the University of Southern California. His work mainly concerns the psychometric study of intelligence. He has carried out research for developing classification testing for the U.S. Air Force. He was the President of the Psychometric Society.

[33] Guilford's model is three-dimensional. Apart from the process dimension it contains a content dimension (types of information which are used by processes) and a product dimension (results of applying processes to specific contents). Later, Guilford extended his model, cf. Guilford J.P.: Some changes in the structure of intellect model. *Educational and Psychological Measurement 48* (1988), 1–4.

[34] In cognitive psychology, the concept of knowledge is related to the content of long-term memory.

from these categories, so-called *metacognitive factors* which are responsible for planning, supervising, and controlling mental processes are distinguished[35] in cognitive psychology. The *triarchic theory of intelligence* introduced by Robert J. Sternberg[36] in 1980 [289] is one of the best-known models based on mental (cognitive) processes.

Summing up, on the basis of a survey of important psychological theories we have identified specific mental/cognitive abilities which can be simulated in AI systems. They include perception and pattern recognition, knowledge representation, problem solving, reasoning, decision making, planning, natural language processing, learning, manipulation and locomotion, social/emotional intelligence and creativity. The issue of simulating these abilities is discussed in the next chapter.

Bibliographical Note

A good introduction to philosophy and the history of philosophy can be found in [59, 155, 255]. Epistemological issues are discussed in [33, 143, 178].

Theories of intelligence in psychology are presented in [31, 108, 290, 291].

[35] If controlling is performed with feedback, then we deal with learning.

[36] Robert Jeffrey Sternberg—a professor of psychology at Yale University, Tufts University, and Oklahoma State University, a Ph.D. student of Gordon H. Bower at Stanford University. He was the President of the American Psychological Association. Sternberg is also known for the *triangular theory of love*.

Chapter 16
Application Areas of AI Systems

Before we discuss the issue of the possibility of constructing an intelligent artificial system in the last chapter, we now summarize practical results concerning application areas of AI systems.[1] As we have mentioned in a previous chapter, designers of such systems do not model a *general intelligence*, rather they focus on methods simulating particular human cognitive/mental abilities and corresponding constructs such as knowledge representation models. Application areas of AI systems will be discussed on the basis of human cognitive abilities identified after the analysis of psychology models of intelligence discussed in the previous chapter.

16.1 Perception and Pattern Recognition

Intelligent behavior depends on *perception* of the external world to some extent. Although a human being perceives with the help of five senses, i.e., sight, hearing, taste, smell, and touch, only the first two senses are simulated in most AI systems. From a technical point of view, both sound and image are treated one- or two-dimensional signals. (Sometimes 3D signals, if a spatial model of the world is defined in the system.)

In AI systems the task of perceiving sound or image is divided into two main phases. The first phase concerns of receiving a corresponding signal with the help of a sensory device (e.g., a camera or a microphone), its preprocessing, and its coding in a certain format. The methods used in this phase belong to conventional[2] areas of computer science (also automatics and electronics) such as *signal processing theory* and *image processing theory*. Both theories were developed remarkably in

[1] In the monograph we do not present specific AI systems, because they are continuously being introduced in the software market. So, this chapter had to be updated each year.

[2] *Conventional* means here that they do not need the support of AI techniques.

M. Flasiński, *Introduction to Artificial Intelligence*,
DOI 10.1007/978-3-319-40022-8_16

the second half of the twentieth century. They allow us to implement systems which surpass human beings in some aspects of sensory perception.[3]

In the second phase of perception, in which sensory information is ingested thoroughly, AI systems can use methods belonging to three groups of models that have been introduced in the monograph, namely *pattern recognition*, *neural networks*, and *syntactic pattern recognition*. Let us notice that also in these areas a lot of efficient techniques have been developed. Optical Character Recognition (OCR) systems, vision systems of industrial robots, optical quality control systems in industry, analysis of satellite images, military object identification systems, and medical image systems are some examples of practical applications of such systems. Recently research has been carried out into constructing systems which are able not only to identify objects, but also to understand (interpret) them [297]. Image understanding is especially useful in the area of advanced medical diagnostics [214, 296].

There are, however, still challenges in this area. Automatic learning is a crucial functionality of pattern recognition systems. In the case of classical pattern recognition and neural networks, models contain adaptive techniques of learning. On the other hand, in syntactic pattern recognition the issue of a system self-learning is more difficult, since it relates to the problem of formal grammar induction. In this area research is still in a preliminary phase.

The second problem concerns *intelligent* integration of information sent by various sensory devices at the same time (e.g., a camera and a microphone) in order to obtain a synthetic sensation.[4] We will return to this problem in the next chapter.

16.2 Knowledge Representation

The problem of adequate *knowledge representation* has been crucial since the very beginning of developments in the AI area. An intelligent system should be able to adapt to its environment, according to our definition formulated in Sect. 15.2. Thus, it should be able to acquire knowledge which describes this environment (*declarative knowledge*), then to store this knowledge in a form allowing a quick and adequate (intelligent) response to any stimulus generated by the environment. Patterns of such responses, represented as *procedural knowledge*, should be stored in the system as well.

A taxonomy of knowledge representation models can be defined according to two basic criteria: the form of knowledge representation and the way of acquiring knowledge.

According to the first criterion, knowledge representation models can be divided into the following three groups.

[3]Certainly the reader has seen crime films in which a blurry photograph made while moving has been processed by a computer system in order to restore a sharp image of a killer.

[4]Such a functionality in the system corresponds to St. Thomas Aquinas' *sensus communis* introduced in the previous chapter.

- *Models of symbolic knowledge representation formulated in an explicit way*. Let us notice that the basic models of this group, i.e., *conceptual dependency graphs*, *semantic networks*, and *scripts* have been introduced by psychologists Roger Schank, Allan M. Collins, and Robert P. Abelson, respectively. Where procedural knowledge is concerned, *rule-based systems* are the most popular representation model. In this book we have also introduced other specific representations for such knowledge, e.g., formal grammars, representations based on mathematical logic, models in reasoning systems, and schemes in Case-Based Reasoning systems.
- *Models of symbolic-numeric knowledge representation formulated in an explicit way*. These models are used if the notions which are the basis for the representation model are fuzzy, i.e., they are ambiguous or imprecise. Bayesian networks, models based on fuzzy sets, and models based on rough sets introduced in Chaps. 12 and 13 are good examples of such models.
- *Models of knowledge representation formulated in an implicit way*. This form is applied if knowledge is represented in a numeric way. It is typical for pattern recognition methods and neural networks. Such representations are of the form of clusters consisting of vectors, sets of parameters (in pattern recognition), and weight vectors (in NNs). Here, representation in an *implicit way* means not only that we lack access to these vectors or parameters. Even if we read these strings of numbers, we could not relate them to the meaning of knowledge coded in such a way. In other words, we are not able to interpret them in terms of the problem description.

Where the second criterion, i.e., the way of acquiring knowledge, is concerned, representation models can be divided into the following two groups.

- *Models in which knowledge can be acquired by the system automatically*. First of all, models of knowledge representation formulated in an implicit way belong to this group. Both pattern recognition methods and neural networks can be self-learning in the case of unsupervised learning techniques. For pattern recognition we use cluster analysis. In the case of symbolic representations such learning is performed via induction, for example grammatical induction in syntactic pattern recognition.
- *Models in which knowledge representation is defined and entered into the system by a knowledge engineer*. Most models of knowledge representation formulated in an explicit way belong to this group.

Summing up, automatic acquisition of knowledge in models based on symbolic knowledge is the crucial issue in this area. An automatic conceptualization is the main problem here and it has not been solved in a satisfactory way till now. Learning methods will be discussed in a more detailed way in Sect. 16.8.

16.3 Problem Solving

We define the area of *problem solving* as research into constructing *generic* methods that can be used for solving general problems. *General Problem Solver, GPS*, constructed by Allen Newell and Herbert A. Simon described in Chap. 1 is a good example of this area of Artificial Intelligence. The dream of AI researchers to construct such a system has not come true yet. Therefore, this problem has been divided into a variety of subproblems such as reasoning, decision making, planning, etc., which are discussed in the next sections.

Returning to the problem of constructing a *general problem solver*, let us notice that *heuristic search methods* and their extension in the form of *evolutionary computing*[5] are good candidates for such a purpose.

Nevertheless, systems based on a search strategy do not solve problems in an autonomous way, but in *cooperation* with a human designer. Let us notice that there are two phases of problem solving with the help of a search strategy, namely:

- a phase of constructing an *abstract model of the problem*, which is the basis for defining states in the state space (cf. Sect. 4.1) and
- a phase of searching the state space.

Methods of searching a state space concern only the second phase. The first phase is performed by a human designer. The development of methods which allow an AI system to autonomously construct an abstract model of a problem on the basis of perception (observation) of the problem seems to be one of the biggest challenges in the area of simulating cognitive/mental abilities.

16.4 Reasoning

Artificial Intelligence systems work perfectly, where *deductive reasoning* is concerned.[6] Deductive reasoning is a type of reasoning in which on the basis of a certain general rule (rules) and a premise, we infer a conclusion (cf. Appendix F.2). Systems based on mathematical logic are the best examples of such reasoning. In Chap. 6 we have introduced two basic models for constructing such systems, namely *First-Order Logic* and *lambda calculus*.

Rule-based systems presented in Chap. 9 are one of the most popular types of reasoning systems. They are applied in business, medicine, industry, communications, transport, etc. In case we deal with imperfect knowledge or fuzzy notions, AI systems based on *non-monotonic logic* introduced in Chap. 12 are constructed or we apply *fuzzy logic* presented in Chap. 13.

---

[5]In fact *evolutionary computing* can be treated as an efficient version of a search strategy.

[6]In this section by reasoning we mean deductive reasoning, whereas later when we discuss the area of machine learning we discuss both deductive and inductive inference.

Thus, in Artificial Intelligence in the area of deductive reasoning we are able to simulate human abilities better than in the case of the remaining cognitive/mental abilities. This results from the dynamic development of mathematical and logic models in the period preceding the birth of Artificial Intelligence. This concerns especially the excellent development of mathematical logic in the first half of the twentieth century. Its models have been used successfully for defining effective algorithms of reasoning.

16.5 Decision Making

Supporting a process of *decision making* was one of the first applications of AI systems. The natural approach based on a simulation of succeeding steps of a decision process performed by a human expert is used in *expert rule-based systems*. A simple example of a simulation of such a decision process has been presented in Sect. 9.2. Let us notice that in order to apply such an approach, *explicit* knowledge in the form of rules representing partial decisions which can be used in any reasoning scenario should be delivered by a human expert. These rules are the basis for constructing an expert system.

In case such knowledge is unavailable an approach based on *pattern recognition* or *neural networks* can be used. Then we build a general specification of a problem with the help of numerical features, as has been presented in Chap. 10. The problem is characterized by a vector of numerical values. The possibility of equating the set of possible decisions with the set of classes determined by the vectors of a learning set is a condition of using such an approach.

If we apply a pattern-recognition-based approach for constructing a system which supports decision making, then *statistical pattern recognition* using the Bayes classifier (cf. Sect. 10.5) can be especially convenient. In such a case a system does not propose one decision in a deterministic way, but it suggests several possible decisions, assigning probability measures to them. As we have mentioned in Sect. 10.5, we can generalize the Bayes classifier by assuming that in case of making erroneous decisions there are various consequences with various costs of an error. The function of the cost of an error together with the *a posteriori* probability is used for defining the function of the risk corresponding to various decisions. Of course, the Bayes classifier tries to minimize the risk function.

If a decision process can be divided into stages, then we can apply a classifier based on *decision trees*, which has been presented in Sect. 10.6.

In case we have to solve a decision problem on the basis of knowledge which is uncertain, imprecise, or incomplete, we should use the methods introduced in Chap. 12, i.e., *Bayes networks* or *Dempster-Shafer Theory*. If a decision problem is described with fuzzy notions, then *fuzzy rule-based systems*, introduced in Chap. 13, or hybrid systems based on fuzzy set theory and model-based reasoning [152], introduced in Chap. 9, can be used.

At the end of the twentieth century effective methods of decision making were developed on the basis of advanced models of decision theory, game theory, and utility theory. Decision support systems are applied in many application areas. Typical application areas include, e.g., economics, management, medicine, national defence, and industrial equipment control.

16.6 Planning

Planning consists of defining a sequence[7] of activities which should result in achieving a predefined target. Simulation of this mental ability seems to be very difficult. It contains a crucial element of *predicting* consequences (results) of taking certain actions. This task is especially difficult if it is performed in a real-time mode and in a changing environment, which is typical in practical applications. Then, a system has to modify (very quickly) a plan which has been already generated, in order to keep up with the changing environment.

Planning methods can be based on a scheme of *state space search*,[8] which has been introduced in Chap. 4. The final state represents a goal which should be achieved as a result of a sequence of activities. Possible activities are defined by transition operators in the state space, and states represent the results of performing these activities. However, defining these intermediate results is a crucial problem. Let us notice that, for example, in case of using a search strategy for problems concerning artifacts, like various games, predicting results of activities is trivial. For example, if a system playing chess makes a decision to make the move *Ra5-h5*, then the result of this activity is obvious, i.e., *Rook* moves from *a5* to *h5*.[9] In this case the predictability of the result of the activity arises from the precise rules in the "world of chess". However, if a system functions in the real world, then the consequences of performing an activity sometimes cannot be determined. For example, if somebody has said something unpleasant to me, I can plan the activity of making a joke of it. Such an activity can result in easing the tension, which is the goal of my activity. However, it can also result in further verbal aggression, if my joke is treated as showing disrespect to my opponent. Thus, predicting consequences of planned activities is a very difficult issue.

Planning in the real world is sometimes connected with some circumstances, facts, or situations which limit the possibility of our activity in the sense of time, space, other conditions related to the physicality of the world, preferences concerning the way of achieving a goal, etc. Then, a planning problem can be expressed as a

[7] We use the term *sequence* in the definition of a planning task. However, it can be a *complex* of activities, which consists of many activity sequences that are performed in a parallel way.

[8] This scheme can be extended to *evolutionary computing* introduced in Chap. 5.

[9] Unless the opponent has been irritated and he/she has knocked the chessboard down from the table. However, if we assume that our opponent is well-bred, we can eliminate such a "result" from our considerations.

Constraint Satisfaction Problem, CSP, which has been introduced in Sect. 4.5. In such a case a planning strategy can be based on one of the CSP search methods.

In the area of Artificial Intelligence planning problems are very important, because of their various practical applications [311]. Therefore, many advanced methods which are based of such models as *temporal logic*, *dynamic logic*, *situation calculus*, and *interval algebra* have been defined in this area recently.

16.7 Natural Language Processing (NLP)

The research area of *Natural Language Processing, NLP*,[10] should be divided into two subareas. The first subarea includes problems which can be solved by an analysis of a language on the syntactic (and lexical) level. For example, text proofreading, extraction of information from a text, automatic summarizing, Optical Character Recognition (OCR), speech synthesis (on the basis of a text), simple *question-answer* dialogue systems, etc. belong to this group. The second subarea contains problems which can be solved by analysis of a language on the semantic level. For example, automatic translation from a natural language into another natural language, speech/text understanding, systems of human-computer verbal communication, etc. belong to this group. This division has been introduced because nowadays only problems belonging to the second group are challenging in Artificial Intelligence.

The Chomsky theory of *generative grammar* introduced in Chap. 8 is a referential model in this area. Although the Chomsky model is sometimes criticized in the area of NLP, since it has not fulfilled all the expectations of NLP researchers, it is usually the point of departure for defining models of NLP such as, e.g., *metamorphosis grammars* [58], *Definite Clause Grammars, DCGs* [225], and *Augmented Transition Networks, ATNs* [318].

At the end of the twentieth century a statistical approach to language analysis was developed. It makes use of *text corpora*, which are large referential sets of texts in a given language. A system refers to a text corpus during a text analysis with the help of stochastic models in order to determine statistical characteristics of the text, which relate to, e.g., possible contexts in which a word occurs, possible uses of a given phrase in the text corpus, etc.

Another approach consists of the use of the generative grammar model together with probability theory, which results in defining *stochastic grammars* and *stochastic automata* introduced in Chap. 8. Such a model is equivalent to the *Markov chain model* (cf. Appendix B.2), which is also used in advanced methods of NLP.

In the models mentioned above a syntax is assumed as a point of departure for language analysis. Such an approach is sometimes not sufficient in case of problems

[10]The notion of *natural language* is used in computer science in relation to such languages as English, German, Chinese, etc. in order to distinguish the issue of computer processing of such languages from the problem of computer processing of *artificial languages*, which is much easier. Artificial languages include, for example, programming languages, and formal languages, which have been presented in Chap. 8.

in which concept understanding is necessary. Then, in order to interpret the semantics of sentences in a proper way, an AI system should have additional knowledge in the form of a world model. This problem can be solved by defining an *ontology*, which has been introduced in Chap. 7. Let us recall that *semantic networks* are one of the most popular formalisms for defining ontologies.

In computerized semantic analysis of spoken language we cope with a much more difficult problem. Communication is the main function of spoken language. From the point of view of this function *non-verbal aspects of a language*[11] such as intonation, stress, etc. are essential. For example, the sentence: "I did not testify under oath that I had seen Cain killing Abel." can be interpreted in a number of ways, depending on which phrase is stressed. Possible interpretations include (the stressed phrase is marked):

- "*I did not testify* under oath that I had seen Cain killing Abel."—the basic interpretation,
- "I did not testify *under oath* that I had seen Cain killing Abel."—I testified, but not under oath,
- "I did not testify under oath that I *had seen* Cain killing Abel."—I overheard the event,
- "I did not testify under oath that I had seen *Cain* killing Abel."—I saw somebody killing Abel, but it was not Cain,
- "I did not testify under oath that I had seen Cain killing *Abel.*"—I saw Cain killing somebody, but it was not Abel.

Passing a message in one specific sense reveals the intention of its sender. He/she passes this sense by stressing the proper phrase. However, the ability to understand the correct sense of the message on the basis of stress, intonation, etc. relates to social intelligence. Although in this case we mean elementary social intelligence, it is very difficult to embed this kind of intelligence in an AI system.

Summing up, Natural Language Processing can be considered a well-developed area of AI. *Chatbots*, mentioned in Chap. 1, where we have presented *ELIZA* designed by Joseph Weizenbaum, simulating human speakers are good examples of successes in NLP. On one hand, some chatbots simulate an intelligent conversation quite well. On the other hand, they still cannot pass the *Turing test*.

16.8 Learning

Learning models in Artificial Intelligence can be divided into two basic groups:

1. experience generalization models,
2. models transforming a representation of a problem domain.

---

[11]Here we distinguish *non-verbal aspects of a language* from *non-verbal communication*, which includes, e.g., body language and facial expression.

In *experience generalization models* we assume the availability of a *learning set* $U = ((\mathbf{X}^1, \mathbf{u}^1), (\mathbf{X}^2, \mathbf{u}^2), \ldots, (\mathbf{X}^M, \mathbf{u}^M))$, where a pair $(\mathbf{X}^j, \mathbf{u}^j)$, $j = 1, \ldots, M$, consists of a *stimulus* $\mathbf{X}^j$, which represents a certain fact occurring in a system environment, and a *response* $\mathbf{u}^j$, which should be generated by the system as a result of receiving this stimulus. In other words, a learning set represents experience gained. An AI system is confronted with this experience, which is formalized in such a way. As a result it should define, via induction (generalization), a response function f such that for each $\mathbf{X}^j$, $j = 1, \ldots, M$, the following rule holds: $f(\mathbf{X}^j) = \mathbf{u}^j$. We say that the response function f is a generator of a proper reaction of the system for the observation (stimulus).

Generalized learning in AI is connected with the behavioral approach in psychology. Such learning is treated as gaining experience, which is done in order to modify the system's behavior. Following this approach, we have trained *neural networks* in Chap. 11 and *classifiers* for pattern recognition in Chap. 10. In both cases a stimulus $\mathbf{X}^j$ is of the form of a vector of numbers, which is used for coding a problem. The scheme for constructing a classifier based on a *decision tree* introduced in Sect. 10.6 belongs to this approach as well.

In the case of neural networks and classifiers we have also discussed models of *unsupervised learning*. Then, a learning set is of the form $U = (\mathbf{X}^1, \mathbf{X}^2, \ldots, \mathbf{X}^M)$. This means that the required reaction is not determined. The system should divide a set of stimuli into groups (subsets) itself. *Cluster analysis* introduced in Sect. 10.7 and *Hebbian learning* presented in Sect. 11.1 are good examples of such learning.

In *experience generalization learning* a stimulus $\mathbf{X}^j$ usually consists of a complex of parameters. However, such learning can also be applied to symbolic representations. The scheme *grammar induction—automaton synthesis* is a good example here. Then the response function f takes the form of a formal automaton.

Models transforming a representation of a problem domain correspond instead to models of *cognitive psychology*. In this case, an AI system should construct a world representation, i.e., an *ontology* introduced in Chap. 7. Then the system should *transform* it on the basis of the new knowledge gained. Thus, a learning process can be divided into two phases, ontology construction and ontology transformation.

In order to *construct* an ontology the system should, firstly, define *concepts* on the basis of observations of the world. Then, it should define structures which describe semantic relations among these notions. Unfortunately, AI systems are not able to perform such a task nowadays.[12]

However, AI systems are able to learn by transforming ontologies predefined by a human designers. In this case the system extracts knowledge from the ontology by transformation operations. For example, if there are the following two rules in our knowledge base[13]:

---

[12]It seems that in order to perform such a task, a system should be able to learn via insight into the heart of the matter. Such a way of learning, however, results from *understanding* the heart of the matter.

[13]In the example we assume that the ontology is constructed with the help of First-Order Logic.

$$C = aunt(A) \quad \Leftrightarrow \quad [\, \exists B \; B = mother(A) \;\wedge\; C = sister(B) \,] \\ \vee \; [\exists B \; B = father(A) \;\wedge\; C = sister(B)] \,,$$

and

$$B = parent(A) \quad \Leftrightarrow \quad [\, B = mother(A) \;\vee\; B = father(A) \,] \,,$$

then the system can infer a new rule:

$$C = aunt(A) \quad \Leftrightarrow \quad [\, \exists B \; B = parent(A) \;\wedge\; C = sister(B) \,] \,.$$

In fact, systems based on models which transform a representation of a problem domain simulate a cognitive activity. However, referring to the St. Thomas Aquinas definition of three generic operations of the intellect (cf. Sect. 15.1), this cognitive activity is limited to the third one, i.e., *reasoning* only.[14] In other words, the system extracts new knowledge from knowledge which is already stored in its knowledge base. Nevertheless, a human designer has to construct an ontology with the help of the two remaining cognitive operations, i.e., defining concepts (*simple apprehension*) and pronouncing judgments. Let us notice that system learning via an ontology transformation is possible only if the human designer is able to *encode* semantic knowledge into its syntax in a precise and unambiguous way.[15] Unfortunately, at present this is impossible for many application areas.

Syntactic pattern analysis systems introduced in Chap. 8 are AI systems which are able to generate a *structural representation* of some aspects of the world in an automatic way. However, such a representation is limited to *physical objects* which are extracted from an image and to *spatial-topological relations* among them. These systems perform neither abstraction processes nor conceptualization. Thus, there is no ontology construction in this case either.

Learning models which transform a problem domain have been developed dynamically in AI since the 1980s. The most popular methods include *Explanation-Based Learning, EBL* [205], *Relevance-Based Learning, RBL* [3], and *Inductive Logic Programming, ILP* [207].

16.9 Manipulation and Locomotion

As we have discussed in the previous chapter, *kinesthetic intelligence* related to both manipulation and locomotion abilities has been identified in the *sensimotor stage* of cognitive development of an infant (from birth to about age two) by Jean Piaget. Since we do not remember this stage of our life well, we do not realize the difficulty of acquiring these abilities. The simulation of these abilities is one of the most difficult problems in Artificial Intelligence, strictly speaking in *robotics*, which

[14] In the sense of proceeding from one proposition to another according to logical rules.

[15] As has been done in our genealogy example above.

is an interdisciplinary research area making use of models of automatic control, mechatronics, mechanics, electronics, cybernetics, and computer science.

Firstly, manipulation and locomotion abilities of robots (or similar devices) depend strongly on functionalities of other systems such as perception/pattern recognition systems, problem-solving systems, or planning systems. Successes and challenges in these research areas have been discussed in previous sections.

Secondly, manipulation and locomotion abilities of robots also depend on the technological possibilities of execution devices, such as effectors, actuators, etc. Let us notice that in this case sometimes we do not want to simulate human abilities. For example, where locomotion is concerned, some animals have a clear advantage over humans. Therefore, mobile robots for military or search-and-rescue applications are often constructed on the basis of the locomotion abilities of insects (hexapod robots), snakes (snakebots), or four-limbed animals (e.g., the BigDog quadruped robot), not to mention intelligent aerial mobile robots (drones) and underwater drones. Generally, in the area of locomotion constructors of mobile robots and devices have achieved amazing achievements recently.

Manipulation abilities of robots surpass those of humans in certain applications, especially if high precision, manual dexterity, or high resistance to tiredness are required. Manipulation microsurgical robots and robots aiding microbiology experiments are good examples here. Of course, these robots are telemanipulators (or remote telemanipulators) which are controlled by operators (e.g., surgeons). Summing up, there have been remarkable results in the area of intelligent manipulators and one can expect further successes in this field.

In spite of the fact that there are some interesting and usually spectacular results in the area of humanoid/android robotics, we still await robots which can simulate a violin virtuoso or a prima ballerina.

16.10 Social Intelligence, Emotional Intelligence and Creativity

At the end of the twentieth century research into simulating both social intelligence and emotional intelligence in AI systems began. This has concerned synthetic aspects of the problem, e.g., expression of emotions by a robot face, as well as analytic aspects, e.g., recognizing human mood on the basis of speech intonation. Simulating human abilities in the analytic aspect is, of course, more difficult. In order to analyze facial expression and features of speech (intonation, stress, etc.) advanced pattern recognition methods are applied. Rule-based systems are used for the purpose of integrating vision and sound. Surely, research in this area is very important, since its results, together with achievements in robotics, can be applied in medicine, social security, etc. *Distinct* emotional messages sent by humans via, e.g., facial expressions are recognized quite well nowadays by AI systems. Will robots be able to recognize them in case these messages are not clear? We must await for the answer.

In 2010 the first International Conference on Computer Creativity was organized at the prestigious University of Coimbra, which was established in 1290. The issue of the possibility of simulating human creativity discussed during the conference is really controversial. It seems that a view of Margaret Boden,[16] who distinguishes two types of creativity, can be helpful in this discussion [32]. *Exploratory creativity* consists of searching a predefined conceptual space.[17] However, if we deliberately transform or transcend a conceptual space, then we deal with *transformational creativity*. Simulation of transformational creativity in artificial systems is a really challenging task in the AI area.

Creative AI systems are implemented for solving general problems, generating music and visual art, etc. Various AI methods such as state space search, neural networks, genetic algorithms, semantic networks, and reasoning by analogy are used for these purposes.

Bibliographical Note

The issue of simulation of various human mental/cognitive abilities is usually discussed in fundamental books on Artificial Intelligence. The following monographs are recommended [18, 19, 55, 147, 189, 211, 241, 256, 261, 262, 273, 315].

[16] Margaret Boden—a professor of cognitive science at the University of Sussex. Her work concerns the overlapping fields of: psychology, philosophy, cognitive science, and AI. She was the Vice-President of the British Academy.

[17] Let us notice that this type of creativity can be simulated via cognitive simulation, i.e., searching a state space.

Chapter 17
Prospects of Artificial Intelligence

In Chap. 15 philosophical (epistemological) approaches to issues of mind, cognition, knowledge, and human intelligence have been presented. In the first section of this chapter contemporary views concerning the essence of artificial intelligence are discussed. As one can easily notice these views result from epistemological assumptions.[1] Philosophical assumptions also influence the views of authorities in the AI field concerning the a possibility of constructing intelligent systems. Since this monograph is an introduction to the field, its author tries not to take part in the discussion on AI, but just presents various ideas in the theory of mind.

Potential barriers which are also challenges in the AI field are discussed in the second section. Research areas which are crucial for the further development of Artificial Intelligence are presented in the third section.

17.1 Issues of Artificial Intelligence

Let us begin our considerations with an analysis of the term *artificial intelligence*. In fact, it has two basic meanings. Firstly, it means a *common* research field of computer science and robotics,[2] in which development of systems performing tasks which require intelligence when performed by humans is a research goal.

Secondly, it means a *feature* of artificial systems which allows them to perform tasks that require intelligence, when made by humans. Thus, in this meaning artificial

[1]Therefore, the reader is recommended to recall the considerations contained in Sect. 15.1.

[2]Usually it is assumed that *Artificial Intelligence* is a subfield of computer science. However, in this case we exclude from AI studies such important issues as, e.g., manipulation and locomotion performed by certain AI systems.

M. Flasiński, *Introduction to Artificial Intelligence*,
DOI 10.1007/978-3-319-40022-8_17

intelligence is not a thing, but a property of certain systems, just as mobility is a property of mobile robots[3] which allows them to move.

Let us notice that *artificial intelligence* in the second meaning is the subject of research in a discipline called cognitive science rather than computer science or robotics.

Cognitive science is a new interdisciplinary research field on mind and cognitive processes concerning not only humans, but also artificial systems. Its research focuses on issues which belong to philosophy, psychology, linguistics, neuroscience, computer science, logic, etc.

In the first chapter we have discussed the *Chinese room* thought experiment, which was introduced by Searle [269]. On the basis of this experiment views concerning artificial intelligence can be divided into the following two groups:

- *Strong Artificial Intelligence*, which claims that a properly programmed computer is equivalent to a human brain and its mental activity,
- *Weak Artificial Intelligence*, in which a computer is treated as a device that can simulate the performance of a brain. In this approach a computer is also treated as a convenient tool for testing hypotheses concerning brain and mental processes.

According to Searle the *Chinese room* shows that simulation of human mental activity with the help of a computer (Weak AI) does not mean that these activities take place in a computer in the same way as they do in a human brain. In other words, the brain is not a computer, and computing processes performed according to computer programs should not be treated as equivalent to mental processes in a human brain.

The term *computationalism* is related to Strong AI. It means the view that a human brain is a computer and any mental process is a form of computing.[4] So a mind can be treated as an information processing system. Computationalists assume that information processed in both kinds of system is of a symbolic form.[5]

Let us begin our presentation of views in the modern theory of mind with those which are close to Strong AI. They relate to the Cartesian *mind-body problem*, which has been presented in Chap. 15. This problem can be described as the issue of the place of mental processes in the physical (material) word.

Adherents of *analytical (logical) behaviorism* treat the mind-body problem with reserve, treating it as an unscientific pseudo-problem [258]. If we talk about mental states, we use a kind of metaphor. In fact, we just want to describe human behavior.

---

[3]In fact, we could say *artificial mobility*, since a robot is an *artefact*, i.e., an artificial object, which does not exists in nature, so its properties are also "artificial". While we, as *Homo sapiens* do not object a term *robot mobility*, in case of a term *computer intelligence* we prefer to add *artificial*. Of course, somebody could say that in this case a term *artificial* means *imperfect*. In the author's opinion, it is not a good interpretation. Does the reader think that some day we construct a mobile robot, which can dance Odette in *Swan Lake* like Sylvie Guillem?.

[4]Let us notice that this view is consistent with the philosophical views of T. Hobbes and G.W. Leibniz, which have been presented in Chap. 15.

[5]An assumption of *computationalists* about the symbolic form of information processed by intelligent systems triggered a discussion with followers of *connectionism* in the 1980s and 1990s (cf. Sect. 3.1).

Therefore, instead of using concepts related to mind, i.e., using an inadequate language, we should use terms describing behavioral patterns.[6] Gilbert Ryle, who has been mentioned in Sect. 2.4 (structural models of knowledge representation) is one of the best-known logical behaviorists.

Physicalism is also a theory which can be used to defend Strong AI, since mental phenomena are treated here as identical to physiological phenomena which occur in a brain. There are two basic approaches in physicalism, and assuming one of them has various consequences in a discussion about Artificial Intelligence.

Type-identity theory (type physicalism) was introduced by John J.C. Smart[7] [279] and Ullin Place[8] in the 1950s [227]. It asserts that mental states of a given type are identical to brain (i.e., physical) states of a certain type. If one assumes this view, then the following holds. Let us assume that in the future we will define a mapping between types of brain state and types of mental state. Then, our discussion of concepts and the nature of intelligence in philosophy and in psychology (in Chap. 15) could be replaced by a discussion in the fields of neuroscience and neurophysiology (*reductive physicalism*). Thus, the construction of artificial systems will depend only on the state of knowledge in these fields and technological progress in the future. This theory has been further developed by David M. Armstrong[9] [11], among others.

The weaker assumption is the basis of *token-identity theory (token physicalism)*. Although any mental state is identical to a certain brain state, mental states of a given type need not necessarily be identical to brain states of a certain type. *Anomalous monism* was introduced by Donald Davidson[10] in 1970 [64]. It is a very interesting theory from the point of view of a discussion about Artificial Intelligence. According to this theory, although mental events are identical to brain (physical) events, there are no deterministic principles which allow one to predict mental events. Davidson also assumed that mental phenomena *supervene*[11] on brain (physical) phenomena. For example, if the brains of two humans are in the indistinguishable states, then their

[6] We have introduced the issue of inadequate language in Chap. 15, presenting the views of William of Ockham and Ludwig Wittgenstein. Analytical behaviorism has been introduced on the basis of the views of the *Vienna Circle*.

[7] John Jamieson Carswell Smart—a professor of philosophy at the University of Adelaide and Monash University (Australia). His work concerns metaphysics, theory of mind, philosophy of science, and political philosophy.

[8] Ullin T. Place—a professor at the University of Adelaide and the University of Leeds. His work concerns the philosophy of mind and psychology. According to his will, his brain is located in a display case at the University of Adelaide with the message: *Did this brain contain the consciousness of U.T. Place?*

[9] David Malet Armstrong—a professor of philosophy at the University of Sydney, Stanford University, and Yale University. His work concerns theory of mind and metaphysics.

[10] Donald Herbert Davidson—a professor of philosophy at the University of California, Berkeley and also other prestigious universities (Stanford, Harvard, Princeton, Oxford). He significantly influenced philosophy of mind, epistemology, and philosophy of language. He was known as an indefatigable man who had a variety of interests, such as playing piano, flying aircraft, and mountain climbing.

[11] We say that a set of properties M *supervene* on a set of properties B if and only if any two beings which are indistinguishable w.r.t. the set B are also indistinguishable w.r.t. the set M.

mental states are also indistinguishable. Theory of supervenience has been further developed by Jaegwon Kim[12] [157]. Consequently, one can conclude that mental phenomena cannot be reduced to physical phenomena, and laws of psychology cannot be reduced to principles of neuroscience (*non-reductive physicalism*).

In order to preserve a chronology (at least partially), let us consider now certain views which relate to Weak AI. In 1961 John R. Lucas[13] formulated the following argument against the possibility of constructing a cybernetic machine that is equivalent to a mathematician on the basis of *Gödel's limitation (incompleteness) theorem* [188]. A human mind can recognize the truth of the Gödel sentence, whereas a machine (as a result of Gödel's incompleteness theorem) cannot, unless it is inconsistent. However, if a machine is inconsistent, it is not equivalent to a human mind. Let us notice that the Lucas argument concerns the intelligence of an outstanding human being, who is capable of developing advanced theories in logic. In 1941 Emil Post[14] had similar objections to machine intelligence, when he wrote in [230]:

> "We see that a machine would never give a complete logic; for once the machine is made we could prove a theorem it does not prove."

Although some logicians do not agree with this view of J.R. Lucas, modified versions of it appear in the literature from time to time, such as, e.g., an idea of Roger Penrose[15] presented in *"The Emperor's New Mind"* [224].

In 1972 Hubert Dreyfus[16] expressed his criticism of Strong AI in a monograph entitled *"What Computers Can't Do: The Limits of Artificial Intelligence"* [74]. He has presented four, in his opinion, unjustified assumptions defined by adherents of Strong AI. The *biological assumption* consists of treating the brain as a kind of digital machine, which processes information by discrete operations.[17] Viewing the mind as a system which processes information according to formal rules is the *psychological assumption*. The conviction that knowledge of any kind can be defined with a formal representation is the *epistemological assumption*. Finally, adherents of Strong AI are convinced that the world consists of independent beings, their properties, relations among beings, and categories of beings. Consequently all of them can be described

[12] Jaegwon Kim—a professor of philosophy at Brown University, Cornell University, and the University of Notre Dame. His work concerns philosophy of mind, epistemology, and metaphysics.

[13] John Randolph Lucas—a professor of philosophy of Merton College, University of Oxford, elected as a Fellow of the British Academy. He is known for a variety of research interests, including philosophy of science, philosophy of mind, business ethics, physics, and political philosophy.

[14] Emil Leon Post—a professor of logic and mathematics at the City University of New York (CUNY). His pioneering work concerns fundamental areas of computer science such as computability theory and formal language theory.

[15] Roger Penrose—a professor of the University of Oxford, mathematician, physicist, and philosopher. In 1988 he was awarded the Wolf Prize (together with Stephen Hawking) for a contribution to cosmology.

[16] Hubert Lederer Dreyfus—a professor of philosophy at the University of California, Berkeley. His work concerns phenomenological and existentialist philosophy, and philosophical foundations of AI.

[17] Let us notice that H. Dreyfus formulated this argument when the study of neural networks was beyond the research mainstream in AI.

adequately with formal models, e.g., by representing them by constant symbols, predicate (relation) symbols, function symbols, etc. in FOL. Dreyfus calls this view the *ontological assumption* and he claims that there is also an unformalizable aspect of our knowledge which results from our body, our culture, etc. Therefore, this kind of (unconscious) knowledge cannot be represented with the help of formal (symbolic) models, because it is stored in our brains in an *intuitive form*.[18]

The first version of *functionalism*, which is one of the most influential theories in Artificial Intelligence, was formulated by Hilary Putnam[19] in 1960 [232]. According to this theory, mental states are connected by causal relations in an analogous way to formal automata states, which have been discussed in Chap. 8. Similarly as automaton states are used for defining its behavior via the transition function, mental states play a functional role in the mind. Additionally, mental states are in causal relationships with mental system inputs (sensors) and outputs (effectors).[20] In early *machine functionalism*[21] the following computer analogy was formulated: *brain = hardware and mind = software*. Consequently, mental states can be represented by various physical media (e.g., a brain, a computer, etc.) similarly as software can be implemented by various computers.[22] The Turing machine is especially attractive as a mind model in functionalism.[23]

John R. Searle has criticized functionalism on the basis of his *Chinese room* thought experiment [269], which has been introduced in Chap. 1. In this experiment he tries to show that a system can behave as if it had *intentional states*[24] if we deliver a set of instructions[25] allowing it to perform such a simulation. J. Searle calls such intentionality *"as-if intentionality"* [270]. However, this does not mean that the system really has *intrinsic intentionality*.[26] Thus, in functionalism, which equates an

---

[18]Dreyfus represents here the *phenomenological* point of view, which has been introduced in Sect. 15.1. Especially this relates to the work of Martin Heidegger.

[19]Hilary Whitehall Putnam—a professor of philosophy at Harvard University. He is known for a variety of research interests, including philosophy of mind, philosophy of language, philosophy of science and mathematics, and computer science (the Davis-Putnam algorithm). A student of H. Reichenbach, R. Carnap, and W.V.O. Quine. Due to his scientific achievements, he has been elected a fellow of the American Academy of Arts and Sciences and the British Academy, and he was the President of American Philosophical Association.

[20]Analogously to the way we have defined *transducers* in Chap. 8.

[21]At the end of the twentieth century H. Putnam weakened his *orthodox* version of functionalism and in 1994 he published a paper *"Why Functionalism Didn't Work"*. Nevertheless, new theories (e.g., psychofunctionalism represented by Jerry Fodor and Zenon Pylyshyn) were developed on the basis of his early model.

[22]This thesis was formulated by H. Putnam in the late 1960s as an argument against type-identity theory. It is called *multiple realizability*.

[23]Since the Turing machine is an automaton of the greatest computational power (cf. Appendix E).

[24]The concept of intentionality has been introduced in Sect. 15.1, when the views of Franz Brentano have been presented.

[25]For example, a computer program is such a set of instructions.

[26]In other words, a computer does not want to translate a story, does not doubt whether it has translated a story properly, is not curious to know how a story ends, etc.

information system with a human being, there is no difference between something that is really intentional and something that is apparently intentional.

Daniel Dennett[27] proposed another approach to the issue of intentionality in 1987 [68]. The behavior of systems can be explained on three levels of abstraction. At the lowest level, called the *physical stance* and concerning both the physics and chemistry domains, we explain the behavior of a system in a causal way with the help of the principles of science. The intermediate level, called the *design stance*, includes biological systems and systems constructed in engineering. We describe their behavior in a functional way.[28] Minds and software belong to the highest level, called the *intentional stance*. Their behavior can be explained using concepts of intentionality, beliefs, etc.[29]

The argument of the *Chinese room* can be challenged if one assumes the most extreme view which supports Strong AI, namely *eliminative materialism (eliminativism)* introduced by Patricia Smith Churchland[30] and Paul M. Churchland[31] [49]. According to this view psychical phenomena do not exist. Concepts such as intentionality, belief, and mind do not explain anything. So they should be removed from science and replaced with terms of biology and neuroscience.

Researchers who develop AI systems also take part in the discussion about Strong AI. Similarly to the case of philosophers and cognitivists, views on this matter are divided. For some of them, successes in constructing AI systems show that in the future the design of an "artificial brain" will be possible. Hans Moravec[32] and Raymond Kurzweil[33] are the most notable researchers who express such a view.

17.2 Potential Barriers and Challenges in AI

In Sect. 15.2 we have introduced a psychological definition of intelligence as a set of abilities which allow one firstly, to adapt to a changing environment, and secondly, a cognitive activity consisting of creating and operating abstract structures. After

[27]Daniel Clement Dennett III—a professor of philosophy at Tufts University. His work concerns philosophy of mind and philosophy of science. He was a student of G. Ryle and W.V.O. Quine.

[28]For example, if a fish moves its fins, then it swims; if a thermometer senses that it is too cold, then a thermostat turns up the heat.

[29]Of course, according to Searle, Dennett does not make a distinction between *as-if intentionality* and *intrinsic intentionality*.

[30]Patricia Smith Churchland—a professor of philosophy at the University of California, San Diego and University of Manitoba. Her work concerns philosophy of mind, neurophilosophy, and medical ethics.

[31]Paul M. Churchland—a professor of philosophy at the University of California, San Diego and University of Manitoba. His work concerns philosophy of mind, neurophilosophy, and epistemology.

[32]Hans Moravec—a researcher at Carnegie Mellon University. In 1980 he constructed a TV-equipped robot at Stanford University. He was a co-founder of Seegrid Corporation, which is a company developing autonomous robots.

[33]Raymond "Ray" Kurzweil—an inventor and a futurist. A specialist in computer recognition of characters and speech.

the analysis of AI achievements made in Chap. 16 we can conclude that potential barriers to AI development concern the second component of this definition. We will try to identify these barriers on the basis of the classification of cognitive operations introduced by St. Thomas Aquinas, because in our opinion it specifies the essence of generic cognitive processes adequately. Let us recall that he distinguished three acts of the intellect, namely concept comprehension, pronouncing a judgment, and reasoning.

Let us begin with the third cognitive operation, because the greatest achievements in AI have been obtained in this area. As we have discussed in Chap. 16, *reasoning* is defined as proceeding from one proposition to another according to reliable rules of deduction. In the first half of the twentieth century sound theoretical foundations and effective methods of deductive reasoning were developed in mathematical logic. These methods are used in Artificial Intelligence successfully. Since a simulation of human reasoning should be performed according to logical principles, we use them for designing AI systems.

In the case of a simulation of *concept comprehension* research results are not so impressive. The standard Aristotelian approach to concept definition, which consists of giving its nearest genus and its specific difference, is used in formal sciences (e.g., in mathematics) successfully, but it is not so effective in other sciences, and it is usually inadequate in everyday life. There are two reasons for the difficulty of applying this approach in AI. Firstly, the Aristotelian rule is a very general principle. Therefore, defining an effective method (algorithm) on the basis of such a general principle is troublesome. Secondly, the Aristotelian creation of concepts by abstracting is based on the assumption of the existence of crisp categories. However, modern psycholinguistics claims that categories are of a fuzzy and radial nature, as we have discussed when presenting Lakoff cognitive linguistics in Chap. 1. Consequently, a process of abstracting is treated as an intrinsic intellectual process of comprehending the heart of the matter. However, modern science does not answer the question: How does a concept comprehension process, interpreted in such a way, proceed?

Sometimes, concept comprehension is considered equivalent to cluster analysis. This is a remarkable simplification of the problem. Let us notice that in the case of a *concept*, two its aspects are distinguished, namely its *intension*, which is the internal content, i.e., the set of properties that characterize objects falling within this concept, and its *extension*, which defines its range of applicability by designating objects falling within the concept. In cluster analysis there is a designer of the system, who has to define the feature space (the intensional aspect). The system only groups objects in clusters (the extensional aspect). We could talk about a system which comprehends concepts if it generates a feature space on the basis of *observation* of example objects.

The process of *pronouncing a judgment* is a generic cognitive process, which is little understood in psychology and philosophy. In order to discuss the possibility of a simulation of this process in AI we use the Kantian taxonomy of propositions, which has been discussed in Chap. 15. Simulation of a generation of *analytic propositions* a priori is performed in AI systems. Let us recall that such propositions concern knowledge already existing in our minds. In AI systems in the case of such propositions we refer to a knowledge base directly, e.g., we find the correct part of a semantic network, or we derive a required proposition with the resolution method.

Synthetic propositions, which expand our knowledge, are divided into two groups. *Synthetic propositions a posteriori* are derived on the basis of experience gained. In AI such propositions are obtained by generalized learning, which has been discussed in Sect. 16.8. Unsupervised learning of neural networks and cluster analysis are the best examples of such learning. Although, as we have mentioned in Sect. 16.8, there are a lot of open problems in this area, we use here the paradigm of inductive reasoning used successfully in empirical sciences.

Unfortunately, we are still unable to simulate the process of generating mathematical theorems, which is a fundamental process of mathematical development. According to I. Kant, such theorems correspond to *synthetic propositions* a priori. Let us analyze this problem in a more detailed way.

Mathematical theories are axiomatic-deductive systems. Firstly, basic notions and axioms[34] are defined. Then, a theory is developed by *deductive reasoning*,[35] which is based on the *modus ponendo ponens* rule. This rule is interpreted in the following way:

If the expression: $If\ A, then\ B$ is true
and the expression: A is true,
then the expression: B is true as well.

When we develop axiomatic-deductive systems, we can apply this rule in two ways, namely as a progressive deduction or a regressive deduction [30].

In a *progressive deduction* we start from a true premise and we try to infer a conclusion. Thus, such a process is a kind of *symbolic computing*. This computing consists of manipulating symbolic expressions in order to generate new expressions. The system *Logic Theorist*, which has been presented in Chap. 1, is based on this method of deductive reasoning.

On the other hand, in the case of a *regressive deduction*, we first formulate a conclusion and then we try to justify it via pointing out expressions of the system which can be used to derive this conclusion.

A remarkable expansion of axiomatic-deductive systems is obtained with the help of regressive deduction. Important research results in mathematics have been achieved in this way [30, 224]. Let us notice that formulating a conclusion, whose truth has not been proved at the moment of formulation is a crucial moment in this method. And again, modern science does not answer the question: How does the process of formulating such conclusions proceed? This phenomenon is usually described with such terms as insight, inspiration, or intuition [224]. Of course, such a description does not allow us to define algorithms which simulate this cognitive process.

---

[34] *Axioms* are propositions that are assumed to be true.

[35] Concepts related to deductive reasoning are contained in Appendix F.2.

Two problems identified above, which are fundamental barriers to AI development, result in more specific key problems that have been discussed in Chap. 16. The fact that we do not know the mechanisms of concept comprehension results in difficulty with developing satisfactory methods of automatic generation of ontologies in the area of knowledge representation and learning, automatic construction of abstract models of problems in the area of problem solving, and semantic analysis in Natural Language Processing.

The lack of models which describe the process of pronouncing a judgment is the main barrier in the areas of planning, automatic learning (the problem of formulating hypotheses), social intelligence, and creativity.

These barriers should not be used as an argument against the possibility of the development of intelligent systems in the future. They constitute, in the author's opinion, the main challenge for research in Artificial Intelligence.

17.3 Determinants of AI Development

Let us notice that most AI models presented in the second part of the book have been defined on the basis of ideas which are outside computer science. Cognitive simulation, semantic networks, frames, scripts, and cognitive architectures have been developed on the basis of psychological theories. The models of standard reasoning, and non-monotonic reasoning are logical theories. Genetic algorithms, evolution strategies, evolutionary programming, genetic programming, swarm intelligence, and artificial immune systems are inspired by biological models. Mathematics has contributed to Bayes networks, fuzzy sets, rough sets, and standard pattern recognition. Theories of linguistics have influenced the development of syntactic pattern recognition. Artificial neural networks simulate models of neuroscience. Physics delivers methods based on statistical mechanics, which make algorithms of problem solving and learning algorithms more efficient. It seems that only rule-based systems have been defined in computer science.

Thus, the development of Artificial Intelligence treated as a research area has been influenced strongly by the theories of the scientific disciplines mentioned above. It seems that AI will be developed in a similar way in the future.

Now, let us try to identify the most important AI prospects of the disciplines mentioned above. The main scheme of AI determinants is shown in Fig. 17.1.

As we have concluded in the previous section, the crucial barriers in the areas of general problem solving, automatic learning, Natural Language Processing, planning, and creativity result from our lack of psychological models of two generic cognitive processes, namely concept comprehension and pronouncing a judgment. Any research result relating to these processes would be very useful as a starting point for studies into a computer simulation of these processes.

Communication between humans and AI systems and between AI systems (multi-agent systems) requires much more effective NLP methods. Advanced models of

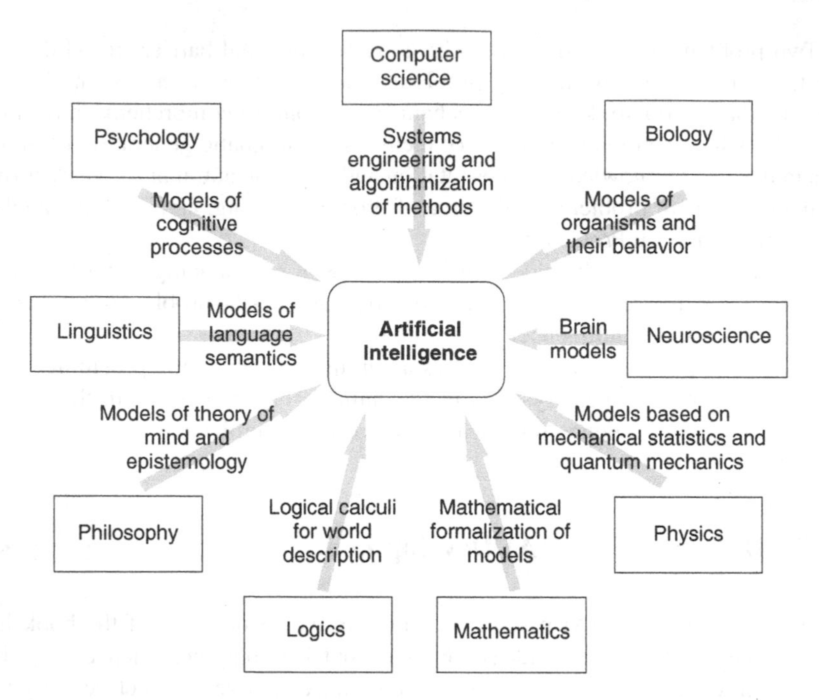

Fig. 17.1 Determinants of AI development

syntax analysis developed in linguistics are successfully used in AI. Let us hope that adequate models of semantic analysis will be defined in linguistics in the very near future.

If advanced neuroimaging and electrophysiology techniques in neuroscience allow us to unravel the mysteries of the human brain, then this will help us to construct more effective connectionist models.

Models of organisms and their physiological processes and evolutionary mechanisms will be an inexhaustible source of inspiration for developing general methods of problem solving.

Further development of new logical calculi of a descriptive power that allows us to represent many aspects of the physical world would allow a broader application of reasoning methods in expert systems. Mathematics should help us to formalize models of biology, psychology, linguistics, etc. that could be used in AI.

As we have discussed in the second part of the monograph, AI methods are very often computationally inefficient. In order to develop efficient AI methods we should use computational models that are based on mechanical statistics or quantum mechanics delivered by modern physics.

New effective techniques of software and system engineering should be developed in computer science. This would allow us to construct hybrid AI systems and multiagent systems. The algorithmization of methods which are based on models developed in various scientific disciplines is the second goal of AI research in computer science.

Hopefully, philosophy will deliver modern models of theory of mind and epistemology. As we have seen in Sect. 17.1 they play an important and inspiring role in progress in Artificial Intelligence.

Finally, let us notice that AI researchers should cooperate more strongly, because of the interdisciplinary nature of this research area. What is more, any AI researcher should broaden his/her interests beyond his/her primary discipline. The development of a new discipline, cognitive science, should help us to integrate various scientific disciplines that contribute to progress in Artificial Intelligence.

Bibliographical Note

Fundamental issues of theory of mind are presented in [45, 120, 134, 158, 186, 224, 292].

Appendix A
Formal Models for Artificial Intelligence Methods: Formal Notions for Search Methods

As we mentioned in Chap. 2, search methods are based on a concept of *cognitive simulation* developed in psychology and mind theory. In this appendix, firstly we introduce fundamental notions of state space and search tree in a formal way. Then, we present properties of heuristic function [256]. A constraint satisfaction problem (CSP) is formalized at the end [305].

A.1 State Space, Search Tree and Heuristic Function

Definition A.1 Let P be a problem, Σ a set of representations of all the configurations of the problem P, $\Sigma_S \subset \Sigma$ a set of representations of initial (start) configurations of P, $\Sigma_F \subset \Sigma$ a set of representations of final (goal) configurations of P, Γ a finite set of operators that can be used for solving P. *A state space* of the problem P is a node- and edge-labelled directed graph

$$G = (V, E, \Sigma, \Gamma, \phi), \text{ where}$$

V is a set of graph nodes corresponding to states of the problem P,
E is a set of graph edges of the form (v, λ, w), $v, w \in V, \lambda \in \Gamma$; an edge $(v, \lambda, w) \in E$ represents a transition from a state v to a state w as a result of an application of an operator λ,
$\phi : V \to \Sigma$ is the node labeling function.

A node $v \in V$ such that $\phi(v) \in \Sigma_S$ is called a *start node* of the state space and it represents an *initial state of the problem P*, and a node $v \in V$ such that $\phi(v) \in \Sigma_F$ is called a *goal node* of the state space and it represents a *goal state of the problem P*. A *solution of the problem P* is any path in the graph G that begins at a start node and ends in a goal node.[1]

[1]In fact, solving a problem means an application of a sequence of operators that ascribe labels to edges of the path.

M. Flasiński, *Introduction to Artificial Intelligence*,
DOI 10.1007/978-3-319-40022-8

State spaces usually are too big to be searched directly. Therefore, searching for a solution, we treat them as a problem domain. Then we successively span a tree structure on such a domain. Let us introduce the following definition.

Definition A.2 Let G be a state space of a problem P. A *search tree*[2] T is a directed tree spanned on G (on a part of G) such that the root v of T is a start node of G.

As we discussed in Chap. 4, if we construct heuristic methods of a state space search, we should define a heuristic function determining (for each state) a distance to a goal state. Now, we introduce formal notions related to this function [256].

Definition A.3 Let $G = (V, E, \Sigma, \Gamma, \phi)$ be a state space of a problem P. A *heuristic function* is a function

$$h : V \to \mathbb{R}_+ ,$$

where $\mathbb{R}_+$ is a set of non-negative real numbers, such that it ascribes a cost of reaching a goal node of G for any node $v \in V$.

We will assume a cost-based interpretation of a heuristic function in the definitions below. It means that we will try to move from a given node to a node with a minimum value of a heuristic function. If we assumed a quality-based interpretation of a heuristic function we would try to go to a node with a maximum value (quality) of a heuristic function. If one uses the second interpretation, the symbol $\leq$ should be replaced with the symbol $\geq$.

Definition A.4 A heuristic function h is *admissible* if for each node v of a state space G (or: of a search tree) the following condition holds:

$$h(v) \leq C(v),$$

where $C(v)$ is an actual cost of a path from the node v to the goal node.

Definition A.5 A heuristic function h is *consistent* (*monotone*) (cf. Fig. A.1) if for each node v of a state space G (or: of a search tree) and for each child node $S(v)$ of the node v the following conditions holds.

If $C(v, S(v))$ is the cost of moving from the node v to the node $S(v)$, then

$$h(v) \leq C(v, S(v)) + h(S(v)) .$$

Of course,[3] any consistent heuristic function is admissible.

[2] A search tree defined in such a way is called a *partial search tree*.

[3] Sometimes, we claim an additional condition to be fulfilled. If k is the goal node of a state space G, then $h(k) = 0$.

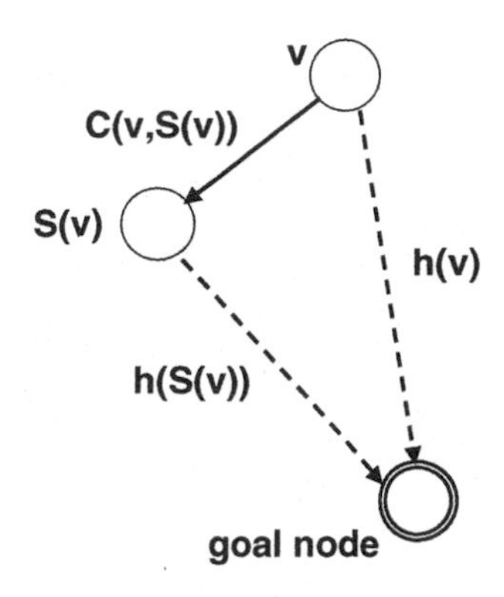

Fig. A.1 Consistency of a heuristic function

Definition A.6 Let h_1, h_2 be admissible heuristic functions. The heuristic function h_2 *dominates* the heuristic function h_1 if for each node v of a state space G the following condition holds:

$$h_2(v) \geq h_1(v).$$

As we discussed in Chap. 4, having two admissible functions, one should use a dominating function.

A.2 Constraint Satisfaction Problem

Definition A.7 *Constraint Satisfaction Problem, CSP* is a triple

$$\mathcal{P} = (Z, D, C),$$

where $Z = \{x_1, x_2, \ldots, x_n\}$ is a finite set of variables,
$D = (D_{x_1}, D_{x_2}, \ldots, D_{x_n})$, where D_{x_i} is a set of possible values that can be ascribed to x_i, called a *domain* of x_i,
C is a finite set of constraints (restrictions, conditions) imposed on a subset of Z.

Let $x_1, x_2, \ldots, x_n \in Z, \alpha_1 \in D_{x_1}, \alpha_2 \in D_{x_2}, \ldots, \alpha_n \in D_{x_n}$.

With $(\langle x_1, \alpha_1 \rangle, \langle x_2, \alpha_2 \rangle, \ldots, \langle x_n, \alpha_n \rangle)$ we denote ascribing a value α_i for each variable x_i, $i = 1, 2, \ldots, n$.

Definition A.8 Let $\mathcal{P} = (Z, D, C)$ be a constraint satisfaction problem. $\mathcal{P}$ is *satisfiable* if there exists $(\langle x_1, \alpha_1 \rangle, \langle x_2, \alpha_2 \rangle, \ldots, \langle x_n, \alpha_n \rangle)$ such that:

$$\forall c \in C : (\langle x_1, \alpha_1 \rangle, \langle x_2, \alpha_2 \rangle, \ldots, \langle x_n, \alpha_n \rangle) \text{ satisfies } c.$$

$(\langle x_1, \alpha_1 \rangle, \langle x_2, \alpha_2 \rangle, \ldots, \langle x_n, \alpha_n \rangle)$ is called a *solution* of the problem $\mathcal{P}$.

Appendix B
Formal Models for Artificial Intelligence Methods: Mathematical Foundations of Evolutionary Computation

When we discuss evolutionary computation methods in Chap. 5, we make use of some notions of probability theory, like standard deviation, normal distribution, etc. In Sect. B.1 we introduce them [118, 119] on the basis of a *probabilistic space* presented in Appendix I.[4]

Pioneering works on genetic algorithms were published in the 1960s. Since then researchers have tried to formalize this approach, which is based on a biological metaphor, with various mathematical models. At the end of the twentieth century, it turned out that the Markov chain, in which populations are represented with a state space, is a convenient model for this purpose [310]. Notions concerning the Markov chain [115, 118, 119, 293] are presented in Sect. B.2.

B.1 Selected Notions of Probability Theory

A notion of σ-algebra generated by a family of sets is included in Appendix I. In order to introduce notions of probability theory used in our considerations, firstly we present a definition of a special σ-algebra generated by a family of open sets.[5]

Definition B.1 Let X be a topological space. A σ-algebra generated by a family of open sets of the space X is called the *Borel σ-algebra*. Any element of a *Borel σ-algebra* is called a *Borel set*. A family of all Borel sets on X is denoted with $\mathcal{B}(X)$.

After defining a Borel set, we can introduce the following notions: probability distribution, random vector (random variable), and distribution of random vector (distribution of random variable).

In Appendix I a probability space $(\Omega, \mathcal{F}, P)$ is introduced. In fact, we are often interested in a space Ω equivalent to $\mathbb{R}^n$ ($\mathbb{R}$ denotes a set of real numbers), and

[4] Appendix I contains basic notions of probability theory that are used for probabilistic reasoning in intelligent systems.

[5] Open set and topological space are defined in Appendix G.1.

M. Flasiński, *Introduction to Artificial Intelligence*,
DOI 10.1007/978-3-319-40022-8

consequently in a σ-algebra $\mathcal{F}$ being a family of Borel sets $\mathcal{B}(\mathbb{R}^n)$. Let us introduce a definition of probability distribution.

Definition B.2 A probability measure P such that the triple $(\mathbb{R}^n, \mathcal{B}(\mathbb{R}^n), P)$ is a probability space is called an *n-dimensional probability distribution.*

Definition B.3 An n-dimensional probability distribution P is called a *discrete distribution* iff there exists a Borel set $S \subset \mathbb{R}^n$ such that:

$$P(S) = 1 \text{ and } s \in S \Rightarrow P(\{s\}) > 0\,.$$

If $S = \{s_i : i = 1, \ldots, m\}$, where $m \in \mathbb{N}$ (i.e. m is a natural number) or $m = \infty$ and $P(\{s_i\}) = p_i$, then for each Borel set $A \subset \mathbb{R}^n$ the following formula holds:

$$P(A) = \sum_{i\,:\,s_i \in A} P(\{s_i\}) = \sum_{i\,:\,s_i \in A} p_i\,,$$

and in addition:

- $p_i > 0$, for each $i = 1, \ldots, m$,
- $\sum_{i=1}^{m} p_i = 1$.

Definition B.4 An n-dimensional probability distribution P is called a *continuous distribution* iff there exists an integrable function $f : \mathbb{R}^n \longrightarrow \mathbb{R}$ such that for each Borel set $A \subset \mathbb{R}^n$ the following formula holds:

$$P(A) = \int_A f(x)\, dx\,,$$

where $\int_A f(x)\, dx$ denotes a multiple integral of a function f over A. A function f is called a *probability density function*, and in addition:

- $f(x) \geq 0$, for each $x \in \mathbb{R}^n$,
- $\int_{\mathbb{R}^n} f(x)\, dx = 1$.

Definition B.5 Let $(\Omega, \mathcal{F}, P)$ be a probability space. A function $X : \Omega \longrightarrow \mathbb{R}^n$ is called a *random vector* iff:

$$X^{-1}(B) = \{\omega \in \Omega : X(\omega) \in B\} \in \mathcal{F}$$

for each Borel set $B \in \mathcal{B}(\mathbb{R}^n)$.

A one-dimensional random vector is called a *random variable*.

Definition B.6 Let $(\Omega, \mathcal{F}, P)$ be a probability space, $X : \Omega \longrightarrow \mathbb{R}^n$ a random vector. A distribution P_X defined by a formula:

$$P_X(B) = P(X^{-1}(B)),$$

for $B \in \mathcal{B}(\mathbb{R}^n)$, is called a *distribution of a vector* X.

A distribution of a random variable is defined in an analogous way (we have $\mathbb{R}$, instead of $\mathbb{R}^n$).

Now, we introduce definitions of basic parameters of random variable distributions: expected value, variance, and standard deviation.

Definition B.7 Let $(\Omega, \mathcal{F}, P)$ be a probability space, $X : \Omega \longrightarrow \mathbb{R}$ a random variable. An *expected value* of X is computed as:

$$m = \mathbb{E}(X) = \sum_{i=1}^{m} x_i p_i \, ,$$

if X is a random variable of a discrete distribution: $P(X = x_i) = p_i, i = 1, \ldots, m$, $m \in \mathbb{N}$ or $m = \infty$, or it is computed as:

$$m = \mathbb{E}(X) = \int_{-\infty}^{+\infty} x f(x) \, dx \, ,$$

if X is a random variable of a continuous distribution with a probability density function f.

Definition B.8 Let $(\Omega, \mathcal{F}, P)$ be a probability space, $X : \Omega \longrightarrow \mathbb{R}$ a random variable with a finite expected value $m = \mathbb{E}(X)$. A *variance* of X is computed as:

$$\sigma^2 = \mathbb{D}^2(X) = \mathbb{E}((X - m)^2) = \sum_{i=1}^{m} (x_i - m)^2 p_i,$$

if X is a random variable of a discrete distribution: $P(X = x_i) = p_i, i = 1, \ldots, m$, $m \in \mathbb{N}$ or $m = \infty$, or it is computed as:

$$\sigma^2 = \mathbb{D}^2(X) = \mathbb{E}((X - m)^2) = \int_{-\infty}^{+\infty} (x - m)^2 f(x) \, dx,$$

if X is a random variable of a continuous distribution with a probability density function f. A *standard deviation* of a random variable X is computed as:

$$\sigma = \sqrt{\mathbb{D}^2(X)}.$$

At the end of this section we introduce a notion of the normal distribution, called also the Gaussian distribution.

Definition B.9 A distribution P is called the *normal* (*Gaussian*) distribution iff there exist numbers: $m, \sigma \in \mathbb{R}, \sigma > 0$ such that a function $f : \mathbb{R} \longrightarrow \mathbb{R}$, given by a formula:

$$f(x) = \frac{1}{\sigma\sqrt{2\pi}} e^{-\frac{1}{2}(\frac{x-m}{\sigma})^2}, x \in \mathbb{R},$$

is a probability density function of a distribution P. The normal distribution with parameters: m (an expected value) and σ (a standard deviation) is denoted with $N(m, \sigma)$.

B.2 Markov Chain Model for Genetic Algorithm

Discussing an idea of a genetic algorithm in Sect. 5.1, we noticed that a state space (in the sense used in search methods) is created by consecutive populations (strictly speaking: by their unique representations) generated with the algorithm. Thus, processing in a genetic algorithm can be treated as a process of generating consecutive populations corresponding to states of this space in successive genetic epochs $t \in T = \{0, 1, \ldots\}$. These populations: $P_0, P_1, \ldots$ define a sequence $\{P_t : t \in T\}$. Such a sequence can be treated formally as a stochastic process.

Definition B.10 Let $(\Omega, \mathcal{F}, P)$ be a probability space. A family of random variables $\{X_t : t \in T\}$ on Ω is called a *stochastic process*. A set T is called the index set.

Let us denote a set of real numbers with $\mathbb{R}$, and a set of natural numbers with $\mathbb{N}$. If $T = \mathbb{R}$, then we say $\{X_t : t \in T\}$ is a *continuous-time process*. If $T = \mathbb{N}$, then we say $\{X_t : t \in T\}$ is a *discrete-time process*.

Operators of genetic algorithm (crossover, mutation, etc.) applied for a given population do not depend on previous populations. It means that a sequence of random variables $\{X_t : t \in T\}$, being a discrete-time stochastic process fulfills the Markov condition, i.e., it is a Markov chain.

Definition B.11 A discrete-time process $\{X_t : t \in \mathbb{N}\}$ having values in at most countable set $\{x_0, x_1, \ldots\}$, which fulfills the Markov condition, i.e.: for each $n \in \mathbb{N}, j \in \mathbb{N}, m \in \mathbb{N}, m < n$, for each $j_1 < j_2 < \cdots < j_m < n$, for each $i_1, \ldots, i_m \in \mathbb{N}$ the following condition holds:

$$P(X_n = x_j | X_{j_m} = x_{i_m}, \ldots, X_{j_1} = x_{i_1}) = P(X_n = x_j | X_{j_m} = x_{i_m})$$

is called a *Markov chain*.

The set $\{x_0, x_1, \ldots\}$ is called a *state space*.[6]
A probability $P\left(X_n = x_j | X_{n-1} = x_i\right)$ of going from a state x_i to a state x_j in a time step from $n - 1$ to n is denoted with p^n_{ij}.

[6]Each state x_i, $i = 0, 1, \ldots$ represents a population P_i.

Probabilities of transitions in at most a countable number of states form a *transition matrix* defined as:

$$\mathbf{M^n} = p_{ij}^n.$$

If a probability p_{ij}^n of transition from a state x_i to a state x_j is independent of time n, then a Markov chain is called *homogeneous*. We define this property in the following way.

Definition B.12 A Markov chain $\{X_t : t \in \mathbb{N}\}$ is called a *homogeneous Markov chain* iff it fulfills the following condition:

$$P(X_n = x_j | X_{n-1} = x_i) = P(X_m = x_j | X_{m-1} = x_i),$$

for any n and m.

Parameters of genetic operators used in a genetic algorithm are constant.[7] It means that a transition from a population x_i to a population x_j in one time step is the same for any time step. Thus, a genetic algorithm can be modeled as a homogeneous Markov chain that allows us to define a transition matrix for a state space and to analyze many properties of this algorithm.

[7] This condition does not hold for, for example, evolution strategies, cf. Sect. 5.2.

Appendix C
Formal Models for Artificial Intelligence Methods: Selected Issues of Mathematical Logic

Formal notions used in Chap. 6 for discussing methods based on a mathematical logic are introduced in this appendix. Basic definitions of First-Order Logic, a resolution method of automated theorem proving, abstract rewriting systems, and the lambda calculus are introduced in successive sections [16, 17, 20, 24, 46, 81, 177, 181, 244, 317].

C.1 First-Order Logic

As we mentioned in Chap. 6, *First-Order Logic* (*FOL*) is the main formal model used for constructing intelligent systems based on mathematical logic. Notions concerning the syntax of FOL that are necessary for introducing a resolution method and basic notions concerning the semantics of FOL are presented in this section.

A set of symbols of a FOL language consists of the following elements:

- constant symbols,
- variable symbols,
- predicate (relation) symbols,
- function symbols,
- logical symbols: $\neg$, $\wedge$, $\vee$, $\Rightarrow$, $\Leftrightarrow$,
- quantifiers: $\forall$, $\exists$,
- the equality symbol: $=$,
- auxiliary symbols (e.g., parentheses).

Examples of the use of these symbols for describing the real world were presented in Sect. 6.1.

First of all, let us introduce the syntax of FOL, beginning with the notions of signature and a set of terms.

Definition C.1 Let $\Sigma_n^F, n \geq 0$ be a set family of n-ary function symbols, $\Sigma_n^P, n \geq 1$ a set family of n-ary predicate symbols. A pair $\Sigma = (\Sigma_n^F, \Sigma_n^P)$ is called a *signature* Σ.

M. Flasiński, *Introduction to Artificial Intelligence*,
DOI 10.1007/978-3-319-40022-8

Sometimes we differentiate constant symbols in a definition of signature, because of their specific interpretation. Then a signature is defined as a triple $\Sigma = (\Sigma^C, \Sigma_n^F, \Sigma_n^P)$, where Σ^S is a set family of constant symbols, and Σ_n^F is defined for $n \geq 1$.

Definition C.2 Let X be an infinite countable set of variable symbols. A set of *terms* $\mathcal{T}_\Sigma(X)$ over the signature Σ and the set X is defined inductively in the following way.

- If $x \in X$, then $x \in \mathcal{T}_\Sigma(X)$.
- For each $n \geq 1$ and for each $f \in \Sigma_n^F$, if $t_1, \ldots, t_n \in \mathcal{T}_\Sigma(X)$, then $f(t_1, \ldots, t_n) \in \mathcal{T}_\Sigma(X)$.

Now, we introduce notions of: atomic formula and formula (well-formed formula).

Definition C.3 An *atomic formula* over Σ and X is defined in the following way.

- A false symbol $\perp$ is an atomic formula.
- If $t_1, t_2 \in \mathcal{T}_\Sigma(X)$, then an expression $t_1 = t_2$ is an atomic formula.
- For each $n \geq 1$ and for each $p \in \Sigma_n^P$, if $t_1, \ldots, t_n \in \mathcal{T}_\Sigma(X)$, then an expression $p(t_1, \ldots, t_n)$ is an atomic formula.

Definition C.4 A *formula (well-formed formula, wff)* over Σ and X is defined inductively in the following way.

- Any atomic formula is a formula.
- If φ, ψ are formulas, then $\neg\varphi$, $\varphi \wedge \psi$, $\varphi \vee \psi$, $\varphi \Rightarrow \psi$, $\varphi \Leftrightarrow \psi$ are formulas.
- If φ is a formula and $x \in X$, then $(\forall x)(\varphi)$, $(\exists x)(\varphi)$ are formulas. A formula φ in a formula of the form $(\forall x)(\varphi)$ or in the formula of the form $(\exists x)(\varphi)$ is called a *scope of the quantifier*.

Examples of the use of formulas for representing laws concerning the real world were presented in Sect. 6.1.

Let us introduce notions that allow us to define sentences and propositional functions of FOL.

Definition C.5 A variable x is *bound in its occurrence* in a formula φ iff x occurs directly after a quantifier or x is inside the scope of some quantifier that x occurs directly after.

For example, variables that are bound in their occurrences are underlined in the following formulas:

$(\exists x \forall z (r(\underline{x}, \underline{z}) \Rightarrow s(\underline{x}, \underline{z})))$

$(\exists x (r(\underline{x}, z) \Rightarrow \forall z\ s(\underline{x}, \underline{z})))$

Definition C.6 If a variable x is not bound in its occurrence in a formula φ, then x is *free in this occurrence* in φ.

In the example above, a variable which is not underlined is a free variable in its occurrence in the formula.

Definition C.7 A variable x in a formula φ is *free* in φ iff x is free in φ in at least one occurrence. A variable x in a formula φ is *bound* in φ iff x is bound in every occurrence in φ.

Definition C.8 A formula with no free variables is called a *sentence*. Formulas that are not sentences are called *propositional functions*.

At the end of this section, we introduce notions concerning the semantics of FOL.

Definition C.9 A *structure* $\mathfrak{A}$ over the signature $\Sigma = (\Sigma_n^F, \Sigma_n^P)$, or Σ-*structure*, is a pair $(\mathcal{U}, \mathcal{I})$, where $\mathcal{U}$ is nonempty set called a *universe*, $\mathcal{I}$ is a function, called an *interpretation function* that assigns a function $f^{\mathfrak{A}} : \mathcal{U}^n \longrightarrow \mathcal{U}$ to each function symbol $f \in \Sigma_n^F$, and it assigns a relation $p^{\mathfrak{A}} \subseteq \mathcal{U}^n$ to each predicate symbol $p \in \Sigma_n^P$.

Definition C.10 An *assignment* (*valuation*) in a Σ-structure $\mathfrak{A}$ is a function $\varrho : X \longrightarrow \mathcal{U}$. Additionally, for an assignment ϱ, a variable $x \in X$ and an element $a \in \mathcal{U}$ let us define an assignment $\varrho_x^a : X \longrightarrow \mathcal{U}$ in $\mathfrak{A}$ that maps x to a in such a way that ϱ_x^a is equal to ϱ for all the variables that are different from x:

$$\varrho_x^a(y) = \begin{cases} \varrho(y), & \text{if } y \neq x, \\ a, & \text{otherwise.} \end{cases}$$

Definition C.11 An *interpretation* (*value*) of a term $t \in \mathcal{T}_\Sigma(X)$ in a Σ-structure $\mathfrak{A}$ under an assignment ϱ, denoted $[\![t]\!]_\varrho^{\mathfrak{A}}$, is defined inductively in the following way:

- $[\![x]\!]_\varrho^{\mathfrak{A}} = \varrho(x), x \in X$.
- $[\![f(t_1, \ldots, t_n)]\!]_\varrho^{\mathfrak{A}} = f^{\mathfrak{A}}([\![t_1]\!]_\varrho^{\mathfrak{A}}, \ldots, [\![t_n]\!]_\varrho^{\mathfrak{A}}), f \in \Sigma_n^F, t_1, \ldots, t_n \in \mathcal{T}_\Sigma(X), n \geq 1$.

Definition C.12 Let there be given a formula φ and a Σ-structure $\mathfrak{A}$ over the same signature, and an assignment ϱ. We define what it means for a formula φ to be *satisfied* in the structure $\mathfrak{A}$ under the assignment ϱ, denoted

$$(\mathfrak{A}, \varrho) \models \varphi,$$

in the following way.

- $(\mathfrak{A}, \varrho) \models \bot$ never holds.
- For any $n \geq 1, p \in \Sigma_n^P$ and for any $t_1, \ldots, t_n \in \mathcal{T}_\Sigma(X)$, we assume that: $(\mathfrak{A}, \varrho) \models p(t_1, \ldots, t_n)$ iff $([\![t_1]\!]_\varrho^{\mathfrak{A}}, \ldots, [\![t_n]\!]_\varrho^{\mathfrak{A}}) \in p^{\mathfrak{A}}$.
- $(\mathfrak{A}, \varrho) \models t_1 = t_2$ iff $[\![t_1]\!]_\varrho^{\mathfrak{A}} = [\![t_2]\!]_\varrho^{\mathfrak{A}}$.
- $(\mathfrak{A}, \varrho) \models \varphi \wedge \psi$ iff $(\mathfrak{A}, \varrho) \models \varphi$ and $(\mathfrak{A}, \varrho) \models \psi$.
- $(\mathfrak{A}, \varrho) \models \varphi \vee \psi$ iff $(\mathfrak{A}, \varrho) \models \varphi$ or $(\mathfrak{A}, \varrho) \models \psi$ or both.
- $(\mathfrak{A}, \varrho) \models \varphi \Rightarrow \psi$ iff either not $(\mathfrak{A}, \varrho) \models \varphi$ or else $(\mathfrak{A}, \varrho) \models \psi$.
- $(\mathfrak{A}, \varrho) \models \varphi \Leftrightarrow \psi$ iff $(\mathfrak{A}, \varrho) \models \varphi$ and $(\mathfrak{A}, \varrho) \models \psi$ if and only if $(\mathfrak{A}, \varrho) \models \varphi$ and $(\mathfrak{A}, \varrho) \models \psi$.

- $(\mathfrak{A}, \varrho) \models (\forall x)(\varphi)$ iff for each $a \in \mathcal{U}$, $(\mathfrak{A}, \varrho_x^a) \models \varphi$.
- $(\mathfrak{A}, \varrho) \models (\exists x)(\varphi)$ iff there exists an $a \in \mathcal{U}$ such that $(\mathfrak{A}, \varrho_x^a) \models \varphi$.

Definition C.13 We say that a formula φ is *satisfiable in a* Σ*-structure* $\mathfrak{A}$ iff there exists an assignment ϱ in $\mathfrak{A}$ such that $(\mathfrak{A}, \varrho) \models \varphi$.

Definition C.14 We say that a formula φ is *satisfiable* iff there exists a Σ- structure $\mathfrak{A}$, in which φ is satisfiable.

Definition C.15 A formula φ is *valid in a* Σ*-structure* $\mathfrak{A}$, if for each assignment ϱ in $\mathfrak{A}$ $(\mathfrak{A}, \varrho) \models \varphi$. We say that $\mathfrak{A}$ is a *model* for the formula φ, denoted

$$\mathfrak{A} \models \varphi.$$

Definition C.16 Let Φ be a set of formulas. A Σ-structure $\mathfrak{A}$ is a *model* for Φ, denoted $\mathfrak{A} \models \Phi$ iff for each formula $\varphi \in \Phi$, $\mathfrak{A} \models \varphi$.

Definition C.17 A formula φ is *valid*, or it is a *tautology*, denoted $\models \varphi$, iff φ is valid in every Σ-structure.

C.2 Resolution Method

In Sect. 6.2 we introduced a resolution method as a basic inference method by theorem proving. For this purpose we used notions of literal, clause, and Horn clause. Let us introduce formal definitions of these notions.

Definition C.18 A *literal* is either an atomic formula (a *positive literal*) or the negation of an atomic formula (a *negative literal*).

Definition C.19 A *clause* is a finite disjunction of literals, i.e.,

$$L_1 \vee L_2 \vee \cdots \vee L_n,$$

where L_i, $1 \leqslant i \leqslant n$ is a literal.

Definition C.20 A *Horn clause* is a clause containing at most one positive literal.

Now, we introduce a resolution rule in a formal way.

Definition C.21 Let $A \vee B_1 \vee \cdots \vee B_n$ and $\neg A \vee C_1 \vee \cdots \vee C_k$ be clauses, where A is an atomic formula, $B_1, \ldots, B_n$ and $C_1, \ldots, C_k$ are literals. A *resolution rule* is an inference rule of the form

$$\frac{A \vee B_1 \vee \cdots \vee B_n, \neg A \vee C_1 \vee \cdots \vee C_k}{B_1 \vee \cdots \vee B_n \vee C_1 \vee \cdots \vee C_k}.$$

A formula $B_1 \vee \cdots \vee B_n \vee C_1 \vee \cdots \vee C_k$ is called a *resolvent*, formulas $A \vee B_1 \vee \cdots \vee B_n$ and $\neg A \vee C_1 \vee \cdots \vee C_k$ are called *clashing formulas*.

In Sect. 6.2, after presenting a resolution rule, we discussed an issue of matching rules in an inference system with operations of substitution and unification. Let us introduce these operations in a formal way.

Definition C.22 A *substitution* σ is a set of replacements of variables by terms

$$\sigma = \{t_1/x_1, \dots, t_n/x_n\},$$

where $x_1, \cdots, x_n$ are variables different one from another, $t_1, \dots, t_n$ are terms.

If W is an expression,[8] and $\sigma = \{t_1/x_1, \dots, t_n/x_n\}$ is a substitution, then $W[\sigma]$ denotes an expression resulting from W by substituting all free occurrences of variables $x_1, \dots, x_n$ with terms $t_1, \dots, t_n$, respectively.[9]

Definition C.23 Let $W_1, W_2, \dots, W_n$ be expressions. A substitution σ is called a *unifier* of these expressions iff

$$W_1[\sigma] = W_2[\sigma] = \dots = W_n[\sigma].$$

A procedure of applying for expressions a substitution being their unifier is called a *unification of these expressions.*

The second issue concerning an inference by resolution is connected with representing formulas in some standard (normal) forms for the purpose of efficiency. Now, we introduce these forms.

Definition C.24 A formula φ is in *negation normal form* iff negation symbols occur only immediately before atomic formulas.

Definition C.25 A formula φ is in *prenex normal form* iff φ is of the form

$$Q_1x_1Q_2x_2 \dots Q_nx_n\psi,$$

where Q_i is $\forall$ or $\exists$, and ψ is an open formula (i.e., it does not contain quantifiers).

Definition C.26 A formula φ is in *conjunctive normal form* (*CNF*) iff φ is a finite conjunction of clauses, i.e., φ is of the form

$$C_1 \wedge C_2 \wedge \dots \wedge C_k = (L_1^1 \vee \dots \vee L_1^{n_1}) \wedge (L_2^1 \vee \dots \vee L_2^{n_2}) \wedge \dots \wedge (L_k^1 \vee \dots \vee L_k^{n_k}),$$

where $C_i = L_i^1 \vee \dots \vee L_i^{n_i}$, $1 \leqslant i \leqslant k$ is a clause consisting of literals $L_i^1, \dots, L_i^{n_i}$.

Definition C.27 Let a formula φ be in conjunctive normal form

$$C_1 \wedge C_2 \wedge \dots \wedge C_k = (L_1^1 \vee \dots \vee L_1^{n_1}) \wedge (L_2^1 \vee \dots \vee L_2^{n_2}) \wedge \dots \wedge (L_k^1 \vee \dots \vee L_k^{n_k}).$$

---

[8] An expression means here a formula or a term.

[9] In fact, we are interested in *allowable substitutions*, i.e., such that any variable contained in terms t_i does not become a bound variable. In other words, a substitution is allowable if any occurrence of a variable x_i in W is not inside the scope of any quantifier that bounds the variable included in t_i.

A *clausal normal form* of φ is a set

$$\{C_1, C_2, \ldots, C_k\} = \{\{L_1^1, \ldots, L_1^{n_1}\}, \{L_2^1, \ldots, L_2^{n_2}\}, \ldots, \{L_k^1, \ldots, L_k^{n_k}\}\}.$$

C.3 Abstract Rewriting Systems and the Lambda Calculus

As we mentioned in Sect. 6.5, Abstract Rewriting Systems (ARS) are one of the best formal exemplifications of a concept of a physical symbol system, introduced by A. Newell and H.A. Simon. Such systems can be divided into Term Rewriting Systems (TRS) presented in this section, String Rewriting Systems (SRS), and Graph Rewriting Systems (GRS).[10] The lambda calculus, being a special kind of TRS, plays an important role in Artificial Intelligence (cf. Sect. 6.5).

Definition C.28 An *Abstract Rewriting System, ARS* is a pair

$$ARS = (A, \{\rightarrow_\alpha : \alpha \in I\}),$$

where A is a set, $\rightarrow_\alpha$ is a set of binary relations, called *rewrite relations*, on A that are indexed by a set I.

Definition C.29 Let $ARS = (A, \{\rightarrow_\alpha : \alpha \in I\})$ be an abstract rewriting system and $\alpha \in I$.

(a) If $(a, b) \in \rightarrow_\alpha$, for $a, b \in A$, then we talk about a *direct step of rewriting* a into b, denoted $a \rightarrow_\alpha b$ (b is also called a (direct) *reduct* of a).
(b) A *rewriting sequence* (or a *rewriting*) with $\rightarrow_\alpha$ is a finite or infinite sequence $a_0 \rightarrow_\alpha a_1 \rightarrow_\alpha a_2 \rightarrow_\alpha \ldots$.

The transitive and reflexive closure of a relation $\rightarrow_\alpha$ is denoted with $\twoheadrightarrow_\alpha$.

Thus, $a \twoheadrightarrow_\alpha b$, if there exists a finite (also: empty) rewriting sequence $a \equiv a_0 \rightarrow_\alpha a_1 \rightarrow_\alpha \cdots \rightarrow_\alpha a_n \equiv b$, where $\equiv$ denotes an identity of elements belonging to the set A.

An inverse relation to a relation $\rightarrow_\alpha$ ($\twoheadrightarrow_\alpha$) is denoted with ${}_\alpha\leftarrow$ (${}_\alpha\twoheadleftarrow$).

Definition C.30 A relation $\rightarrow_\alpha$ in a set A is *weakly confluent*, in other words, has the *weak Church-Rosser property* (cf. Fig. C.1a) iff the following condition is fulfilled

$$\forall a, b, c \in A \exists d \in A (b_\alpha \leftarrow a \rightarrow_\alpha c \Rightarrow b \twoheadrightarrow_\alpha d_\alpha \twoheadleftarrow c).$$

Definition C.31 A relation $\rightarrow_\alpha$ in a set A is *confluent*, in other words, has the *Church-Rosser property* (cf. Fig. C.1b) iff the following condition is fulfilled

$$\forall a, b, c \in A \exists d \in A \ (b_\alpha \twoheadleftarrow a \twoheadrightarrow_\alpha c \Rightarrow b \twoheadrightarrow_\alpha d_\alpha \twoheadleftarrow c).$$

[10] Notions concerning SRS and GRS are presented in Appendix E.

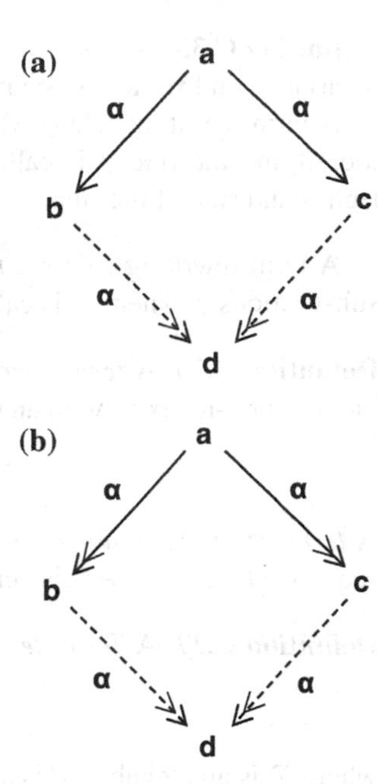

Fig. C.1 Church-Rosser property

Definition C.32 Let $ARS = (A, \rightarrow_\alpha)$ be an abstract rewriting system.

(a) $a \in A$ is a *normal form* iff there does not exist $b \in A$ such that $a \rightarrow_\alpha b$.
(b) If $a \twoheadrightarrow_\alpha b$ and $b \in A$ is a normal form, then we say that $a \in A$ *has a normal form* and b is a normal form (for) a.

Now, we introduce basic notions for term rewriting systems.

Definition C.33 An *alphabet* Σ contains:

- a countable infinite *set of variables* $\mathcal{V} = \{a_0, b_0, c_0, \ldots, z_0, a_1, b_1, c_1, \ldots, z_1, a_2, b_2, \ldots\}$
- a nonempty set of *function symbols*: $f, g, \ldots$

A *context*, denoted $C[\ ]$, is a term[11] that includes a single occurrence of a symbol $\square$, which means an empty place. A substitution of a term t ($t \in \mathcal{T}_\Sigma(\mathcal{V})$) in $\square$ results in $C[t] \in \mathcal{T}_\Sigma(\mathcal{V})$. We say that a term t is a subterm of $C[t]$, denoted $t \subseteq C[t]$.

Definition C.34 Mapping from $\mathcal{T}_\Sigma(\mathcal{V})$ to $\mathcal{T}_\Sigma(\mathcal{V})$ fulfilling the following condition:

$$\sigma(f(t_1, \ldots, t_n)) = f(\sigma(t_1), \ldots, \sigma(t_n)), \text{ where } f \text{ is a function symbol,}$$

is called a *substitution*. A substitution $\sigma(t)$ is often denoted with t^σ.

[11] A notion of term is introduced in the first section of this appendix.

Definition C.35 A *term rewriting rule* is a pair (t, s), $t, s \in \mathcal{T}_\Sigma(\mathcal{V})$ such that t is not a variable, and variables occurring in s are included in t.

A term rewriting rule (t, s) is often denoted with $r : t \to s$, where r is an index identifying the rule, t is called the left-hand side of the rule, and s is called the right-hand side of the rule.

A term rewriting rule $r : t \to s$ defines a set of rewrites $t^\sigma \to_r s^\sigma$ for all the substitutions σ. Then, t^σ is called an *r-redex* and s^σ is called an *r-contractum.*

Definition C.36 A *term rewriting step* according to a rewriting rule r is a replacement of an r-redex t^σ with an r-contractum s^σ inside a context $C[\ ]$, denoted

$$C[t^\sigma] \to_r C[s^\sigma].$$

A *term rewriting sequence* is a finite or infinite sequence $t_0 \to t_1 \to t_2 \to \cdots$ A sequence $t_0 \to \cdots \to t_n$ is denoted with $t_0 \twoheadrightarrow t_n$.

Definition C.37 A *Term Rewriting System, TRS* is a pair

$$ARS = (\Sigma, R),$$

where Σ is an alphabet, R is a set of term rewriting rules.

At the end of this section, we present basic definitions of the lambda calculus for notions introduced informally in Chap. 6.

Definition C.38 Let $\mathcal{V}$ be a countable infinite set of variables, $\Sigma = \mathcal{V} \cup \{(,), \lambda\}$, an alphabet. A *set of* Λ *expressions* over Σ is defined inductively in the following way.

- If $x \in \mathcal{V}$, then $x \in \Lambda$.
- If $M, N \in \Lambda$, then $(MN) \in \Lambda$ (*application*).
- If $M \in \Lambda$ and $x \in \mathcal{V}$, then $(\lambda x.M) \in \Lambda$ (λ-*abstraction*).

The following simplifying notation is used in the lambda calculus.

- Outmost parentheses can be omitted.
- It is assumed that an application is left-associative, i.e. instead of $(MN)P$ one can write MNP.
- Instead of $\lambda x_1(\lambda x_2(\ldots(\lambda x_n M)\ldots))$ one can write $\lambda x_1 x_2 \ldots x_n M$.

An operator of a lambda abstraction λ binds variables in such a way that all the occurrences of a variable x in an expression $\lambda x.M$ are *bound*. Let us introduce a definition of free variables in the lambda calculus.

Definition C.39 *A set of free variables* of a lambda expression M, denoted $\mathcal{FV}(M)$, is defined inductively in the following way.

$\mathcal{FV}(x) = \{x\}$,
$\mathcal{FV}(MN) = \mathcal{FV}(M) \cup \mathcal{FV}(N)$,
$\mathcal{FV}(\lambda x.M) = \mathcal{FV}(M) \setminus \{x\}$.

A variable is a *free variable* of an expression M if it belongs to a set $\mathcal{FV}(M)$, otherwise it is a *bound variable*.

Let $M \equiv N$ mean an identity of M and N up to renaming bound variables. A substitution for a free variable is defined in the following way.

Definition C.40 A *substitution* of an expression N for a free occurrence of a variable x in an expression M, denoted $M[x := N]$, is defined in the following way.

$x[x := N] \equiv N$,
$y[x := N] \equiv y$, if $y \not\equiv x$,
$(M_1 M_2)[x := N] \equiv (M_1[x := N])(M_2[x := N])$,
$(\lambda y.M)[x := N] \equiv \lambda y.(M[x := N])$, if $y \not\equiv x$ and $y \notin \mathcal{FV}(N)$.

At the end, let us introduce formal definitions of beta-reduction (β-reduction) and alpha-conversion (α-conversion) which were discussed in Chap. 6.

Definition C.41 A *beta-reduction*, denoted $\rightarrow_\beta$, is the smallest relation in a set Λ fulfilling the following conditions.

- $(\lambda x.M)N \rightarrow_\beta M[x := N]$.
- If $M \rightarrow_\beta M'$, then: $MZ \rightarrow_\beta M'Z$, $ZM \rightarrow_\beta ZM'$, and $\lambda x.M \rightarrow_\beta \lambda x.M'$.

Definition C.42 An *alpha-conversion*, denoted $\stackrel{\alpha}{\equiv}$, is the smallest equivalence relation in a set Λ fulfilling the following condition.

$$\lambda x.M \stackrel{\alpha}{\equiv} \lambda y.(M[x := y]) \text{ for any } y \notin \mathcal{FV}(M).$$

Lambda expressions up to an alpha conversion are called *lambda terms*.

Appendix D
Formal Models for Artificial Intelligence Methods: Foundations of Description Logics

Description logics is the family of formal systems that are based on mathematical logic and are used for inferring in ontologies. As we mentioned in Chap. 7, ontology is a model of a conceptual knowledge concerning a specific application domain. In 1979 Patrick J. Hayes discussed in [130] a possible use of First-Order Logic (FOL) semantics for Minsky frame systems. Since the beginning of the 1980s many description logics ($\mathcal{AL}$, $\mathcal{FL}^-$, $\mathcal{FL}_0$, $\mathcal{ALC}$, etc.) have been developed. In general, each of these logics can be treated as a certain subset of FOL.[12] So, one can ask: "Why do we not use just FOL for representing ontologies?". There are two reasons pointed out as an answer to this question in the literature. First of all, the use of FOL without some restrictions does not allow us to take into account the structural nature of ontologies. (And a structural aspect of ontologies is vital in an inference procedure.) Secondly, we demand an *effective* inference procedure. Therefore, description logics are defined on the basis of decidable subsets of FOL.[13]

In 1991 Manfred Schmidt-Schauß and Gert Smolka defined in [266] one of the most popular description logics, namely logic $\mathcal{ALC}$. Its syntax and semantics is introduced in the first section, whereas a formal notion of knowledge base defined with this logic is presented in the second section [15, 266].

D.1 Syntax and Semantic of Logic $\mathcal{ALC}$

In Chap. 7 we introduced basic elements useful for defining structural models of knowledge representation, i.e.: *objects*, *concepts*, and *roles* representing relations between objects. We discriminate additionally *atomic concepts* relating to basic (elementary) notions of a given domain. Let us introduce these elements in a formal way.

[12]There are also description logics based on second-order logic.

[13]FOL is not decidable in general.

M. Flasiński, *Introduction to Artificial Intelligence*,
DOI 10.1007/978-3-319-40022-8

Definition D.1 Let N_C be a set of atomic concept names, N_R be a set of role names, N_O be a set of object names. A triple (N_C, N_R, N_O) is called a *signature*.

Instead of *atomic concept name, role name, object name* we will say *atomic concept, role, object*.

Definition D.2 Let (N_C, N_R, N_O) be a *signature*. A set of (*descriptions of*) $\mathcal{ALC}$-*concepts* is the smallest set defined inductively as follows.

1. The following constructs are $\mathcal{ALC}$-concepts:
   (a) $\top$, the universal concept,
   (b) $\bot$, the empty concept,
   (c) every atomic concept $A \in N_C$.
2. If C and D are $\mathcal{ALC}$-concepts, $R \in N_R$, then the following constructs are $\mathcal{ALC}$-concepts:
   (a) $C \sqcap D$,
   (b) $C \sqcup D$,
   (c) $\neg C$,
   (d) $\forall R \cdot C$,
   (e) $\exists R \cdot C$.

Before we introduce a formal characterization of semantics of logic $\mathcal{ALC}$, we interpret elements defined above in an intuitive way. The *universal notion* corresponds to the whole domain that an ontology is constructed for, whereas the *empty concept* represents a concept that has no instances. Elements defined in points 2(a), 2(b), and 2(c) correspond to the intersection of two concepts, the union of two concepts, and the complement of a concept, respectively. A universal quantification, 2(d), determines a set of objects, for which all the relations with the help of a role R concern objects that fall within a concept C. An existential quantification, 2(e), determines a set of objects that are at least once in a relation represented by a role R with an object that falls within a concept C.

Definition D.3 Let (N_C, N_R, N_O) be a signature. An *interpretation* is a pair

$$\mathcal{I} = (\Delta^{\mathcal{I}}, \cdot^{\mathcal{I}}), \text{ where}$$

$\Delta^{\mathcal{I}}$ is a nonempty set called the domain of $\mathcal{I}$,
$\cdot^{\mathcal{I}}$ is an interpretation function, which maps every $\mathcal{ALC}$-concept into a subset $\Delta^{\mathcal{I}}$, and every role into a subset $\Delta^{\mathcal{I}} \times \Delta^{\mathcal{I}}$,

such that for each C and D being $\mathcal{ALC}$-concepts, $R \in N_R$, the following conditions hold:

1(a) $(\top)^{\mathcal{I}} = \Delta^{\mathcal{I}}$,
1(b) $(\bot)^{\mathcal{I}} = \emptyset$,
2(a) $(C \sqcap D)^{\mathcal{I}} = C^{\mathcal{I}} \cap D^{\mathcal{I}}$,

2(b) $(C \sqcup D)^{\mathcal{I}} = C^{\mathcal{I}} \cup D^{\mathcal{I}}$,
2(c) $(\neg C)^{\mathcal{I}} = \Delta^{\mathcal{I}} \backslash C^{\mathcal{I}}$,
2(d) $(\forall R \cdot C)^{\mathcal{I}} = \{x \in \Delta^{\mathcal{I}} : \forall y \in \Delta^{\mathcal{I}} (x, y) \in R^{\mathcal{I}} \Rightarrow y \in C^{\mathcal{I}}\}$,
2(e) $(\exists R \cdot C)^{\mathcal{I}} = \{x \in \Delta^{\mathcal{I}} : \exists y \in \Delta^{\mathcal{I}} (x, y) \in R^{\mathcal{I}} \wedge y \in C^{\mathcal{I}}\}$.

$C^{\mathcal{I}}$ ($r^{\mathcal{I}}$) is called the *extension* of the concept C (the role r) in the interpretation $\mathcal{I}$. If $x \in C^{\mathcal{I}}$, then x is called an *instance* (*object*) of the notion C in the interpretation $\mathcal{I}$.

Additionally, it is assumed that a concept C is *included* in a concept D, denoted $C \sqsubseteq D$ iff for any $\Delta^{\mathcal{I}}$, $\cdot^{\mathcal{I}}$ the following condition holds: $C^{\mathcal{I}} \subseteq D^{\mathcal{I}}$.

D.2 Definition of Knowledge Base in Logic $\mathcal{ALC}$

In Chap. 7 we defined a knowledge base as a system (structure) of frames consisting of *class frames* and *object frames*. A set of class frames constitutes a terminological knowledge, and a set of object frames corresponds to knowledge about specific objects belonging to a domain. These sets are defined in logic $\mathcal{ALC}$ with the help of notions of *TBox* (*Terminological part of knowledge base*) and *ABox* (*Assertional part of knowledge base*), respectively. Both TBox and ABox contain knowledge in the form of axioms. Axioms of TBox are defined with a *general concept inclusion*.

Definition D.4 A *general concept inclusion* is of the form $C \sqsubseteq D$, where C, D are $\mathcal{ALC}$-concepts.

A *TBox* is a finite set of general concept inclusions.

An interpretation $\mathcal{I}$ is a *model of a general concept inclusion* $C \sqsubseteq D$, if $C^{\mathcal{I}} \subseteq D^{\mathcal{I}}$.

An interpretation $\mathcal{I}$ is a *model of a TBox* $\mathcal{T}$, if it is a model of every general concept inclusion $\mathcal{T}$.

$C \sqsubseteq D$ and $D \sqsubseteq C$ is denoted as $C \equiv D$. An axiom of a TBox can be of the form of a *definition*, i.e., $A \equiv D$, where A is a unique concept name.

Now, we characterize an ABox. It can contain axioms of two types. The first type relates to assertions describing a fact that an object is an instance of a given concept, i.e., an object belongs to a given class, denoted $C(a)$, for example: *Polish(John-Kowalski)*. The second type includes statements representing a fact that a pair of objects constitutes an instance of a role, denoted $R(a, b)$, for example: *Married-couple(John-Kowalski, Mary-Kowalski)*. Let us formalize our considerations.

Definition D.5 An *assertional axiom* is of the form $C(a)$ or $R(a, b)$, where C is an $\mathcal{ALC}$-concept, $R \in N_R$, $a, b \in N_O$.

An *ABox* is a finite set of assertional axioms.

An interpretation $\mathcal{I}$ is called a *model of an assertional axiom* $C(a)$ iff $a^{\mathcal{I}} \in C^{\mathcal{I}}$. An interpretation $\mathcal{I}$ is called a *model of an assertional axiom* $R(a, b)$ iff $(a^{\mathcal{I}}, b^{\mathcal{I}}) \in R^{\mathcal{I}}$.

An interpretation $\mathcal{I}$ is called a *model of an ABox* $\mathcal{A}$ iff $\mathcal{I}$ is a model of every assertional axiom of $\mathcal{A}$.

At the end of this appendix, let us introduce a formal definition of a knowledge base constructed with the help of logic $\mathcal{ALC}$.

Definition D.6 A pair $\mathcal{K} = (\mathcal{T}, \mathcal{A})$, where $\mathcal{T}$ is a TBox, $\mathcal{A}$ is an ABox is called a *knowledge base*.

An interpretation $\mathcal{I}$ is a *model of a knowledge base* $\mathcal{K}$ iff $\mathcal{I}$ is a model of $\mathcal{T}$, and $\mathcal{I}$ is a model of $\mathcal{A}$.

Appendix E
Formal Models for Artificial Intelligence Methods: Selected Notions of Formal Language Theory

Selected notions that were used for discussing syntactic pattern recognition in Chap. 8 are introduced in this appendix. Definitions of Chomsky's (string) generative grammars and the LL(k) subclass of context-free grammars are introduced in the first section. Notions of finite-state automaton, pushdown automaton and Turing machine are presented in the second section. The last section includes definitions of edNLC and ETPL(k) graph grammars.

E.1 Chomsky's String Grammars

An *alphabet* Σ is a finite nonempty set of symbols.

A *string* (*word*) over an alphabet Σ is any string consisting of symbols of an alphabet Σ that is of a finite length.

A string that does not include any symbol is called the *empty word* and it is denoted with λ.

A set of all the strings over an alphabet Σ that are of a finite nonzero length is denoted with Σ^+.

A set including all the strings over an alphabet Σ that are of a finite length and the empty word is called the *Kleene closure*, and it is denoted with Σ^*. It can be defined as: $\Sigma^* = \Sigma^+ \cup \{\lambda\}$.

Let S_1, S_2 be sets of strings. $S_1 S_2$ denotes a set of strings: $S_1 S_2 = \{\alpha\beta : \alpha \in S_1, \beta \in S_2\}$, i.e. the set consisting of strings that are *catenations* of strings belonging to S_1 with strings belonging to S_2.

Now, we introduce four classes of grammars of the Noam Chomsky model [141, 250].

Definition E.1 A *phrase-structure grammar* (*unrestricted grammar, type-0 grammar*) is a quadruple

$$G = (\Sigma_N, \Sigma_T, P, S), \text{ where}$$

M. Flasiński, *Introduction to Artificial Intelligence*,
DOI 10.1007/978-3-319-40022-8

Σ_N is a set of nonterminal symbols,
Σ_T is a set of terminal symbols, $\Sigma = \Sigma_N \cup \Sigma_T$,
P a set of productions (rewriting rules) of the form: $\alpha \rightarrow \gamma$, in which $\alpha \in \Sigma^* \Sigma_N \Sigma^*$ is called the left-hand side of the production, and $\gamma \in \Sigma^*$ is called the right-hand side of the production,
S is the start symbol (axiom), $S \in \Sigma_N$.

We assume that $\Sigma_N \cap \Sigma_T = \emptyset$.

Definition E.2 Let $\beta, \delta \in \Sigma^*$. We denote

$$\beta \underset{G}{\Longrightarrow} \delta (\text{or } \beta \Longrightarrow \delta, \text{ if } G \text{ is assumed})$$

iff $\beta = \eta_1 \alpha \eta_2$, $\delta = \eta_1 \gamma \eta_2$ and $\alpha \rightarrow \gamma \in P$, where P is a set of productions of the grammar G.

We say that β directly derives δ in the grammar G, and we call such direct deriving a *derivational step* in the grammar G.

The reflexive and transitive closure of the relation $\Longrightarrow$, denoted with $\Longrightarrow^*$, is called a *derivation* in the grammar G.

Definition E.3 The *language* generated by the grammar G is a set

$$L(G) = \{\phi \in \Sigma_T^* : S \overset{*}{\Longrightarrow} \phi\}.$$

Definition E.4 A *context-sensitive grammar* (*type*-1 *grammar*) is a quadruple

$$G = (\Sigma_N, \Sigma_T, P, S), \text{ where}$$

Σ_N, Σ_T, S are defined as in Definition E.1,
P is a set of productions of the form: $\eta_1 A \eta_2 \rightarrow \eta_1 \gamma \eta_2$,
in which $\eta_1, \eta_2 \in \Sigma^*, A \in \Sigma_N, \gamma \in \Sigma^+$. Additionally we assume that a production of the form $A \rightarrow \lambda$ is allowable, if A does not occur in any production of P in its right-hand side.

Definition E.5 A *context-free grammar* (*type*-2 *grammar*) is a quadruple

$$G = (\Sigma_N, \Sigma_T, P, S), \text{ where}$$

Σ_N, Σ_T, S are defined as in Definition E.1,
P is a set of productions of the form: $A \rightarrow \gamma$, in which $A \in \Sigma_N, \gamma \in \Sigma^*$.

Definition E.6 A *regular* (*or right-regular*) *grammar* (*type*-3 *grammar*) is a quadruple

$$G = (\Sigma_N, \Sigma_T, P, S), \text{ where}$$

Σ_N, Σ_T, S are defined as in Definition E.1,
P is a set of productions of the form: $A \rightarrow \gamma$, in which $A \in \Sigma_N$, $\gamma \in \Sigma_T \cup \Sigma_T \Sigma_N \cup \{\lambda\}$.

As we discussed in Chap. 8, a context-free grammar is of a sufficient descriptive power for most applications of syntactic pattern recognition systems. Unfortunately, a pushdown automaton that analyzes context-free languages is inefficient in the sense of computational complexity. Therefore, there have been defined certain subclasses of context-free grammars such that corresponding automata are efficient. LL(k) grammars, introduced in an intuitive way in Chap. 8, are one of the most popular such subclasses. Let us characterize them in a formal way [180].

Definition E.7 Let $G = (\Sigma_N, \Sigma_T, P, S)$ be a context-free grammar defined as in Definition E.4, $\eta \in \Sigma^*$, and $|x|$ denotes the length (a number of symbols) of a string $x \in \Sigma^*$. $FIRST_k(\eta)$ denotes a set of all the terminal prefixes of strings of the length k (or of the length less than k, if a terminal string shorter than k is derived from α) that can be derived from η in the grammar G, i.e.

$$FIRST_k(\eta) = \{x \in \Sigma_T^* : (\eta \overset{*}{\Longrightarrow} x\beta \wedge |x| = k) \vee (\eta \overset{*}{\Longrightarrow} x \wedge |x| < k), \beta \in \Sigma^*\}.$$

Let $G = (\Sigma_N, \Sigma_T, P, S)$ be a context-free grammar. $\Longrightarrow_L^*$ denotes a *leftmost derivation* in the grammar G, i.e. a derivation such that a production is always applied to the leftmost nonterminal.

Definition E.8 Let $G = (\Sigma_N, \Sigma_T, P, S)$ be a context-free grammar defined as in Definition E.4. A grammar G is called an *LL(k) grammar* iff for every two leftmost derivations

$$S \underset{L}{\overset{*}{\Longrightarrow}} \alpha A \delta \underset{L}{\overset{*}{\Longrightarrow}} \alpha \beta \delta \underset{L}{\overset{*}{\Longrightarrow}} \alpha x$$

$$S \underset{L}{\overset{*}{\Longrightarrow}} \alpha A \delta \underset{L}{\overset{*}{\Longrightarrow}} \alpha \gamma \delta \underset{L}{\overset{*}{\Longrightarrow}} \alpha y,$$

where $\alpha, x, y \in \Sigma_T^*$, $\beta, \gamma, \delta \in \Sigma^*$, $A \in \Sigma_N$, the following condition holds

$$\text{If } FIRST_k(x) = FIRST_k(y), \text{ then } \beta = \gamma.$$

The condition formulated above for a grammar G means that for any derivational step of a derivation of a string w that is derivable in G, we can choose a production in an unambiguous way on the basis of an analysis of some part of w that is of length k. We say that the grammar G has the *property of an unambiguous choice of a production with respect to the k-length prefix in a leftmost derivation*.

E.2 Formal Automata

In this section we present definitions of two types of automata that are useful in syntactic pattern recognition and a Turing machine (because of its meaning in discussions concerning AI, cf. Chap. 17) [141, 250].

Definition E.9 A (*deterministic*) *finite-state automaton* is a quintuple

$$A = (Q, \Sigma_T, \delta, q_0, F), \text{ where}$$

Q is a finite nonempty set of states,
Σ_T is a finite set of input symbols,
$\delta : Q \times \Sigma_T \longrightarrow Q$ is the state-transition function,
$q_0 \in Q$ is the initial state,
$F \subseteq Q$ is a set of final states.

Now, we introduce notions allowing us to describe a computation performed by a finite-state automaton.

Let a current situation in an automaton be represented with a pair $(q, \alpha) \in Q \times \Sigma_T^*$, called an *automaton instantaneous configuration*. The first element represents a state of the automaton, the second a part of an input string that has not been read till now. Let $\vdash$ denote a *direct step of an execution of an automaton*.

A finite-state automaton analyzes an input string β according to the following scheme. At the beginning the automaton is in a state q_0, and an input string β is at its input, so (q_0, β) is the *initial configuration* of the automaton. The automaton reads symbols of the input string one by one, and it performs succeeding steps according to the following rule.

$$(q_i, a\gamma) \vdash (q_k, \gamma) \Leftrightarrow \delta(q_i, a) = q_k,$$

where δ is the state-transition function $q_i, q_k \in Q$, $a \in \Sigma_T, \gamma \in \Sigma_T^*$.

The automaton stops if it reaches a configuration (q_m, λ) such that $q_m \in F$ (q_m is a final state) and there is an empty word λ at its input.

Definition E.10 A (*deterministic*) *pushdown automaton* is a seven-tuple

$$A = (Q, \Sigma_T, \Phi, \delta, q_0, Z_0, F), \text{ where}$$

Q is a finite nonempty set of states,
Σ_T is a finite set of input symbols,
Φ is a finite set of stack symbols,
$\delta : Q \times (\Sigma_T \cup \{\lambda\}) \times \Phi \longrightarrow Q \times \Phi^*$ is the transition function,
$q_0 \in Q$ is the initial state,
$Z_0 \in \Phi$ is the initial stack symbol,
$F \subseteq Q$ is a set of final states.

An instantaneous configuration of a pushdown automaton is represented with a triple $(q, \alpha, \phi) \in Q \times \Sigma_T^* \times \Phi^*$, where the first two elements are the same as for a finite-state automaton, and the third element represents the content of the stack (the first symbol corresponds to the topmost symbol of the stack).

A pushdown automaton analyzes an input string β according to the following scheme. (q_0, β, Z_0) is the initial configuration. The automaton reads symbols of the input string one by one, and it performs succeeding steps according to the following rule.

$$(q_i, a\gamma, Z\phi) \vdash (q_k, \gamma, \eta\phi) \Leftrightarrow \delta(q_i, a, Z) = (q_k, \eta),$$

where δ is the transition function, $q_i, q_k \in Q$, $a \in \Sigma_T \cup \{\lambda\}$, $\gamma \in \Sigma_T^*$, $\eta, \phi \in \Phi^*$.

There are two (various) definitions of finishing a computation in case of pushdown automata. In the first one (an acceptance by a final state) (q_m, λ, ξ), where $q_m \in F$, λ the empty word, is the final configuration. In the second definition (an acceptance by an empty stack) (q_m, λ, λ), where $q_m \in Q$, is the final configuration.

Definition E.11 A (*one-tape*) *Turing machine* is a seven-tuple

$$A = (Q, \Phi, B, \Sigma_T, \delta, q_0, F), \text{ where}$$

Q is a finite nonempty set of states,
Φ is a finite nonempty set of tape symbols,
$B \in \Phi$ is the blank symbol,
$\Sigma_T \subseteq \Phi \backslash \{B\}$ is a set of input symbols,
$\delta : (Q \backslash F) \times \Phi \longrightarrow Q \times \Phi \times \{L, R\}$ is the transition function,
$q_0 \in Q$ is the initial state,
$F \subseteq Q$ is a set of final states.

A Turing machine consists of a finite control (defined by the transition function) and an infinite tape divided into cells. An input word is placed on the tape, each symbol in one cell, and all the cells on the left and on the right of the word are marked with the blank symbol B. At the beginning a reading/writing head of the control is set over a cell containing the first symbol of the word to be analyzed. It is represented with the *initial configuration*: $q_0 X_1 X_2 \dots X_n$, where q_0 is the initial state, $X_1 X_2 \dots X_n$ denotes the content of the cells (the input word). In general, an instantaneous configuration of a Turing machine is of the form $X_1 X_2 \dots X_{i-1} q X_i X_{i+1} \dots X_n$, which means placing of the head over the i-th cell, and the machine is in state q.

A Turing machine analyzes an input word according to the following scheme.

$$X_1 \dots X_{i-1} q_k X_i X_{i+1} \dots X_n \vdash X_1 \dots X_{i-1} Y q_m X_{i+1} \dots X_n \,,$$

if $\delta(q_k, X_i) = (q_m, Y, R)$ (a change of a state from q_k to q_m a change of a symbol on the tape from X_i to Y, and moving the head to the right), and

$$X_1 \dots X_{i-1} q_k X_i X_{i+1} \dots X_n \vdash X_1 \dots X_{i-2} q_m X_{i-1} Y X_{i+1} \dots X_n \,,$$

if $\delta(q_k, X_i) = (q_m, Y, L)$.

A Turing machine is described in a more detailed way in [141].

E.3 Graph Grammars

Graph grammars, which can be treated as systems of graph rewriting (*Graph Rewriting System, GRS*), are the third type, apart from term rewriting systems (cf. Appendix C.3) and string rewriting systems,[14] of Abstract Rewriting Systems (ARS). A variety of classes have been defined in the theory of graph grammars. In order to discuss an issue of syntax analysis of multi-dimensional patterns in Chap. 8, we introduced one of the most popular classes, namely edNLC graph grammars. Now, we define them in a formal way [149].

Definition E.12 A *directed node- and edge-labeled graph, EDG graph* over Σ and Γ is a quintuple

$$H = (V, E, \Sigma, \Gamma, \phi), \text{ where}$$

V is a finite nonempty set of nodes,
Σ is a finite nonempty set of node labels,
Γ is a finite nonempty set of edge labels,
E is a set of edges of the form (v, γ, w), in which $v, w \in V, \gamma \in \Gamma$,
$\phi : V \longrightarrow \Sigma$ is the node-labeling function.

A set of all the EDG graphs over Σ and Γ is denoted with $EDG_{\Sigma,\Gamma}$. EDG graphs can be generated with edNLC graph grammars.

Definition E.13 An *edge-labeled directed node-label controlled graph grammar, edNLC graph grammar* is a quintuple

$$G = (\Sigma, \Delta, \Gamma, P, Z), \text{ where}$$

Σ is a finite nonempty set of node labels,
$\Delta \subseteq \Sigma$ is a set of terminal node labels,
Γ is a finite nonempty set of edge labels,
P is a finite set of productions of the form (l, D, C), where

$l \in \Sigma$,

$D \in EDG_{\Sigma,\Gamma}$,

$C : \Gamma \times \{in, out\} \longrightarrow 2^{\Sigma \times \Sigma \times \Gamma \times \{in,out\}}$ is the embedding transformation,

$Z \in EDG_{\Sigma,\Gamma}$ is the start graph, called the axiom.

[14]The Chomsky string grammar introduced in Sect. E.1 is a specific case of a string rewriting system.

A derivational step for string grammars is simple (cf. Definition E.2). Unfortunately, as we have seen for a genealogy example in Sect. 8.5, in the case of graph grammars the derivational step is complex, mainly because of the form of the embedding transformation. Let us formalize it with the following definition.

Definition E.14 Let $G = (\Sigma, \Delta, \Gamma, P, Z)$ be an edNLC graph grammar.

Let $H, \overline{H} \in EDG_{\Sigma,\Gamma}$. Then H *directly derives* $\overline{H}$ in G, denoted $H \underset{G}{\Longrightarrow} \overline{H}$, if there exists a node $v \in V_H$ and a production $(l, D, C) \in P$ such that the following holds.

(1) $l = \phi_H(v)$.

(2) There exists an isomorphism from $\overline{H}$ onto a graph $X \in EDG_{\Sigma,\Gamma}$ constructed as follows. Let $\overline{D}$ be a graph isomorphic to D such that $V_H \cap V_{\overline{D}} = \emptyset$, and let h be an isomorphism from D onto $\overline{D}$. Then

$$X = (V_X, E_X, \Sigma, \Gamma, \phi_X), \text{ where}$$

$V_X = (V_H \setminus \{v\}) \cup V_{\overline{D}}$,

$$\phi_X(y) = \begin{cases} \phi_H(y), & \text{if } y \in V_H \setminus \{v\}, \\ \phi_{\overline{D}}(y), & \text{if } y \in V_{\overline{D}}, \end{cases}$$

$E_X = (E_H \setminus \{(n, \gamma, m) : n = v \text{ or } m = v\}) \cup \{(n, \gamma, m) : n \in V_{\overline{D}}, m \in V_{X \setminus \overline{D}}$ and there exists an edge $(m, \lambda, v) \in E_H$ such that $(\phi_X(n), \phi_X(m), \gamma, out) \in C(\lambda, in)\}$ $\cup \{(m, \gamma, n) : n \in V_{\overline{D}}, m \in V_{X \setminus \overline{D}}$ and there exists an edge $(m, \lambda, v) \in E_H$ such that $(\phi_X(n), \phi_X(m), \gamma, in) \in C(\lambda, in)\} \cup \{(n, \gamma, m) : n \in V_{\overline{D}}, m \in V_{X \setminus \overline{D}}$ and there exists an edge $(v, \lambda, m) \in E_H$ such that $(\phi_X(n), \phi_X(m), \gamma, out) \in C(\lambda, out)\}$ $\cup \{(m, \gamma, n) : n \in V_{\overline{D}}, m \in V_{X \setminus \overline{D}}$ and there exists an edge $(v, \lambda, m) \in E_H$ such that $(\phi_X(n), \phi_X(m), \gamma, in) \in C(\lambda, out)\}$.

A graph grammar of the edNLC class is of a very big descriptive power [149]. Unfortunately, a membership problem for this grammar is non-polynomial. Therefore, similarly as for string context-free grammars, its subclass, namely the ETPL(k) graph grammar, with a polynomial membership problem has been defined. For ETPL(k) grammars an efficient graph automaton has been constructed. Now, we introduce definitions concerning ETPL(k) grammars [93, 94].

Definition E.15 Let H be an EDG graph. H is called an *IE graph* iff the following conditions hold.

(1) A graph H contains a directed tree T such that nodes of T have been indexed according to Breadth-First Search (BFS).
(2) Nodes of a graph H are indexed in the same way as nodes of T.
(3) Every edge in a graph H is directed from a node having a lower index to a node having a greater index.

Definition E.16 Let G be an edNLC graph grammar defined as in Definition E.13. G is called a *TLPO graph grammar* iff it fulfills the following conditions.

(1) The start graph Z and graphs D of the right-sides of all the productions are IE graphs.
(2) For each graph of the right-hand side D, a directed spanning tree T is of at most two levels, and a node indexed with 1 is labeled with a terminal symbol.
(3) Each graph belonging to a derivation in G is an IE graph.
(4) For each derivational step, a production is applied to a node with the least index.
(5) Node indices do not change during a derivation.

A derivation fulfilling conditions (4) and (5) is called a *regular leftmost derivation.*

The next definition recalls the idea that was applied for LL(k) grammars in Definition E.8. We demand an unambiguity of a production choice during a regular left-hand side derivation. It makes the computation of an automaton efficient. For a string LL(k) grammar such an unambiguity concerns the k-length prefix of a word. In the case of IE graphs it concerns a subgraph. Such a subgraph contains a node v having an index determining the position of a production application and its k successors. Such a subgraph is called a *k-successors handle*. If for every derivational step in a grammar G we can choose a production in an unambiguous way on the basis of an analysis of a k-successors handle, then we say that G has the *property of an unambiguous choice of a production with respect to the k-successors handle in a regular leftmost derivation.*

Definition E.17 Let G be a TLPO graph grammar. G is called a *PL(k) graph grammar* iff G has the property of an unambiguous choice of a production with respect to the k-successors handle in a regular leftmost derivation.

Definition E.18 Let G be a PL(k) graph grammar. G is called an *ETPL(k) graph grammar* iff the following condition is fulfilled. If (v, λ, w), where $\phi(v) \in \Delta$, is an edge of an IE graph H belonging to a certain regular leftmost derivation, then this edge is preserved by all the embedding transformations applied in succeeding steps of the derivation.

For ETPL(k) graph grammars there have been defined both a polynomial graph automaton and a polynomial algorithm of grammatical inference [93, 94, 96].

Appendix F
Formal Models for Artificial Intelligence Methods: Theoretical Foundations of Rule-Based Systems

Definitions that allow one to describe a rule-based system in a formal way [52] are introduced in the first section. An issue of reasoning in logic is presented in the second section. There are many approaches to this issue in modern logic, e.g., introduced by Kazimierz Ajdukiewicz, Jan Łukasiewicz, Charles Sanders Peirce, Willard Van Orman Quine. In this monograph we present a taxonomy of reasoning according to Józef Maria Bocheński[15] [30].

F.1 Definition of Generic Rule-Based Systems

There are several definitions of generic rule-based systems in the literature. However, most of them relate to rule-based systems of the specific form. In our opinion, one of the most successful trials of constructing such a formal model is the one developed by a team of Claude Kirchner (INRIA). We present a formalization of a rule-based system according to this approach [52].

Let us assume definitions of a signature $\Sigma = (\Sigma^C, \Sigma_n^F, \Sigma_n^P)$, a set of terms $\mathcal{T}_\Sigma(X)$, a substitution σ, and semantics of FOL, as in Appendix C. Additionally, let $\mathcal{F}_\Sigma(X)$ denote a set of formulas, $Var(t)$ a set of variables occurring in a term (a set of terms) t, $FV(\phi)$ a set of free variables occurring in a formula ϕ, $Dom(\sigma)$ the domain of a substitution σ, $\mathcal{R}$ a set of labels. A theory is a set of formulas $\mathcal{T}$ that is closed under a logical consequence, i.e., for each formula φ the following holds: if $\mathcal{T} \models \varphi$, then $\varphi \in \mathcal{T}$.

[15] Józef Maria Bocheński, OP, a professor and the rector of the *Université de Fribourg*, a professor of *Pontificia Studiorum Universitas a Sancto Thoma Aquinate* (Angelicum) in Rome, a logician and philosopher, Dominican. He was known as an indefatigable man having a good sense of humor, e.g., he gained a pilot's licence while in his late sixties.

M. Flasiński, *Introduction to Artificial Intelligence*,
DOI 10.1007/978-3-319-40022-8

Definition F.1 A term t is called a *ground term*, if $Var(t) = \emptyset$.

Definition F.2 The *Herbrand universe* for any $\mathcal{F}_\Sigma(X)$ is a set $\mathcal{H}$ that is defined inductively in the following way.

- If $a \in \Sigma^C$ occurs in a formula belonging to $\mathcal{F}_\Sigma(X)$, then $a \in \mathcal{H}$. (If there is no constant in formulas of $\mathcal{F}_\Sigma(X)$, then we add any constant to $\mathcal{H}$.)
- For every $n \geq 0$ and for every $f \in \Sigma_n^F$, if $t_1, \ldots, t_n$ are terms belonging to $\mathcal{H}$, then $f(t_1, \ldots, t_n) \in \mathcal{H}$.

Thus, the Herbrand universe for a given set of formulas $\mathcal{F}_\Sigma(X)$ is a set of all the ground terms that have been defined with the help of function symbols out of constants occurring in formulas of $\mathcal{F}_\Sigma(X)$.

Definition F.3 A *fact* is a ground term.

Definition F.4 A *working memory* $\mathcal{WM}$ is a *set of facts*. In other words, a working memory $\mathcal{WM}$ is a subset of the Herbrand universe $\mathcal{H}$.

Definition F.5 A *(positive) pattern* is a term $p \in \mathcal{T}_\Sigma(X)$, and a *negative pattern* is a term of the form $\neg p$. A set of positive and negative patterns is denoted as $P = P^+ \cup P^-$ and it is called the *set of patterns*.

Definition F.6 Let S be a set of facts. A *pattern of updating* PU of a set of facts S is a pair $PU = (rem, add)$, where $rem = \{r : r \in \mathcal{T}_\Sigma(X)\}$ is a set containing patterns of the terms that should be removed from the set of facts, $add = \{a : a \in \mathcal{T}_\Sigma(X)\}$ is a set containing patterns of the terms that should be added to the set of facts.

Definition F.7 A *rule* is a triple

$$(R, COND, ACT), \text{ where}$$

$R \in \mathcal{R}$ is the *rule label*,
$COND$ is of the form (P, ϕ), where P is the set of patterns, ϕ is a formula such that $FV(\phi) \subseteq Var(P)$,
$ACT = (rem, add)$ is a pattern of updating a working memory $\mathcal{WM}$ such that: $Var(rem) \subseteq Var(P^+)$ and $Var(add) \subseteq Var(P^+)$.

COND is called a *condition (antecedent) of the rule*, and *ACT* an *action (consequent) of the rule*. A rule can be written also in the form:

$$R : \text{IF } COND \text{ THEN } ACT.$$

Definition F.8 Let S be a set of facts, $P = P^+ \cup P^-$ be a set of patterns. P^+ *matches* S according to a theory $\mathcal{T}$ and a substitution σ, denoted $P^+ \ll^\sigma_\mathcal{T} S$ iff the following condition is fulfilled.

$$\forall p \in P^+ \exists t \in S \sigma(p) =_\mathcal{T} t.$$

P^- *mismatches* S according to a theory $\mathcal{T}$, denoted $P^- \not\ll_{\mathcal{T}} S$ iff the following condition is fulfilled.

$$\forall \neg p \in P^- \forall t \in S \forall \sigma\ \sigma(p) \neq_{\mathcal{T}} t.$$

Definition F.9 Let σ be a substitution, $\mathcal{WM}$ be a working memory, $\mathcal{WM}' \subseteq \mathcal{WM}$. A rule $(R, COND, ACT)$, where $COND = (P, \phi)$, $ACT = (rem, add)$, $(\sigma, \mathcal{WM}')$-*matches* the working memory $\mathcal{WM}$ iff the following conditions are fulfilled.

- $P^+ \ll^{\sigma}_{\mathcal{T}} \mathcal{WM}'$,
- $P^- \not\ll_{\mathcal{T}} \mathcal{WM}$,
- $\mathcal{T} \models \sigma(\phi)$, where $\mathcal{WM}'$ is the minimal subset of $\mathcal{WM}$.

Definition F.10 Let a rule $(R, COND, ACT)$, $COND = (P, \phi)$, $ACT = (rem, add)$, $(\sigma, \mathcal{WM}')$-*matches* the working memory $\mathcal{WM}$. An *application* of this rule is a modification of the working memory $\overline{\mathcal{WM}}$ defined in the following way.

$$\overline{\mathcal{WM}} = (\mathcal{WM} \setminus \sigma(rem)) \cup \sigma(add).$$

An application of a rule is denoted by $\mathcal{WM} \Rightarrow \overline{\mathcal{WM}}$. A *sequence of rule applications* is denoted by $\mathcal{WM}_0 \Rightarrow \mathcal{WM}_1 \Rightarrow \cdots \Rightarrow \mathcal{WM}_n$.

Definition F.11 Let $\mathcal{WM}$ be a working memory, $\mathcal{R}$ be a set of rules, $\mathcal{WM}' \subseteq \mathcal{WM}$. A set

$$\mathcal{CS} = \{(R, \mathcal{WM}') : \exists (R, COND, ACT) \in \mathcal{R} \text{ such that } (\sigma, \mathcal{WM}')\text{-matches } \mathcal{WM}\}$$

is called a *set of conflicting rules* of $\mathcal{R}$ for the working memory $\mathcal{WM}$.

Definition F.12 A *conflict resolution method* $\mathcal{CRM}$ is an algorithm, which, for a set of rules $\mathcal{R}$ and a sequence of rule applications,

$$\mathcal{WM}_0 \Rightarrow \mathcal{WM}_1 \Rightarrow \cdots \Rightarrow \mathcal{WM}_n,$$

computes a unique element of a set of conflicting rules of $\mathcal{R}$ for the working memory $\mathcal{WM}_n$.

Definition F.13 A *generic rule-based system* $\mathcal{GRBS}$ is a quadruple:

$$\mathcal{GRBS} = (\mathcal{WM}, \mathcal{R}, \mathcal{CRM}, \mathcal{T}), \text{ where}$$

$\mathcal{WM}$ is a working memory,
$\mathcal{R}$ is a set of rules, called a *rule base*,
$\mathcal{CRM}$ is a conflict resolution method,
$\mathcal{T}$ is a pattern matching theory.

One can easily notice that the model presented above [52] is very formal. In practice, rule-based systems are constructed according to such a formalized model very rarely. Firstly, conflict resolution methods usually concern a single application of a rule. Secondly, facts belonging to a working memory $\mathcal{WM}$ and patterns belonging to a set P are of the form $f(t_1, \ldots t_n)$, where t_i, $i = 1, \ldots, n$ are symbols of variables or constants (for facts, only constants), f is a function symbol. Moreover, a set of patterns P is usually not defined explicitly in a rule antecedent $COND = (P, \phi)$. Instead, patterns *occur* in a formula ϕ and a module of rules' matching extracts them from ϕ. A formula ϕ is a conjunction of literals.[16] A matching process is usually of the form of syntactic pattern matching, which means that a theory $\mathcal{T}$ is empty. Frequently, an application of a rule action ACT is defined as a replacement of a constant c_k by a constant $\overline{c_k}$ in a ground term $f_m(c_1, \ldots, c_k, \ldots, c_n)$, which results in obtaining a ground term $f_m(c_1, \ldots, \overline{c_k}, \ldots, c_n)$.[17] Thus, we can simply define an action ACT as an assignment $f_m^k := \overline{c_k}$. Of course, an action can be a sequence of such assignments.

Taking into account such simplifications,[18] a *rule-based system* can be defined as a triple:

$$\mathcal{RBS} = (\mathcal{WM}, \mathcal{R}, \mathcal{CRM}),$$

where $\mathcal{WM}, \mathcal{R}, \mathcal{CRM}$ are of the simplified form discussed above.

Our considerations above concerned rules of a *declarative type*. Such rules only modify the working memory. If we construct a *control* (steering) rule-based system, we define also *reactive rules* that influence the system external environment.[19] Modeling such a system with the formalism presented above, we can assume that an execution of a reactive rule consists in changing a "control (steering) term" in the working memory. Then, a specialized module of the system interprets such a term and calls a proper procedure contained in a library of control procedures.

F.2 Logical Reasoning—Selected Notions

An *inference* is a reasoning that consists in acknowledging a statement to be true assuming some statements are true. In logic, an inference is made with the help of *rules of inference*.

[16] Notions concerning FOL are introduced in Appendix C.

[17] According to the formalism presented, we could simulate such an operation as removing the first ground term from the working memory and adding the second one.

[18] Of course, some rule-based systems cannot be simplified in such a way. For example, an expert system mentioned in Chap. 9 and developed by the author operates on graphs. Thus, in this case assuming the "shallow" (one-level) form of terms is impossible.

[19] An action of a reactive rule can be of the form of a command, e.g. *open_valve(V34)*.

A rule of inference is usually defined in the following form:

$$\frac{\begin{array}{c}A_1\\ A_2\\ \ldots,\\ A_n\end{array}}{B}$$

where $A_1, A_2, \ldots, A_n$ are statements assumed to be true, whereas B is a statement we acknowledge to be true, or in the form:

$$\frac{A_1, A_2, \ldots, A_n}{B}.$$

A notation introduced above can be interpreted as follows: "*If we assume that statements represented with expressions* $A_1, A_2, \ldots, A_n$ *are true, then we are allowed to acknowledge a statement represented with an expression B is true*".

The three following basic types of reasoning can be distinguished in logic.

Deduction is based on the *modus ponendo ponens* rule, which is of the form:

$$\frac{\begin{array}{l}\mathit{if}\, A,\, \mathit{then}\, B\\ A\end{array}}{B}.$$

Thus, deduction is a type of reasoning, in which on the basis of a certain general rule and a *premise* we infer a *conclusion*.

Abduction is based on the rule of the form:

$$\frac{\begin{array}{l}\mathit{if}\, A,\, \mathit{then}\, B\\ B\end{array}}{A}.$$

In abductive reasoning we use a certain rule and a conclusion (usually a certain (empirical) observation) to derive a premise (usually interpreted as the best explanation of this conclusion).

From the point of view of logic only deduction is reliable reasoning.

(Incomplete) induction[20] (in a sense, induction by incomplete enumeration) can be treated as a special case of abduction. It consists in inferring a certain generalization about a class of objects on the basis of premises concerning some objects belonging to this class. In the simplest case induction can be defined in the following way [2]:

[20] In this appendix we do not introduce all types of induction (e.g. induction by complete enumeration, eliminative induction), but only such a type that relates to methods described in this book.

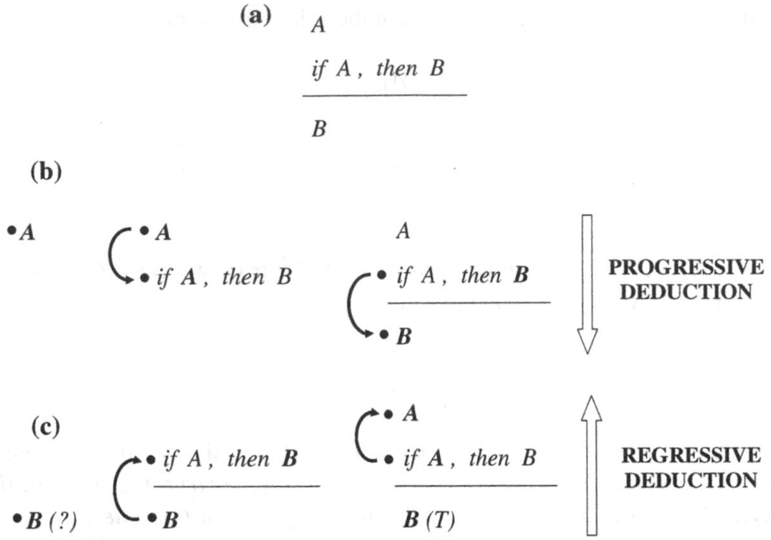

(a)

$$\frac{A \qquad \textit{if } A, \textit{ then } B}{B}$$

(b)

•A

•A
• if A, then B

A
• if A, then B
———
• B

PROGRESSIVE DEDUCTION

(c)

•B (?)

• if A, then B
———
• B

• A
• if A, then B
———
B (T)

REGRESSIVE DEDUCTION

Fig. F.1 Two kinds of deduction: **a** the inference rule for deduction, **b** a progressive deduction, **c** a regressive deduction

$$\frac{\begin{array}{l} S_1 \textit{ is } P \\ S_2 \textit{ is } P \\ \ldots \\ S_n \textit{ is } P \end{array}}{\textit{every } S \textit{ is } P}$$

Now, let us notice that two kinds of deductive reasoning can be distinguished.

- In *progressive deductive reasoning* we start from true premises and we infer on their basis. In other words, a reason is given and a consequent is to be derived (cf. Fig. F.1b). Every computation (including a symbolic computation) is of the form of progressive deduction. (The final conclusion being the result of the computation is formulated at the end.)
- In *regressive deductive reasoning* we start from a consequent to be inferred and we look for true premises, which can be used for proving the consequent. As is shown in Fig. F.1c, a consequent is given and a reason is looked for.

Appendix G
Formal Models for Artificial Intelligence Methods: Mathematical Similarity Measures for Pattern Recognition

As we discussed in Chap. 10, the idea of similarity of two objects (phenomena) is a fundamental one in the area of pattern recognition and cluster analysis. Firstly we introduce mathematical foundations for defining similarity measures, then we survey the most popular measures.

G.1 Metric and Topological Spaces

Let us introduce basic notions concerning metric and topological spaces [294].

Definition G.1 Let X be a nonempty set. A *metric* on a set X is any function[21] $\rho : X \times X \longrightarrow \mathbb{R}_+$ fulfilling the following conditions.

1. $\forall x \in X : \rho(x, y) = 0$ iff $x = y$.
2. $\forall x, y \in X : \rho(x, y) = \rho(y, x)$.
3. $\forall x, y, z \in X : \rho(x, z) \leq \rho(x, y) + \rho(y, z)$.

Definition G.2 If ρ is a metric on a set X, then a pair (X, ρ) is called a *metric space*. Elements of a metric space (X, ρ) are called *points*. For any $x, y \in X$, a value $\rho(x, y)$ is called a *distance between points* x and y.

Definition G.3 Let (X, ρ) be a metric space. A *ball* (*open ball*) of a radius $r > 0$ and centered at a point $a \in X$ is a set:

$$K(a, r) = \{x \in X : \rho(a, x) < r\}.$$

Definition G.4 Let (X, ρ) be a metric space. A set $U \subset X$ is called an *open set* iff every point of a set U is included in a set U together with some ball centered at this point, i.e.

$$\forall x \in U \, \exists r > 0 : K(x, r) \subset U.$$

[21] $\mathbb{R}_+ = [0, +\infty)$.

M. Flasiński, *Introduction to Artificial Intelligence*,
DOI 10.1007/978-3-319-40022-8

Definition G.5 Let X be a nonempty set, $\mathcal{T}$ be a family of subsets of X. The family $\mathcal{T}$ is called a *topology* for X, if it fulfills the following conditions.

- $\emptyset, X \in \mathcal{T}$.
- A finite intersection of elements of $\mathcal{T}$ is the element of $\mathcal{T}$.
- An arbitrary union of elements of $\mathcal{T}$ is the element of $\mathcal{T}$.

Definition G.6 If $\mathcal{T}$ is a topology for a set X, then a pair $(X, \mathcal{T})$ is called a *topological space*. The members of $\mathcal{T}$ are called *open sets* in $(X, \mathcal{T})$.

G.2 Metrics Used in Pattern Recognition

In pattern recognition and cluster analysis selection of an adequate metric is essential for the effectiveness of the method constructed. Now we present the most popular metrics in this area.

Definition G.7 The *Minkowski metric* ρ_p is given by the formula:

$$\rho_p(x, y) = \left(\sum_{i=1}^{n} |x_i - y_i|^p \right)^{1/p} .$$

For cases $p = 2$ and $p = 1$ of the Minkowski metric, the following metrics are defined.

Definition G.8 The *Euclidean metric* ρ_2 is given by the formula:

$$\rho_2(x, y) = \sqrt{\sum_{i=1}^{n} |x_i - y_i|^2} .$$

Definition G.9 The *Manhattan metric* ρ_1 is given by the formula:

$$\rho_1(x, y) = \sum_{i=1}^{n} |x_i - y_i| .$$

If $p \to \infty$, then the following metric is received.

Definition G.10 The *Chebyshev metric* ρ_∞ is given by the formula:

$$\rho_\infty(x, y) = \max_{1 \le j \le n} \{|x_i - y_i|\} .$$

In order to illustrate the differences among metrics (Euclidean, Manhattan, Chebyshev), balls of radius 1 centered at a point having coordinates (0, 0) (unit balls) are shown in Fig. G.1.

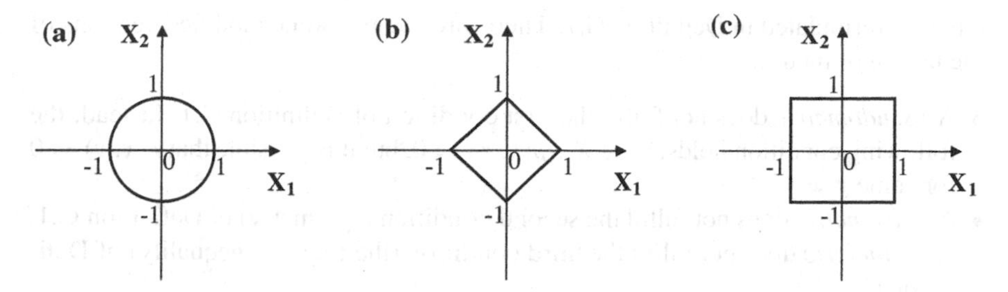

Fig. G.1 Unit balls constructed with various metrics: **a** Euclidean, **b** Manhattan, **c** Chebyshev

The metrics introduced above are used primarily for the recognition of patterns, which are represented by vectors of *continuous* features. If patterns are represented with binary feature vectors or with structural/syntactic descriptions, then metrics of a different nature are applied. Let us present such metrics.

In computer science and artificial intelligence the Hamming metric [124] plays an important role. For example, we introduced this metric discussing Hamming neural networks in Chap. 11. Let Σ_T be a set of terminal symbols (alphabet).[22]

Definition G.11 Let there be given two strings of characters (symbols): $x = x_1 x_2 \dots x_n, y = y_1 y_2 \dots y_n \in \Sigma_T^*$. Let $H = \{x_i, i = 1, \dots, n : x_i \neq y_i\}$. A distance ρ_H between strings x and y in the sense of the *Hamming metric* equals $\rho_H(x, y) = |H|$, where $|H|$ is the number of elements of a set H.

In other words, the Hamming metric defines on how many positions two strings differ one from another.

The Levenshtein metrics [104, 179] are generalizations of the Hamming metric. Let us introduce them.

Definition G.12 Let there be given two strings of characters (symbols): $x, y \in \Sigma_T^*$. A transformation $F : \Sigma_T^* \longrightarrow \Sigma_T^*$ such that $y \in F(x)$ is called a *string transformation*. Let us introduce the following string transformations.

1. A *substitution error transformation* F_S: $\eta_1 a \eta_2 \longmapsto^{F_S} \eta_1 b \eta_2 a, b \in \Sigma_T, a \neq b, \eta_1, \eta_2 \in \Sigma_T^*$.
2. A *deletion error transformation* F_D: $\eta_1 a \eta_2 \longmapsto^{F_D} \eta_1 \eta_2 a \in \Sigma_T, \eta_1, \eta_2 \in \Sigma_T^*$.
3. An *insertion error transformation* F_I: $\eta_1 \eta_2 \longmapsto^{F_I} \eta_1 a \eta_2 a \in \Sigma_T, \eta_1, \eta_2 \in \Sigma_T^*$.

Definition G.13 Let there be given two strings of characters (symbols): $x, y \in \Sigma_T^*$. A distance ρ_L between strings x and y in the sense of the (*simple*) *Levenshtein metric* is defined as the smallest number of string transformations F_S, F_D, F_I required to obtain the string y from the string x.

Before we introduce generalizations of the simple Levenshtein metric, let us notice that in computer science sometimes we do not want to preserve all the properties of

[22]Notions of formal language theory are introduced in Appendix E.

a metric formulated in Definition G.1. Therefore, we use some modified versions of the notion of metric.

- A *pseudometric* does not fulfill the first condition of Definition G.1. Instead, the following condition holds: $\forall x \in X : \rho(x, x) = 0$, but it is possible that $\rho(x, y) = 0$ for some $x \neq y$.
- A *quasimetric* does not fulfill the second condition (symmetry) of Definition G.1.
- A *semimetric* does not fulfill the third condition (the triangle inequality) of Definition G.1.

In the following definitions we call all these modified versions, briefly, a metric [104].

Definition G.14 Let there be given two strings of characters (symbols): $x, y \in \Sigma_T^*$. Let us ascribe weights α, β, γ to string transformations: F_S, F_D, F_I, respectively. Let M be a sequence of string transformations applied to obtain the string y from the string x such that we have used s_M substitution error transformations, d_M deletion error transformations and i_M insertion error transformations.

Then, a distance ρ_{LWTE} between strings x and y in the sense of the *Levenshtein metric weighted according to a type of an error* is given by the following formula:

$$\rho_{LWTE}(x, y) = \min_M \{\alpha \cdot s_M + \beta \cdot d_M + \gamma \cdot i_M\} .$$

Let us note that if the weight of a deletion error transformation β differs from the weight of an insertion error transformation γ, then the Levenshtein metric weighted according to the type of an error is a quasimetric.

Definition G.15 Let $S(a, b)$ denote the cost of a substitution error transformation described as in point 1 of Definition G.12, $S(a, a) = 0$, $D(a)$ denote the cost of a deletion error transformation described as in point 2 of Definition G.12.

Let $I(a, b)$ denote the cost of the insertion of a symbol b before a symbol a, i.e.

$$\eta_1 a \eta_2 \overset{F_I}{\longmapsto} \eta_1 b a \eta_2 a, b \in \Sigma_T, \eta_1, \eta_2 \in \Sigma_T^* ,$$

and, additionally, let $I'(b)$ denote the cost of the insertion of a symbol b at the end of a word.

Let M be a sequence of string transformations applied to obtain the string y from the string x, where $x, y \in \Sigma_T^*$, and $c(M)$ denotes the sum of costs S, D, I, I' of all the transformations of a sequence M.

Then, a distance ρ_{LWE} between strings x and y in the sense of the *Levenshtein metric weighted with errors* is given by the following formula:

$$\rho_{LWE}(x, y) = \min_M \{c(M)\} .$$

Appendix H
Formal Models for Artificial Intelligence Methods: Mathematical Model of Neural Network Learning

When we discussed neural networks in Chap. 11, we introduced the basic model of their learning, namely the back propagation method. In the second section of this appendix we present a formal justification for the principles of the method. In the first section we introduce basic notions of mathematical analysis [121, 122, 237] that are used for this justification.

H.1 Selected Notions of Mathematical Analysis

Firstly, let us introduce notions of vector space and normed vector space.

Definition H.1 Let V be a nonempty set closed under an addition operation $+$, and K be a field. Let $\cdot$ be an external operation of the left-hand side multiplication, i.e. it is a mapping from $K \times V$ to V, where its result for a pair $(a, \mathbf{w}) \in K \times V$ is denoted $a \cdot \mathbf{w}$, briefly $a\mathbf{w}$.

A *vector space* is a structure consisting of the set V, the field K, and operations $+, \cdot$, which fulfils the following conditions.

- The set V with the operation $+$ is the Abelian group.
- $\forall a, b \in K, \mathbf{w} \in V$: $a(b\mathbf{w}) = (ab)\mathbf{w}$.
- $\forall a, b \in K, \mathbf{w} \in V$: $(a + b)\mathbf{w} = a\mathbf{w} + b\mathbf{w}$.
- $\forall a \in K, \mathbf{w}, \mathbf{u} \in V$: $a(\mathbf{w} + \mathbf{u}) = a\mathbf{w} + a\mathbf{u}$.
- $\forall \mathbf{w} \in V$: $1 \cdot \mathbf{w} = \mathbf{w}$, where 1 is the identity element of multiplication in K.

Definition H.2 Let X be a vector space over a field K. A *norm* on X is a mapping $\| \cdot \| : X \longrightarrow \mathbb{R}_+$ fulfilling the following conditions.

- $\forall x \in X$: $\|x\| = 0 \Leftrightarrow x = \mathbf{0}$, where $\mathbf{0}$ is the zero vector in X.
- $\forall x \in X, \lambda \in K$: $\|\lambda x\| = |\lambda| \cdot \|x\|$.
- $\forall x, y \in X$: $\|x + y\| \leq \|x\| + \|y\|$.

M. Flasiński, *Introduction to Artificial Intelligence*,
DOI 10.1007/978-3-319-40022-8

Definition H.3 Let $\|\cdot\|$ be a norm on a vector space X. A pair $(X, \|\cdot\|)$ is called a *normed vector space.*

Further on, we assume X is a normed vector space.

Now, we can define directional derivative, partial derivative and gradient. Let $U \subset X$ be an open subset of X.

Definition H.4 Let there be given a function $f : U \longrightarrow \mathbb{R}$ and $v \neq 0$ the vector in X. If there exists a limit of a difference quotient

$$\lim_{h \to 0} \frac{f(a + hv) - f(a)}{h},$$

then this limit is called a *directional derivative* of the function f along the vector v at the point a, denoted $\partial_v f(a)$.

Let $X = \mathbb{R}^n$, and vectors $e_1 = (1, 0, 0, \dots, 0)$, $e_2 = (0, 1, 0, \dots, 0)$, $\dots$, $e_n = (0, 0, 0, \dots, 1)$ constitute a canonical basis for a space X. Let $U \subset X$ be an open subset of X.

Definition H.5 If there exist directional derivatives $\partial_{e_1} f(a), \partial_{e_2} f(a), \dots, \partial_{e_n} f(a)$ of a function $f : U \longrightarrow \mathbb{R}$ along vectors of the canonical basis $e_1, e_2, \dots, e_n$, then they are called *partial derivatives* of the function f at the point a, denoted $\frac{\partial f}{\partial x_1}(a), \frac{\partial f}{\partial x_2}(a), \dots, \frac{\partial f}{\partial x_n}(a)$.

Let $f : U \longrightarrow \mathbb{R}$ be a function, where the set $U \subset \mathbb{R}^n$ is an open set. Let us assume that there exist partial derivatives: $\frac{\partial f}{\partial x_1}(a), \frac{\partial f}{\partial x_2}(a), \dots, \frac{\partial f}{\partial x_n}(a)$ at the point $a \in U$.

Definition H.6 A vector

$$\nabla f(a) = \left(\frac{\partial f}{\partial x_1}(a), \frac{\partial f}{\partial x_2}(a), \dots, \frac{\partial f}{\partial x_n}(a) \right) \in \mathbb{R}^n.$$

is called a *gradient* of the function f at the point a.

Theorem H.1 *At a given point, a directional derivative has the maximum absolute value in the direction of the gradient vector.*

Thus, a function increases (or decreases) most rapidly in the gradient direction. We will make use of this property in the next section.

H.2 Backpropagation Learning of Neural Networks

In this section we introduce a formalization of the backpropagation method of neural network learning [252], which was presented in an intuitive way in Chap. 11. Firstly, let us discuss its general idea.

We learn a neural network, i.e. we modify its weights, in order to minimize an error function of a classification of vectors belonging to the training set. All the weights of a neural network are variables of this function. Let us denote this function with $E(\mathbf{W})$, where $\mathbf{W} = (W_1, W_2, \ldots, W_N)$ is a vector of weights of all the neurons. At the j-th step of a learning process we have an error $E(\mathbf{W}(j))$, briefly $E(j)$. This error will be minimized with the method of steepest descent, which can be defined in the following way.

$$\mathbf{W}(j+1) = \mathbf{W}(j) - \alpha \nabla E(\mathbf{W}(j)), \tag{H.1}$$

where $\nabla E(\mathbf{W}(j)) = \left(\frac{\partial E(j)}{\partial W_1(j)}, \frac{\partial E(j)}{\partial W_2(j)}, \ldots, \frac{\partial E(j)}{\partial W_N(j)} \right)$ is a gradient of the function E.

Now, let us introduce denotations according to those used in Chap. 11. $N^{(r)(k)}$ denotes the k-th neuron of the r-th layer. Let us assume that a network consists of L layers, and the r-th layer consists of M_r neurons. The output signal of the k-th neuron of the r-th layer at the j-th step of learning is denoted with $y^{(r)(k)}(j)$. The input signal at the i-th input of the k-th neuron of the r-th layer at the j-th step of learning is denoted with $X_i^{(r)(k)}(j)$, and the corresponding weight is denoted with $W_i^{(r)(k)}(j)$.

Let us define a function E as a mean squared error function at the output of the network, i.e.

$$E(j) = \frac{1}{2} \sum_{m=1}^{M_L} (u^{(m)}(j) - y^{(L)(m)}(j))^2, \tag{H.2}$$

where $u^{(m)}(j)$ is a required output signal for the m-th neuron of the L-th layer at the j-th step.

First of all, let us define a formula for a value of the i-th weight of the k-th neuron of the r-th layer at the $(j+1)$-th step of learning. From formulas (H.1) and (H.2) we obtain

$$\begin{aligned} W_i^{(r)(k)}(j+1) &= W_i^{(r)(k)}(j) - \alpha \frac{\partial E(j)}{\partial W_i^{(r)(k)}(j)} \\ &= W_i^{(r)(k)}(j) - \alpha \frac{\partial E(j)}{\partial v^{(r)(k)}(j)} \cdot \frac{\partial v^{(r)(k)}(j)}{\partial W_i^{(r)(k)}(j)} \\ &= W_i^{(r)(k)}(j) - \alpha \frac{\partial E(j)}{\partial v^{(r)(k)}(j)} \cdot X_i^{(r)(k)}(j) . \end{aligned} \tag{H.3}$$

Now, let us introduce the following denotation in the formula (H.3)

$$\delta^{(r)(k)}(j) = - \frac{\partial E(j)}{\partial v^{(r)(k)}(j)} . \tag{H.4}$$

Then, we obtain the following formula.

$$W_i^{(r)(k)}(j+1) = W_i^{(r)(k)}(j) + \alpha \delta^{(r)(k)}(j) X_i^{(r)(k)}(j) . \tag{H.5}$$

The formula (H.5) is analogous to the formula (11.16) in Sect. 11.2 including a description of the back propagation method.[23]

At the end of our considerations, we should derive a formula for $\delta^{(r)(k)}(j)$. Let us determine it, firstly, for neurons of the input layer and hidden layers.

$$\delta^{(r)(k)}(j) = -\frac{\partial E(j)}{\partial v^{(r)(k)}(j)} = -\sum_{m=1}^{M_{r+1}} \frac{\partial E(j)}{\partial v^{(r+1)(m)}(j)} \cdot \frac{\partial v^{(r+1)(m)}(j)}{\partial v^{(r)(k)}(j)}. \tag{H.6}$$

By applying the formula (H.4) and making use of the formula (11.1) introduced in Sect. 11.2, we receive

$$\delta^{(r)(k)}(j) = -\sum_{m=1}^{M_{r+1}} (-\delta^{(r+1)(m)}(j)) \cdot \frac{\partial \sum_{i=1}^{M_{r+1}} W_i^{(r+1)(m)}(j) X_i^{(r+1)(m)}(j)}{\partial v^{(r)(k)}(j)}. \tag{H.7}$$

From the formula (11.12) introduced in Sect. 11.2 we find that

$$\begin{aligned} \delta^{(r)(k)}(j) &= \sum_{m=1}^{M_{r+1}} \delta^{(r+1)(m)}(j) \cdot \frac{\partial \sum_{i=1}^{M_{r+1}} W_i^{(r+1)(m)}(j) y^{(r)(i)}(j)}{\partial v^{(r)(k)}(j)} \\ &= \sum_{m=1}^{M_{r+1}} (\delta^{(r+1)(m)}(j) W_k^{(r+1)(m)}(j)) \cdot \frac{\partial f(v^{(r)(k)}(j))}{\partial v^{(r)(k)}(j)}. \end{aligned} \tag{H.8}$$

Let us note that the derived formula (H.8) is analogous to the formula (11.15) presented in Sect. 11.2.

Deriving the formula (H.8) for the r-th layer, we have made use of parameters of the $(r+1)$-th layer. For the last (L-th) layer we cannot use such a technique. Therefore, $\delta^{(L)(k)}(j)$ is derived directly on the basis of the formula (H.2).

$$\begin{aligned} \delta^{(L)(k)}(j) &= -\frac{\partial E(j)}{\partial v^{(L)(k)}(j)} = -\frac{\partial \frac{1}{2} \sum_{m=1}^{M_L} (u^{(m)}(j) - y^{(L)(m)}(j))^2}{\partial v^{(L)(k)}(j)} \\ &= (u^{(k)}(j) - y^{(L)(k)}(j)) \cdot \frac{\partial y^{(L)(k)}(j)}{\partial v^{(L)(k)}(j)} \\ &= (u^{(k)}(j) - y^{(L)(k)}(j)) \cdot \frac{\partial f(v^{(L)(k)}(j))}{\partial v^{(L)(k)}(j)}. \end{aligned} \tag{H.9}$$

Again, let us notice that the formula (H.9) is analogous to the formula (11.14) introduced in Sect. 11.2.

[23] In Sect. 11.2 we have analyzed (only) two steps of learning. The first step corresponds to the (j)-th step of our considerations in this section. The second step (the "primed" one) corresponds to the $(j+1)$-th step of our considerations here. A parameter α corresponds to a learning rate coefficient η.

Appendix I
Formal Models for Artificial Intelligence Methods: Mathematical Models for Reasoning Under Uncertainty

In the first section fundamental notions of measure theory [251] are introduced in order to define probability space [118, 119], which is a basic definition of probability theory. The Bayesian model [119], which is a basis for constructing probabilistic reasoning systems described in Chap. 12,[24] is introduced in the second section. The third section contains basic notions [67, 271] of the Dempster-Shafer theory.

I.1 Foundations of Measure Theory and Probability Theory

Let us begin with fundamental definitions of measure theory, i.e. σ-algebra, measurable space, measure, and measure space.

Definition I.1 Let Ω be a nonempty set (called a sample space), $\mathfrak{M}$ a family of subsets of Ω. A family $\mathfrak{M}$ is called a *σ-algebra* on the set Ω iff it fulfills the following conditions.

- $\emptyset \in \mathfrak{M}$.
- If $A \in \mathfrak{M}$, then $\Omega \setminus A \in \mathfrak{M}$.
- If $A_1, A_2, A_3, \ldots \in \mathfrak{M}$, then $\bigcup_{i=1}^{\infty} A_i \in \mathfrak{M}$.

Definition I.2 Let $\mathfrak{M}$ be a σ-algebra on Ω. The pair $(\Omega, \mathfrak{M})$ is called a *measurable space*.

If a set A belongs to a σ-algebra $\mathfrak{M}$, then we say that A is *$\mathfrak{M}$-measurable*, or simply *measurable*, if it is clear what the underlying σ-algebra is.

Definition I.3 Let $\mathfrak{M}$ be a family of subsets of a set Ω. Let $\sigma(\mathfrak{M})$ be the intersection of all σ-algebras on Ω containing $\mathfrak{M}$. Then $\sigma(\mathfrak{M})$ is a σ-algebra on Ω containing

[24]This model is also used for defining statistical pattern recognition algorithms, discussed in Chap. 10.

M. Flasiński, *Introduction to Artificial Intelligence*,
DOI 10.1007/978-3-319-40022-8

$\mathfrak{M}$. $\sigma(\mathfrak{M})$ has the following property: if $\mathfrak{N}$ is a σ-algebra on Ω containing $\mathfrak{M}$, then $\sigma(\mathfrak{M}) \subset \mathfrak{N}$. $\sigma(\mathfrak{M})$ is called a *σ-algebra on Ω generated by the family $\mathfrak{M}$*.[25]

Definition I.4 Let $\mathfrak{M}$ be a σ-algebra on a set Ω. A function

$$\mu : \mathfrak{M} \longrightarrow \mathbb{R} \cup \{\infty\}$$

is called a *measure* iff it satisfies the following conditions.

- For each set $A \in \mathfrak{M}$: $\mu(A) \geq 0$.
- $\mu(\emptyset) = 0$.
- If $A_1, A_2, A_3, \ldots \in \mathfrak{M}$ are pairwise disjoint, then $\mu(\bigcup_{i=1}^{\infty} A_i) = \sum_{i=1}^{\infty} \mu(A_i)$.

Definition I.5 Let $(\Omega, \mathfrak{M})$ be a measurable space, μ a measure. The triple $(\Omega, \mathfrak{M}, \mu)$ is called a *measure space*.

Definition I.6 A measure space $(\Omega, \mathcal{F}, P)$ is called a *probability space* iff a measure P fulfills the following condition $P(\Omega) = 1$.

The set Ω is called the *space of elementary events*. The family of sets $\mathcal{F}$ contains sets of events that we want to analyze. Each such a set consists of elementary events. The function (measure) P is called a *probability measure*, and a number $P(A), A \in \mathcal{F}$ is called a *probability of an event A*.

I.2 Bayesian Probability Theory

After defining notions of probability space, space of elementary events and probability of an event, we can introduce the foundations of Bayesian probability theory.

Definition I.7 Let $(\Omega, \mathcal{F}, P)$ be a probability space, $B \in \mathcal{F}$ an event, $P(B) > 0$. A *conditional probability* of an event $A \in \mathcal{F}$, assuming the event B has occurred, is given by a formula

$$P(A|B) = \frac{P(A \cap B)}{P(B)} .$$

Now we introduce *the total probability theorem*.

Theorem I.1 *Let $(\Omega, \mathcal{F}, P)$ be a probability space. Let events $B_1, B_2, \ldots, B_n \in \mathcal{F}$ fulfill the following conditions.*

- *$P(B_i) > 0$, for each $i = 1, 2, \ldots, n$.*
- *$B_i \cap B_j = \emptyset$, for each i and j such that $i \neq j$.*
- *$B_1 \cup B_2 \cup \cdots \cup B_n = \Omega$.*

---

[25] We say that $\sigma(\mathfrak{M})$ is *the smallest σ-algebra on Ω containing the family $\mathfrak{M}$*.

Then for each event $A \in \mathcal{F}$ the following formula holds

$$P(A) = \sum_{i=1}^{n} P(A|B_i) \cdot P(B_i).$$

The following theorem, called Bayes' rule, results from the total probability theorem and a definition of conditional probability.

Theorem I.2 *Let the assumptions of Theorem I.1 be fulfilled. Then the following formula holds*

$$P(B_k|A) = \frac{P(A|B_k) \cdot P(B_k)}{\sum_{i=1}^{n} P(A|B_i) \cdot P(B_i)},$$

for each $k = 1, 2, \ldots, n$.

At the end of this section we define independence of events and conditional independence of events.

Definition I.8 Let $(\Omega, \mathcal{F}, P)$ be a probability space. Events $A, B \in \mathcal{F}$ are *independent* iff the following condition holds

$$P(A \cap B) = P(A) \cdot P(B).$$

If $P(B) > 0$, then this condition is equivalent to

$$P(A|B) = P(A).$$

A notion of independence can be extended to any finite collection of events.

Definition I.9 Let $(\Omega, \mathcal{F}, P)$ be a probability space. Events $A_1, \ldots, A_n \in \mathcal{F}$ are *(mutually) independent* iff for any sub-collection of k events $A_{i_1}, \ldots, A_{i_k}$ the following condition holds

$$P(A_{i_1} \cap \cdots \cap A_{i_k}) = P(A_{i_1}) \cdot \cdots \cdot P(A_{i_k}).$$

Definition I.10 Let $(\Omega, \mathcal{F}, P)$ be a probability space, $A, B, C \in \mathcal{F}, P(C) > 0$. Events A, B are *conditionally independent given an event C* iff the following condition holds

$$P(A \cap B|C) = P(A|C) \cdot P(B|C).$$

This condition is equivalent to

$$P(A|B \cap C) = P(A|C).$$

Similarly to a notion of independence, a conditional independence can be extended to any finite collection of events.

I.3 Basic Notions of Dempster-Shafer Theory

In this section we introduce the basic notions of the theory of belief functions used in Sect. 12.2.

Definition I.11 Let Θ be a set of mutually exclusive events, called a *universe of discourse* (a *frame of discernment*). A function $m : 2^{\Theta} \longrightarrow [0, 1]$ is called a *basic belief assignment* (a *mass assignment function*) iff it fulfills the following conditions.

- $m(\emptyset) = 0$.
- $\sum_{A \subseteq \Theta} m(A) = 1$.

Definition I.12 A function $Bel : 2^{\Theta} \longrightarrow [0, 1]$ is called a *belief function* iff

$$Bel(A) = \sum_{B:B \subseteq A} m(B)$$

for each $A \subseteq \Theta$.

Definition I.13 A function $Pl : 2^{\Theta} \longrightarrow [0, 1]$ is called a *plausibility function* iff

$$Pl(A) = \sum_{B:B \cap A \neq \emptyset} m(B)$$

for each $A \subseteq \Theta$. A plausibility function can be defined on the basis of a belief function in the following way:

$$Pl(A) = 1 - Bel(\bar{A}),$$

where $\bar{A}$ is a complement of the set A.

Appendix J
Formal Models for Artificial Intelligence Methods: Foundations of Fuzzy Set and Rough Set Theories

The basic definitions of fuzzy set theory [321] and rough set theory [216], introduced in Chap. 13 for defining imprecise (vague) notions, are contained in here.

J.1 Selected Notions of Fuzzy Set Theory

Basic notions of fuzzy set theory such as fuzzy set, linguistic variable, valuation in fuzzy logic are introduced. Then selected definitions of the Mamdani fuzzy reasoning [191] (fuzzy rule, fuzzification operator, Mamdani minimum formula of reasoning, center of gravity of membership function) are presented.

Definition J.1 Let U be a nonempty space, called a *universe of discourse*. A set A in the space $U, A \subseteq U$, is called a *fuzzy set* iff

$$A = \{(x, \mu_A(x)) : x \in U\},$$

where

$$\mu_A : U \longrightarrow [0, 1]$$

is a *membership function*, which is defined in the following way

$$\mu_A(x) = \begin{cases} 0, & x \notin A, \\ 1, & x \in A, \\ s, s \in (0, 1), & x \text{ belongs to} A \text{ with a grade of membership } s. \end{cases}$$

M. Flasiński, *Introduction to Artificial Intelligence*,
DOI 10.1007/978-3-319-40022-8

Definition J.2 A *linguistic variable*[26] is a quadruple

$$L = (N, T, U, \mathcal{M}), \text{ where}$$

N is a name of the variable,
T is a set of possible *linguistic values* for this variable,
U is a universe of discourse,
$\mathcal{M}$ is a *semantic function* that ascribes a meaning $\mathcal{M}(t)$ for every linguistic value $t \in T$; a meaning is represented with a fuzzy set $A \in X$.

Definition J.3 Let there be given propositions **P**: "x *is* $\mathcal{P}$" and **Q**: "x *is* $\mathcal{Q}$", where $\mathcal{P}$ and $\mathcal{Q}$ are vague notions with fuzzy sets P and Q ascribed to by a semantic function. Let μ_P and μ_Q be their membership functions, respectively. A *valuation* in fuzzy propositional logic is defined with the help of the *truth degree function* T in the following way.

- $T(\mathbf{P}) = \mu_P(x)$.
- $T(\neg\mathbf{P}) = 1 - T(\mathbf{P})$.
- $T(\mathbf{P} \wedge \mathbf{Q}) = min\{T(\mathbf{P}), T(\mathbf{Q})\}$.
- $T(\mathbf{P} \vee \mathbf{Q}) = max\{T(\mathbf{P}), T(\mathbf{Q})\}$.
- $T(\mathbf{P} \Rightarrow \mathbf{Q}) = max\{T(\neg\mathbf{P}), T(\mathbf{Q})\}$.

Now, we introduce selected notions of the Mamdani model of reasoning[27]

Definition J.4 A rule $(R^k, COND^k, ACT^k)$ is called a *fuzzy rule* iff:

$$COND^k \text{ is of the form : } \quad x_1^k \textsf{ is } A_1^k \wedge \cdots \wedge x_n^k \textsf{ is } A_n^k,$$

where x_i^k, $i = 1, \ldots, n$ is a linguistic variable, A_i^k, $i = 1, \ldots, n$ is a linguistic value,

$$ACT^k \text{ is of the form : } \quad y^k \textsf{ is } B^k,$$

where y^k is a linguistic variable, B^k is a linguistic value.[28]

Definition J.5 Let $X \subseteq \mathbb{R}$ be a domain of a variable x. A *singleton fuzzification operation* of a value $\overline{x}$ of a variable x is a mapping of a value $\overline{x}$ to a fuzzy set $A \subseteq X$ with a membership function given by the following formula:

[26] There are several definitions of *linguistic variable*. In its original version [322] linguistic variable was defined on the basis of context-free grammar. However, because of the complex form of such a definition, nowadays a linguistic variable is defined with a set of linguistic values *explicitly*. We assume such a convention in the definition.

[27] We assume a definition of a rule as in Appendix F.

[28] The definition concerns rules for which a condition is of the form of a conjunction and a consequent is a single element (i.e., of the form of a canonical MISO). In practice, fuzzy rules can be defined in such a form (with certain assumptions).

$$\mu_A(x) = \begin{cases} 1, & x = \overline{x}, \\ 0, & \text{otherwise.} \end{cases}$$

In order to define fuzzy reasoning, we introduce firstly a concept of fuzzy relation.

Definition J.6 Let X, Y be nonempty (non-fuzzy) sets. A *fuzzy relation* R is a fuzzy set defined on the Cartesian product $X \times Y$ such that

$$R = \{((x, y)\, ,\, \mu_R(x, y)) : x \in X, y \in Y,\ \mu_R : X \times Y \longrightarrow [0, 1]\}.$$

Definition J.7 Let X, Y be sets, $A \subseteq X$, $B \subseteq Y$ be fuzzy sets with membership functions μ_A and μ_B. A result of *fuzzy reasoning from A to B in the Mamdani model according to a minimum rule*[29] is a fuzzy relation on $X \times Y$, denoted $A \to B$, in which a membership function is defined with the following formula

$$\mu_{A \to B}(x, y) = \min\{\mu_A(x), \mu_B(y)\}.$$

Definition J.8 Let $A \subseteq X \subseteq \mathbb{R}$ be a fuzzy set with a membership function μ_A. A *defuzzification operation with the help of a center of gravity* consists in ascribing a value $\overline{x}$ to a set A according to the following formula

$$\overline{x} = \frac{\int_X x\mu_A(x)\, dx}{\int_X \mu_A(x)\, dx},$$

assuming the existence of both integrals.

J.2 Selected Notions of Rough Set Theory

Let U be a nonempty space containing objects considered, called a *universe of discourse*, A be a finite nonempty set of attributes describing objects belonging to U. For every attribute $a \in A$ let us define a set of its possible values V_a, called the domain of a. Let $a(x)$ denote a value of an attribute a for an object $x \in U$.

Definition J.9 Let $B \subseteq A$. A relation $I_B \subseteq U \times U$ is called a *B-indiscernibility relation* iff it fulfills the following condition.

$$(x, y) \in I_B \Leftrightarrow \forall a \in B : a(x) = a(y),$$

for every $x, y \in U$. We say that objects x, y are *B-indiscernible*.

[29] Such a somewhat complicated formulation is used by the author in order to avoid naming the formula: *a fuzzy implication* $A \to B$. The Mamdani minimum formula, although very useful in practice, does not fulfil a definition of fuzzy implication.

Since a relation I_B is an equivalence relation, it defines a partition of a universe U into equivalence classes. Let us introduce the following definition.

Definition J.10 Let I_B be a B-indiscernibility relation in a universe U, $x \in U$. A *B-elementary set* $[x]_{I_B}$ is the equivalence class of an object x, i.e.

$$[x]_{I_B} = \{y \in U : (x, y) \in I_B\}.$$

Let $X \subseteq U$.

Definition J.11 A *B-lower approximation* of the set X is a set

$$\underline{B}X = \{x \in U : [x]_{I_B} \subseteq X\}.$$

Definition J.12 A *B-upper approximation* of the set X is a set

$$\overline{B}X = \{x \in U : [x]_{I_B} \cap X \neq \emptyset\}.$$

Definition J.13 A *B-boundary region* of the set X is a set

$$B_{BOUND}X = \overline{B}X \setminus \underline{B}X.$$

Definition J.14 The set X is called a *B-rough set* iff the following condition holds

$$\underline{B}X \neq \overline{B}X.$$

Definition J.15 The set X is called a *B-exact* (*B-crisp*) *set* iff the following condition holds

$$\underline{B}X = \overline{B}X.$$

Definition J.16 Let $card(Z)$ denote the cardinality of a nonempty set Z. A *coefficient of an accuracy of approximation* of the set X with respect to a B-indiscernibility relation is given by the following formula

$$\alpha_B(X) = \frac{card(\underline{B}X)}{card(\overline{B}X)}.$$

Bibliography

1. Aho, A.V.: Indexed grammars an extension of context-free grammars. J. ACM **15**, 647–671 (1968)
2. Ajdukiewicz, K.: Pragmatic Logic. Reidel, Dordrecht (1974)
3. Almuallim, H., Dietterich, T.G.: Learning with many irrelevant features. In: Proceedings of 9th National Conference on Artificial Intelligence AAAI-91, Anaheim, CA, pp. 547–552 (1991)
4. Anderberg, M.R.: Cluster Analysis for Applications. Academic Press, New York (1973)
5. Anderson, J.R.: The Architecture of Cognition. Harvard University Press, Cambridge (1983)
6. Anderson, J.R.: Rules of the Mind. Lawrence Erlbaum, Hillsdale (1993)
7. Apt, K.R.: From Logic Programming to Prolog. Prentice Hall (1997)
8. Apt, K.R.: Principles of Constraint Programming. Cambridge University Press, Cambridge (2003)
9. Aristotle: De anima, Hicks, R.D. (ed.). Cambridge University Press, Cambridge (1907)
10. Aristotle: The Works of Aristotle Translated into English Under the Editorship of WD Ross, 12 vols. Clarendon Press, Oxford (1908–1952)
11. Armstrong, D.: A Materialist Theory of Mind. Routledge and K. Paul, London (1968)
12. Arnold, M.B:. Emotion and Personality. Columbia University Press, New York (1960)
13. Austin, J.L.: How to Do Things with Words. Clarendon Press, Oxford (1962)
14. Baader, F., Nipkow, T.: Term Rewriting and All That. Cambridge University Press, Cambridge (1999)
15. Baader, F., Horrocks, I., Sattler, U.: Description Logics. In: Van Harmelen, F., Lifschitz, V., Porter, B. (eds.) Handbook of Knowledge Representation. Elsevier Science, Amsterdam (2007)
16. Barendregt, H.P.: The Lambda Calculus: its Syntax and Semantics. North-Holland, Amsterdam (1984)
17. Barendregt, H., Barendsen, E.: Introduction to Lambda Calculus (2000)
18. Barr, A., Feigenbaum, E.: The Handbook of Artificial Intelligence, vols. 1–2. HeurisTech Press and Kaufmann, Stanford-Los Altos (1981, 1982)
19. Barr, A., Cohen, P.R., Feigenbaum, E.: The Handbook of Artificial Intelligence, vol. 4. Addison-Wesley, Reading (1989)
20. Barwise, J.: Mathematical Logic. Elsevier Science, Amsterdam (1993)
21. Bechtel, W., Abrahamsen, A.: Connectionism and the Mind: Parallel Processing, Dynamics, and Evolution in Networks. Blackwell, Oxford (2002)
22. Beni, G., Wang, J.: Swarm intelligence in cellular robotic systems. In: Proceedings of the NATO Advanced Workshop on Robots and Biological Systems, Tuscany, Italy (1989)

M. Flasiński, *Introduction to Artificial Intelligence*,
DOI 10.1007/978-3-319-40022-8

23. Beyer, H.G., Schwefel, H.P.: Evolution strategies: a comprehensive introduction. Nat. Comput. **1**, 3–52 (2002)
24. Bezem, M., Klop, J.W., de Vrijer, R.: Term Rewriting Systems. Cambridge University Press, Cambridge (2003)
25. Bielecki, A., Ombach, J.: Dynamical properties of a perceptron learning process–structural stability under numerics and shadowing. J. Nonlinear Sci. **21**, 579–593 (2011)
26. Binet, A., Simon, T.: The Development of Intelligence in Children. Williams and Wilkins, Baltimore (1916)
27. Bishop, C.M.: Neural Networks for Pattern Recognition. Oxford University Press, Oxford (1995)
28. Bishop, C.: Pattern Recognition and Machine Learning. Springer (2006)
29. Bobrowski, L., Niemiro, W.: A method of synthesis of linear discriminant function in the case of nonseparabilty. Pattern Recogn. **17**, 205–210 (1984)
30. Bocheński, J.M.: Die zeitgenossischen Denkmethoden. A. Francke AG Verlag, Bern (1954). (English translation: The Methods of Contemporary Thought. Reidel, Dordrecht (1965))
31. Bock, G.R., Goode, J.A., Webb, K. (eds.): The Nature of Intelligence. Wiley, Chichester (2000)
32. Boden, M.: Computer models of creativity. In: Sternberg, R.J. (ed.) Handbook of Creativity, pp. 351–373. Cambridge University Press, Cambridge (1999)
33. BonJour, L.: Epistemology: Classic Problems and Contemporary Responses. Rowman and Littlefield, Lanham (2002)
34. Booth, T.L.: Probabilistic representation of formal languages. In: Proceedings of 10th Annual Symposium Foundation on Computer Science, Waterloo, Canada, pp. 74–81 (1969)
35. Boser, B.E., Guyon, I.M., Vapnik, V.N.: A training algorithm for optimal margin classifier. In: Proceedings of 5th Annual ACM Workshop COLT, Pittsburgh, USA, pp. 144–152 (1992)
36. Brachman, R.J., Levesque, H.J.: Knowledge Representation and Reasoning. Morgan Kaufmann, San Francisco (2004)
37. Bratko, I.: Programming in Prolog for Artificial Intelligence. Addison-Wesley, Reading (2000)
38. Brentano, F.: Psychology from an Empirical Standpoint. Routledge, London and New York (1995)
39. Brewka, G., Niemelä, I., Truszczyński, M.: Nonmonotonic reasoning. In: van Harmelen, F., et al. (eds.) Handbook of Knowledge Representation, pp. 239–284. Elsevier Science, Amsterdam (2008)
40. Brown, M., Harris, C.: Neuro-fuzzy Adaptive Modeling and Control. Prentice-Hall, Englewood Cliffs, NJ (1994)
41. Bunke, H.O., Sanfeliu, A. (eds.): Syntactic and Structural Pattern Recognition-Theory and Applications. World Scientific, Singapore (1990)
42. Carpenter, G.A., Grossberg, S.: A massively parallel architecture for self-organizing neural pattern recognition machine. Comput. Vis. Graph. Image Process. **37**, 54–115 (1987)
43. DeCastro, L.N., Timmis, J.: Artificial Immune Systems: A New Computational Intelligence Approach. Springer (2002)
44. Cattell, R.B.: Abilities: Their Structure, Growth, and Action. Houghton Mifflin, Boston, MA (1971)
45. Chalmers, D.: Philosophy of Mind: Classical and Contemporary Readings. Oxford University Press, New York (2002)
46. Chang, C.-L., Lee, R.C.-T.: Symbolic Logic and Mechanical Theorem Proving. Academic Press, New York (1997)
47. Chomsky, N.: Aspects of the Theory of Syntax. MIT Press, Cambridge, MA (1965)
48. Church, A.: A note on the Entscheidungs problem. J. Symbolic Logic **1**, 40–41 (1936)
49. Churchland, P.M.: Eliminative materialism and the propositional attitudes. J. Philos. **78**, 67–90 (1981)
50. Cichocki, A., Unbehauen, R.: Neural Networks for Optimization and Signal Processing. Wiley, New York (1993)

51. Cios, K., Pedrycz, W., Swiniarski, R.: Data Mining Techniques. Kluwer Academic Publishers, Dordrecht/London/Boston (1998)
52. Cirstea, H., Kirchner, C., Moossen, M., Moreau, P.-E.: Production systems and rete algorithm formalisation, INRIA Rapport Intermédiaire A04-R-546, Rocquencourt, (2004). http://www.loria.fr/publications/2004/A04-R-546/A04-R-546.ps
53. Clocksin, W.F., Mellish, C.S.: Programming in Prolog. Springer (1994)
54. Cohen, M.A., Grossberg, S.: Absolute stability of global pattern formation and parallel memory storage by competitive neural networks. IEEE Trans. Syst. Man Cybern. **SMC-13**, 815–826 (1983)
55. Cohen, P.R., Feigenbaum, E.: The Handbook of Artificial Intelligence, vol. 3. HeurisTech Press and Kaufmann, Stanford-Los Altos (1982)
56. Collins, A.M., Quillian, M.R.: Retrieval time from semantic memory. J. Verbal Learning Verbal Behav. **8**, 240–248 (1969)
57. Colmerauer, A., Kanoui, P., Pasero, P., Roussel, Ph.: Un système de communication homme-machine en Français. *Internal report. Groupe d'Intelligence Artificielle. Université Aix-Marseille II* (1973)
58. Colmerauer, A.: Les grammaires de métamorphose. Internal report. Groupe d'Intelligence Artificielle. Université Aix-Marseille II (1975). (English translation: metamorphosis grammars. In: Bolc, L. (ed.) Natural Language Communication with Computers, pp. 133–189. Springer (1978))
59. Copleston, F.: A History of Philosophy, vols 1–11. Continuum, London (2003)
60. Cover, T.M., Hart, P.E.: Nearest neighbor pattern classification. IEEE Trans. Inf. Theory **13**, 21–27 (1967)
61. Cramer, N.L.: A representtion for the adaptive generation of simple sequential programs. In: Proceedings of International Conference on Genetic Algorithms Application, Pittsburgh, PA, USA, pp. 183–187 (1985)
62. Crevier, D.: AI: The Tumultous Search for Artificial Intelligence. BasicBooks, New York (1993)
63. Czogała, E., Łęski, J.: Fuzzy and Neuro-fuzzy Intelligent Systems. Springer (2000)
64. Davidson, D.: Mental events. In: Foster, L., Swanson, J.W. (eds.) Experience and Theory. Duckworth Publishers, London (1970)
65. De Jong, K.A.: Evolutionary Computation: A Unified Approach. MIT Press, Cambridge, MA (2006)
66. De Kleer, J.: An assumption based truth maintenance system. Artif. Intell. **28**, 127–162 (1986)
67. Dempster, A.P.: A generalization of Bayesian inference. J. R. Stat. Soc. Ser. B **30**, 205–247 (1968)
68. Dennett, D.: The Intentional Stance. MIT Press, Cambridge, MA (1987)
69. Descartes, R.: Principles of Philosophy. D. Reidel, Dordrecht (1983)
70. Dijkstra, E.W.: A note on two problems in connexion with graphs. Numerische Mathematik **1**, 269–271 (1959)
71. Doherty, P., Łukaszewicz, W., Skowron, A., Szałas, A.: Knowledge Representation Techniques. Springer, A Rough Set Approach (2006)
72. Dorigo, M.: Optimization, learning and natural algorithms, PhD thesis, Politecnico di Milano, Italy (1992)
73. Doyle, J.: A truth maintanence system. Artif. Intell. **12**, 231–272 (1979)
74. Dreyfus, H.L.: What Computers Can't Do. MIT Press, New York (1972)
75. Dubois, D., Prade, H.: Fuzzy Sets and Systems. Academic Press, New York (1988)
76. Duch, W.: What is computational intelligence and where is it going? Stud. Comput. Intell. **63**, 1–13 (2007)
77. Duch, W., Oentaryo, R.J., Pasquier M.: Cognitive architectures. Where do we go from here? In: Proceedings of 1st Conference on Artificial General Intelligence, University of Memphis, TN, pp. 122–136 (2008)
78. Duda, R., Hart, P.: Pattern Classification and Scene Analysis. Wiley-Interscience, New York (1973)

79. Duda, R.O., Hart, P.E., Stork, D.G.: Pattern Classification. Wiley, New York (2001)
80. Dunin-Kęplicz, B., Verbrugge, R.: Teamwork in Multi-Agent Systems: A Formal Approach. Wiley, New York (2010)
81. Ebbinghaus, H.-D., Flum, J., Thomas, W.: Mathematical Logic. Springer (1994)
82. Eberhart, R., Kennedy, J.: A new optimizer using particle swarm theory. In: Proceedings of 6th International Symposium on Micro Machine and Human Science, Nagoya, Japan, pp. 39–43 (1995)
83. Eiben, A.E., Smith, J.E.: Introduction to Evolutionary Computing. Springer (2003)
84. Elman, J.L.: Finding structure in time. Cogn. Sci. **14**, 179–211
85. Everitt, B.: Cluster Analysis. Heinemann Educational Books, London (1977)
86. Farmer, J.D., Packard, N., Perelson, A.: The immune system, adaptation and machine learning. Phys. D **2**, 187–204 (1986)
87. Fausett, L.: Fundamentals of Neural Networks. Prentice-Hall, Englewood Cliffs, New Jersey (1994)
88. Ferber, J.: Multi-Agent Systems: An Introduction to Artificial Intelligence. Addison-Wesley, Reading, MA (1999)
89. Firebaugh, M.W.: Artificial Intelligence. A Knowledge-Based Approach. PWS-Kent Publ. Comp, Boston, MA (1988)
90. Fisher, R.A.: The use of multiple measurements in taxonomic problems. Ann. Eugenics **7**, 179–188 (1936)
91. Fix, E., Hodges, J.L.: Discriminatory analysis, nonparametric discrimination: consistency properties. Technical Report 4, USAF School of Aviation Medicine, Randolph Field, TX (1951)
92. Flasiński, M., Lewicki, G.: The convergent method of constructing polynomial discriminant functions for pattern recognition. Pattern Recogn. **24**, 1009–1015 (1991)
93. Flasiński, M.: On the parsing of deterministic graph languages for syntactic pattern recognition. Pattern Recogn. **26**, 1–16 (1993)
94. Flasiński, M.: Power properties of NLC graph grammars with a polynomial membership problem. Theor. Comput. Sci. **201**, 189–231 (1998)
95. Flasiński, M., Jurek, J.: Dynamically programmed automata for quasi context sensitive languages as a tool for inference support in pattern recognition-based real-time control expert systems. Pattern Recogn. **32**, 671–690 (1999)
96. Flasiński, M.: Inference of parsable graph grammars for syntactic pattern recognition. Fundamenta Informaticae **80**, 379–413 (2007)
97. Flasiński, M.: Syntactic pattern recognition: paradigm issues and open problems. In: Chen, C.H. (ed.) Handbook of Pattern Recognition and Computer Vision, pp. 3–25. World Scientific, New Jersey-London-Singapore (2016)
98. Floyd, R.W.: Syntactic analysis and operator precedence. J. ACM **10**, 316–333 (1963)
99. Fogel, L.J., Owens, A.J., Walsh, M.J.: Artificial Intelligence through Simulated Evolution. Wiley, New York (1966)
100. Fogel, L.J.: Intelligence through Simulated Evolution: Forty Years of Evolutionary Programming. Wiley, New York (1999)
101. Forgy, C.: Rete: a fast algorithm for the many pattern/many object pattern match problem. Artif. Intell. **19**, 17–37 (1982)
102. Fraser, A.: Simulation of genetic systems by automatic digital computers. I. Introduction Aust. J. Biol. Sci. **10**, 484–491 (1957)
103. Fu, K.S., Swain, P.H.: Stochastic programmed grammars for syntactic pattern recognition. Pattern Recogn. **4**, 83–100 (1971)
104. Fu, K.S.: Syntactic Pattern Recognition and Applications. Prentice-Hall, Englewood Cliffs, NJ (1982)
105. Fukami, S., Mizumoto, M., Tanaka, K.: Some considerations on fuzzy conditional inference. Fuzzy Sets Syst. **4**, 243–273 (1980)
106. Fukunaga, K.: Introduction to Statistical Pattern Recognition. Academic Press, Boston, MA (1990)

107. Fukushima, K.: Cognitron: a self-organizing multilayered neural network. Biol. Cybern. **20**, 121–136 (1975)
108. Gardner, H., Kornhaber, M.L., Wake, W.K.: Intelligence: Multiple Perspectives. Harcourt Brace, New York (1996)
109. Glover, F.: Future paths for integer programming and links to artificial intelligence. Comput. Oper. Res. **13**, 533–549 (1986)
110. Gödel, K.: Über formal unentscheidbare Sätze der Principia Mathematica und verwandter Systeme. I. Monatshefte für Mathematik und Physik **38**, 173–198 (1931)
111. Goldberg, D.E.: Genetic Algorithms in Search. Optimization and Machine Learning. Addison-Wesley, Reading (1989)
112. Goleman, D.: Social Intelligence: The New Science of Human Relationships. Bantam Books, New York (2006)
113. Gonzales, R.C., Thomason, M.G.: Syntactic Pattern Recognition: An Introduction. Addison-Wesley, Reading (1978)
114. Gottfredson, L.S.: Why g matters: the complexity of everyday life. Intelligence **24**, 79–132 (1997)
115. Gubner, J.A.: Probability and Random Processes for Electrical and Computer Engineers. Cambridge University Press, Cambridge (2006)
116. Guilford, J.P.: The Nature of Human Intelligence. McGraw-Hill, New York (1967)
117. Gupta, M.M., Jin, L., Homma, N.: Static and Dynamic Neural Networks. Wiley, Hoboken, NJ (2003)
118. Gut, A.: An Intermediate Course in Probability. Springer (1995)
119. Gut, A.: Probability: A Graduate Course. Springer, New York, NY (2005)
120. Guttenplan, S.: A Companion to the Philosophy of Mind. Blackwell, Oxford (1996)
121. Guzman, A.: Continuous Functions of Vector Variables. Birkhäuser, Boston, MA (2002)
122. Guzman, A.: Derivatives and Integrals of Multivariate Functions. Birkhäuser, Boston, MA (2003)
123. Hall, R.P.: Computational approaches to analogical reasoning: a comparative analysis. Artif. Intell. **39**, 39–120 (1989)
124. Hamming, R.W.: Error detecting and error correcting codes. Bell Syst. Tech. J. **29**, 147–160 (1950)
125. Hamscher, W., Console, L., de Kleer, J.: Readings in Model-Based Diagnosis. Morgan Kaufmann, San Francisco, CA (1992)
126. Hart, P.E., Nilsson, N.J., Raphael, B.: A formal basis for the heuristic determination of minimum cost paths. IEEE Trans. Syst. Sci. Cybern. **SSC-4** 100–107 (1968)
127. Hartigan, J.A.: Clustering Algorithms. Wiley, New York (1975)
128. Hassanien, A.E., Suraj, Z., Ślęzak, D., Lingras, P.: Rough Computing: Theories, Technologies, and Applications. Information Science Reference, New York (2008)
129. Hassoun, M.H.: Fundamentals of Artificial Neural Networks. MIT Press, Cambridge, MA (1995)
130. Hayes, P.: The logic of frames. In: Metzing, D. (ed.) Frame. Conceptions and Text Understanding, pp. 46–61. Walter de Gruyter and Co., Berlin (1979)
131. Hayes-Roth, F.: Building Expert Systems. Addison-Wesley, Reading, MA (1983)
132. Hayes-Roth, B.: A blackboard architecture for control. Artif. Intell. **26**, 251–321 (1985)
133. Hebb, D.O.: The Organization of Behavior. Wiley, New York (1949)
134. Heil, J.: Philosophy of Mind. A Contemporary Introduction. Routledge, New York (2004)
135. Hindley, J.R., Seldin, J.P.: Introduction to Combinators and λ-Calculus. Cambridge University Press, Cambridge (1986)
136. Hinton, G.E., Anderson, J.A. (eds.): Parallel Models of Associative Memory. Lawrence Erlbaum, Hillsdale, NJ (1981)
137. Hinton, G.E., Sejnowski, T.J.: Learning and relearning in Boltzmann machines. In: Rumelhart, D.E., et al. (eds.) Parallel Distributed Processing: Explorations in the Microstructure of Cognition, vol. 1, pp. 282–317. Foundations. MIT Press, Cambridge, MA (1986)

138. Hobbes, T.: The Metaphysical System of Hobbes. The Open Court Publishing Company, La Salle, IL (1963)
139. Holland, J.H.: Adaptation in Natural and Artificial Systems. University of Michigan Press, Ann Arbor, MI (1975)
140. Hopfield, J.J.: Neural networks and physical systems with emergent collective computational properties. Proc. Nat. Acad. Sci. USA **79**, 2554–2588 (1982)
141. Hopcroft, J.E., Motwani, R., Ullman, J.D.: Introduction to Automata Theory, Languages, and Computation. Addison-Wesley, Reading, MA (2006)
142. Horn, J.: Understanding human intelligence: where have we come since Spearman? In: Cudeck, R., MacCallum, R.C. (eds.) Factor Analysis at 100: Historical Developments and Future Directions, pp. 230–255. Lawrence Erlbaum, Mahwah, NJ (2007)
143. Horrigan, P.G.: Epistemology: An Introduction to the Philosophy of Knowledge. iUniverse, Lincoln, NE (2007)
144. Hume, D.: An Enquiry concerning Human Understanding. Oxford University Press, Oxford (1999)
145. Husserl, E.: Ideas Pertaining to a Pure Phenomenology and to a Phenomenological Philosophy. Kluwer Academic Publishers, Dordrecht (1998)
146. Jackson, F.: Epiphenomenal qualia. Philos. Q. **32**, 127–136 (1982)
147. Jackson, P.: Introduction to Expert Systems. Addison-Wesley, Harlow-London-New York (1999)
148. Jain, A.K., Duin, R.P.W., Mao, J.: Statistical pattern recogniton: a review. IEEE Trans. Patt. Anal. Mach. Intell. **PAMI-22** 4–37 (2000)
149. Janssens, D., Rozenberg, G., Verraedt, R.: On sequential and parallel node-rewriting graph grammars. Comput. Graph. Image Process. **18**, 279–304 (1982)
150. Johnson, S.C.: Hierarchical clustering schemes. Psychometrika **2**, 241–254 (1967)
151. Jordan, M.I.: Attractor dynamics and parallelism in a connectionist sequential machine. Proceedings of 8th Annual Conference on Cognitive Science Society, Erlbaum, Hillsdale NJ, pp. 531–546 (1986)
152. Kacprzyk, J.: Multistage Fuzzy Control: A Model-Based Approach to Control and Decision-Making. Wiley, Chichester (1997)
153. Kanal, L.N.: On pattern, categories and alternate realities. Pattern Recogn. Lett. **14**, 241–255 (1993)
154. Kant, I.: Critique of Pure Reason. Palgrave Macmillan, Basingstoke (2003)
155. Kenny, A. (ed.): The Oxford Illustrated History of Western Philosophy. Oxford University Press, Oxford (1994)
156. Kilian, J., Siegelmann, H.T.: On the power of sigmoid neural networks. In: Proceedings of 6th Annual Conference on Computation Learning Theory, Santa Cruz, CA, USA, pp. 137–143 (1993)
157. Kim, J.: Supervenience and Mind: Selected Philosophical Essays. Cambridge University Press, Cambridge (1993)
158. Kim, J.: Philosophy of Mind. Westview Press, Boulder (2006)
159. Kirkpatrick, S., Gelatt, C.D., Vecchi, M.P.: Optimization by simulated annealing. Science **220**, 671–680 (1983)
160. Kleene, S.C.: Representation of events in nerve nets and finite automata. In: Shannon, C.E., McCarthy, J. (eds.) Automata Studies, p. 342. Princeton University Press (1956)
161. Kłopotek, M.A., Wierzchoń, S.T.: A new qualitative rough-ret approach to modeling belief functions. In: Polkowski, L., Skowron, A. (eds.) Rough Sets And Current Trends In Computing, pp. 346–353. Springer (1998)
162. Kluźniak, F., Szpakowicz, S.: Prolog for Programmers. Academic Press, New York (1985)
163. Knuth, D.E.: On the translation of languages from left to right. Inf. Control **8**, 607–639 (1965)
164. Kohonen, T.: Correlation matrix memories. IEEE Trans. Comput. **C-21** 353–359 (1972)
165. Kohonen, T.: Self-organized formation of topologically correct feature maps. Biol. Cybern. **43**, 59–69 (1982)
166. Kolodner, J.L.: Case-Based Reasoning. Morgan Kaufmann, San Mateo, CA (1993)

167. Korf, R.E.: Depth-first iterative-deepening: an optimal admissible tree search. Artif. Intell. **27**, 97–109 (1985)
168. Koronacki, J., Raś, Z.W., Wierzchoń, S., Kacprzyk, J. (eds.): Advances in Machine Learning, vol. 1–2. Springer (2010)
169. Kosiński, W. (ed.): Advances in Evolutionary Algorithms. InTech Publishing, Vienna (2008)
170. Kosko, B.: Adaptive bidirectional associative memories. Appl. Opt. **26**, 4947–4960 (1987)
171. Koutroumbas, K., Theodoridis, S.: Pattern Recognition. Academic Press, Boston (2008)
172. Koza, J.R: Genetic Programming: On the Programming of Computers by Means of Natural Selection. MIT Press, Cambridge, MA (1992)
173. Koza, J.R. et al.: Genetic Programming IV: Routine Human-Competitive Intelligence. Springer (2005)
174. Laird, J.E.: Extending the Soar cognitive architecture. In: Proceedings of 1st Conference Artificial General Intelligence, University of Memphis, TN, pp. 224–235 (2008)
175. Lakoff, G.: Women, Fire and Dangerous Things: What Categories Reveal About the Mind. The University of Chicago Press, Chicago, IL (1987)
176. Leibniz, G.W.: Philosophical Texts. Oxford University Press, Oxford (1999)
177. Leitsch, A.: The Resolution Calculus. Springer (1997)
178. Lemos, N.: An Introduction to the Theory of Knowledge. Cambridge University Press, Cambridge (2007)
179. Levenshtein, V.I.: Binary codes capable of correcting deletions, insertions and reversals. Sov. Phys. Dokl. **10**, 707–710 (1966)
180. Lewis II, P.M., Stearns, R.E.: Syntax-directed transductions. J. ACM **15**, 465–488 (1968)
181. Li, W.: Mathematical Logic. Foundations for Information Science. Birkhäuser, Berlin (2010)
182. Ligęza, A.: Logical Foundations for Rule-Based Systems. Springer (2006)
183. Lippmann, R.P.: An introduction to computing with neural nets. IEEE Acoust. Speech Signal Process. **4**, 4–22 (1987)
184. Lloyd, J.W.: Foundations of Logic Programming. Springer (1987)
185. Locke, J.: An Essay Concerning Human Understanding. Hackett Publishing Company, Indianapolis, IN (1996)
186. Lowe, E.J.: An Introduction to the Philosophy of Mind. Cambridge University Press, Cambridge (2000)
187. Lowerre, B.T., Reddy, R.: The HARPY speech understanding system. In: Lea, W.A. (ed.) Trends in Speech Recognition, pp. 340–360. Prentice-Hall, Englewood Cliffs, NJ (1980)
188. Lucas, J.: Minds, machines and Gödel. Philosophy **36**, 112–127 (1961)
189. Luger, G., Stubblefield, W.: Artificial Intelligence: Structures and Strategies for Complex Problem Solving. Benjamin/Cummings, Redwood City, CA (2004)
190. MacQueen, J.: Some methods for classification and analysis of multivariate observations. In: Proceedings of 5th Berkeley Symposium on Mathematical Statistics and Probability, vol. 1. University of California Press, Berkeley, CA, pp. 281–297 (1967)
191. Mamdani, E.H., Assilian, S.: An experiment in linguistic synthesis with a fuzzy logic controller. Int. J. Man Mach. Stud. **7**, 1–13 (1975)
192. Mamdami, E.H.: Application of fuzzy logic to approximate reasoning using linguistic systems. IEEE Trans. Comp. **C-26** 1182–1191 (1997)
193. McCarthy, J.: Programs with common sense. In: Proceedings of Teddington Conference on Mechanization of Thought Processes. Her Majesty's Stationery Office, London, pp. 756–791 (1959)
194. McCarthy, J.: Recursive functions of symbolic expressions and their computation by machine. Commun. ACM **3**, 184–195 (1960)
195. McCarthy, J., Hayes, P.J.: Some philosophical problems from the standpoint of artificial intelligence. Mach. Intell. **4**, 463–502 (1969)
196. McCarthy, J.: Circumscription? A form of non-monotonic reasoning. J. Artif. Intell. **13**, 27–39 (1980)
197. McCorduck, P.: Machines Who Think. A.K. Peters Ltd., Natick, MA (2004)

198. McCulloch, W., Pitts, W.: A logical calculus of the ideas immanent in nervous activity. Bull. Math. Biophys. **7**, 115–133 (1943)
199. McDermott, D., Doyle, J.: Non-monotonic logic I. Artif. Intell. **13**, 41–72 (1980)
200. Michaelson, G.: An Introduction to Functional Programming through Lambda Calculus. Addison-Wesley, Reading, MA (1989)
201. Michalewicz, Z.: Genetic Algorithms + Data Structures = Evolution Programs. Springer (1993)
202. Mill, J.S.: A System of Logic. University Press of the Pacific, Honolulu (2002)
203. Minsky, M.: A framework for representing knowledge. In: Winston, P.H. (ed.) Psychology of Computer Vision. MIT Press, Cambridge, MA (1975)
204. Minsky, M., Papert, S.: Perceptrons–An Introduction to Computational Geometry. MIT Press, Cambridge, MA (1969)
205. Mitchell, T.M., Keller, R.M., Kedar-Cabelli, S.T.: Explanation-based generalization: a unifying view. Mach. Learning **1**, 47–80 (1986)
206. Moore, R.C.: Semantical considerations on nonmonotonic logic. Artif. Intell. **25**, 75–94 (1985)
207. Muggleton, S.H.: Inductive logic programming. New Gener. Comput. **8**, 295–318 (1991)
208. Newell, A., Simon, H.A.: Human Problem Solving. Prentice Hall, Englewood Cliffs, NJ (1972)
209. Newell, A., Simon, H.A.: Computer science as empirical inquiry: symbols and search. Commun. ACM **19**, 113–126 (1976)
210. Newell, A.: Unified Theories of Cognition. Harvard University Press, Cambridge, MA (1990)
211. Nilsson, N.: Artificial Intelligence: A New Synthesis. Morgan Kaufmann, San Francisco, CA (1998)
212. Nilsson, U., Małuszyński, J.: Logic. Programming and Prolog. Wiley, New York (1990)
213. Ockham, W.: Summa logicae. University of Notre Dame Press, Notre Dame (1980)
214. Ogiela, M.R., Tadeusiewicz, R.: Modern Computational Intelligence Methods for the Interpretation of Medical Images. Springer (2008)
215. Pavlidis, T.: Structural Pattern Recognition. Springer, New York (1977)
216. Pawlak, Z.: Rough sets. Int. J. Comput. Inf. Sci. **11**, 341–356 (1982)
217. Pawlak, Z.: Rough Sets-Theoretical Aspects of Reasoning about Data. Kluwer Academic Publishers, Boston (1991)
218. Pawlak, Z., Skowron, A.: Rudiments of rough sets. Inf. Sci. **177**, 3–27 (2007)
219. Pawlak, Z., Skowron, A.: Rough sets: some extensions. Inf. Sci. **177**, 28–40 (2007)
220. Pawlak, Z., Polkowski, L., Skowron, A.: Rough Set Theory. Encyclopedia of Computer Science and Engineering. Wiley, New York (2008)
221. Pearl, J.: Heuristics: Intelligent Search Strategies for Computer Problem Solving. Addison-Wesley, Reading, MA (1984)
222. Pearl, J.: Bayesian networks: a model of self-activated memory for evidential reasoning. In: Proceedings of 7th Conference on Cognitive Science Society, University of California, Irvine, CA, August pp. 329–334 (1985)
223. Pearl, J.: Probabilistic Reasoning in Intelligent Systems: Networks of Plausible Inference. Morgan Kaufmann, San Francisco, CA (1988)
224. Penrose, R.: The Emperor's New Mind. Oxford University Press, Oxford (1989)
225. Pereira, F.C.N., Warren, D.H.D.: Definite clause grammars for language analysis: a survey of the formalism and a comparison with augmented transition networks. Artif. Intell. **13**, 231–278 (1980)
226. Piaget, J.: La Psychologie de l'intelligence. Armand Colin, Paris (1947). (English translation: The Psychology of Intelligence. Routledge, London (2001))
227. Place, U.T.: Is consciousness a brain process? Br. J. Psychol. **47**, 4450 (1956)
228. Plato: Republic, Waterfield, R. (ed.). Oxford University Press, New York (2008)
229. Polkowski, L.: Rough Sets—Mathematical Foundations. Springer (2002)
230. Post, E.L.: Absolutely unsolvable problems and relatively undecidable propositions–account of an anticipation (1941). In: Davis, M. (ed.) The Undecidable, pp. 375–441. Raven Press, New York (1965)

231. Puppe, F.: Systematic Introduction to Expert Systems. Springer (1993)
232. Putnam, H.: Minds and machines. In: Hook, S. (ed.) Dimensions of Mind, pp. 148–180. New York University Press, New York (1960)
233. Quinlan, J.R.: Discovering rules by induction from large collections of examples. In: Michie, D. (ed.) Expert Systems in Micro-Electronic Age, pp. 168–201. Edinburgh University Press, Edinburgh (1979)
234. Rabin, M.O., Scott, D.: Finite automata and their decision problems. IBM J. Res. Dev. **3**, 114125 (1959)
235. Raudys, S.J., Pikelis, V.: On dimensionality, sample size, classification error, and complexity of classification algorithms in pattern recognition. IEEE Trans. Patt. Anal. Mach. Intell. **PAMI-2** 243–251 (1980)
236. Rechenberg, I.: Evolutionsstrategie: Optimierung technischer Systeme nach Prinzipien der biologischen Evolution. Frommann Holzboog, Stuttgart (1973)
237. Reddy, B.D.: Introductory Functional Analysis. Spinger (1998)
238. Reynolds, C.: Flocks, herds and schools: a distributed behavioral model. ACM SIGGRAPH Comput. Graph. **21**, 25–34 (1987)
239. Reiter, R.: A logic for default reasoning. Artif. Intell. **13**, 81–132 (1980)
240. Reiter, R.: On closed world data bases. In: Gaillaire, H., Minker, J. (eds.) Logic and Data Bases, pp. 55–76. Plenum Press, New York (1978)
241. Rich, E., Knight, K., Nair, S.B.: Artif. Intell. Tata McGraw-Hill, New Delhi-New York (1991)
242. Riesbeck, C., Schank, R.: Inside Case-Based Reasoning. Erlbaum, Northvale, NJ (1989)
243. Robinson, J.A.: A machine-oriented logic based on the resolution principle. J. ACM **12**, 23–41 (1965)
244. Robinson, A., Voronkov, A.: Handbook of Automated Reasoning. Elsevier Science, Amsterdam (2006)
245. Rosch, E.: Principles of categorization. In: Rosch, E., Lloyd, B. (eds.) Cognition and Categorization, pp. 27–48. Lawrence Erlbaum, Hillsdale, NJ (1978)
246. Rosenblatt, F.: The perceptron: a probabilistic model for information storage and organization in the brain. Psychol. Rev. **65**, 386–408 (1958)
247. Rosenblatt, F.: Principles of Neurodynamics. Spartan Books, New York (1962)
248. Rosenkrantz, D.J.: Programmed grammars and classes of formal languages. J. ACM **16**, 107–131 (1969)
249. Ross, T.J.: Fuzzy Logic with Engineering Applications. Wiley, Chichester (2004)
250. Rozenberg, G., Salomaa, A. (eds.): Handbook of Formal Languages—I. Springer (1997)
251. Rudin, W.: Real and Complex Analysis. McGraw-Hill, New York (1987)
252. Rumelhart, D.E., Hinton, G.E., Williams, R.J.: Learning representations by back-propagating errors. Nature **323**, 533–536 (1986)
253. Rumelhart, D.E., McClelland, J.L. (eds.): Parallel Distributed Processing: Explorations in the Microstructure of Cognition, vol. 1: Foundations, vol. 2: Psychological and Biological Models. The MIT Press, Cambridge, MA (1986)
254. Rumelhart, D.E.: The architecture of mind: a connectionist approach. In: Posner, M. (ed.) Foundations of Cognitive Science. The MIT Press, Cambridge, MA (1989)
255. Russell, B.: History of Western Philosophy. Routledge, London (1995)
256. Russell, S.J., Norvig, P.: Artif. Intell. A Modern Approach. Prentice Hall, Englewood Cliffs, NJ (2009)
257. Rutkowski, L.: Computational Intelligence. Springer, Methods and Techniques (2008)
258. Ryle, G.: The Concept of Mind. University of Chicago Press, Chicago (1949)
259. Sakai, M., Yoneda, M., Hase, H.: A new robust quadratic discriminant function. In: Procedings of 14th International Conference on Pattern Recognition, Brisbane, Australia, vol. 1, pp. 99–102 (1998)
260. Schaefer, R.: Foundation of Genetic Global Optimization. Springer (2007)
261. Schalkoff, R.J.: Artificial Intelligence: An Engineering Approach. McGraw-Hill, New York (1990)

262. Schalkoff, R.J.: Intelligent Systems: Principles, Paradigms, and Pragmatics. Jones and Barlett, Sudbury, MA (2009)
263. Schank, R.C.: Conceptual dependency: theory of natural language understanding. Cogn. Psychol. **3**, 532–631 (1972)
264. Schank, R.C., Abelson, R.P.: Scripts, Plans, Goals, and Understanding. Lawrence Erlbaum, Hillsdale, NJ (1977)
265. Schmidhuber, J.: Evolutionary principles in self-referential learning. Diploma Thesis. Technische Universität München (1987)
266. Schmidt-Schauß, M., Smolka, G.: Attributive concept descriptions with complements. Artif. Intell. **48**, 1–26 (1991)
267. Schwefel, H.P.: Numerische Optimierung von Computer-Modellen. Birkhäuser, Basel (1977)
268. Schwefel, H.P.: Numerical Optimization of Computer Models. Wiley, New York (1995)
269. Searle, J.R.: Minds, brains and programs. Behav. Brain Sci. **3**, 417–457 (1980)
270. Searle, J.R.: The Rediscovery of the Mind. The MIT Press, Cambridge, MA (1994)
271. Shafer, G.: A Mathematical Theory of Evidence. Princeton University Press, Princeton, NJ (1976)
272. Shannon, C.E.: A mathematical theory of communication. Bell Syst. Tech. J. **27**, 379–423 (1948)
273. Shapiro, S.C.: Encyclopedia of Artificial Intelligence. Wiley, New York (1992)
274. Shoham, Y., Leyton-Brown, K.: Multiagent Systems: Algorithmic, Game-Theoretic, and Logical Foundations. Cambridge University Press, Cambridge (2008)
275. Simon, H.A.: Why should machines learn? In: Michalski, R.S., Carbonell, J.G., Mitchell, T.M. (eds.) Machine Learning: Artificial Intelligence Approach, pp. 25–37. Springer (1984)
276. Skolem, T.: Logisch-kombinatorische Untersuchungen über die Erfüllbarkeit oder Beweisbarkeit mathematischer Sätze nebst einem Theorem über dichte Mengen. Videnskapsselskapets Skrifter. I. Mat.-Naturv. Klasse **4** 1–36 (1920)
277. Skowron, A.: Boolean reasoning for decision rules generation. Lect. Notes Artif. Intell. **689**, 295–305 (1993)
278. Słowiński, R. (ed.): Intelligent Decision Support-Handbook of Applications and Advances of the Rough Sets Theory. Kluwer Academic Publishers, Boston/London/Dordrecht (1992)
279. Smart, J.J.C.: Sensations and brain processes. Philos. Rev. **68**, 141–156 (1959)
280. Sowa, J.F. (ed.): Principles of Semantic Networks: Explorations in the Representation of Knowledge. Morgan Kaufmann, San Mateo, CA (1991)
281. Sowa, J.F.: Knowledge Representation: Logical, Philosophical, and Computational Foundations. Brooks/Cole, New York (2000)
282. Spearman, C.: The Nature of "Intelligence" and the Principles of Cognition. Macmillan, London (1923)
283. Specht, D.F.: Probabilistic neural networks. Neural Netw. **3**, 109–118 (1990)
284. Spinoza, B.: Ethics. Oxford University Press, Oxford (2000)
285. St. Thomas Aquinas: Philosophical Texts, Gilby, T. (ed.). Oxford University Press, New York (1951)
286. Steele Jr., G.L.: Common Lisp. The Language. Digital Press, Bedford, MA (1990)
287. Steinhaus, H.: Sur la division des corps matériels en parties. Bull. Acad. Polon. Sci. **4**, 801–804 (1956)
288. Sterling, L., Shapiro, E.: The Art of Prolog. The MIT Press, Cambridge, MA (1994)
289. Sternberg, R.J.: Sketch of a componential subtheory of human intelligence. Behav. Brain Sci. **3**, 573–614 (1980)
290. Sternberg, R.J., Detterman, D.K. (eds.): What is Intelligence?. Contemporary Viewpoints on its Nature and Definition. Ablex Publishing Corp, Norwood (1986)
291. Sternberg, R.J. (ed.): International Handbook of Intelligence. Cambridge University Press, Cambridge (2004)
292. Stich, S.P., Warfield, T.A. (eds.): The Blackwell Guide to Philosophy of Mind. Blackwell, Oxford (2003)
293. Stirzaker, D.: Elementary Probability. Cambridge University Press, Cambridge (2003)

294. Sutherland, W.A.: Introduction to Metric and Topological Spaces. Oxford University Press, Oxford (2004)
295. Sycara, K.P.: Multiagent systems. AI Mag. **19**, 79–92 (1998)
296. Tadeusiewicz, R., Ogiela, M.R.: Medical Image Understanding Technology. Springer (2004)
297. Tadeusiewicz, R., Ogiela, M.R.: The new concept in computer vision: automatic understanding of the images. Lect. Notes Artif. Intell. **3070**, 133–144 (2004)
298. Tadeusiewicz, R., Lewicki, A.: The ant colony optimization algorithm for multiobjective optimization non-compensation model problem staff selection. Lect. Notes Comput. Sci. **6382**, 44–53 (2010)
299. Tadeusiewicz, R.: Introduction to intelligent systems. In: Wilamowski, B.M., Irvin, J.D. (eds.) The Industrial Electronics Handbook Intelligent Systems, pp. 1–12. CRC Press, Boca Raton (2011)
300. Thorndike, E.L.: Intelligence and its uses. Harper's Mag. **140**, 227–235 (1920)
301. Thorndike, E.L.: The Fundamentals of Learning. Teachers College Press, New York (1932)
302. Thurstone, L.L.: A law of comparative judgment. Psychol. Rev. **34**, 273–286 (1927)
303. Thurstone, L.L.: Primary Mental Abilities. University of Chicago Press, Chicago, IL (1938)
304. Touretzky, D.S.: Common LISP. A Gentle Introduction to Symbolic Computation. Benjamin/Cummings, Redwood City, CA (1990)
305. Tsang, E.: Foundations of Constraint Satisfaction. Academic Press, London (1993)
306. Turing, A.M.: On computable numbers, with an application to the Entscheidungsproblem. Proc. Lond. Math. Soc. Ser. **2**(42), 230–265 (1937)
307. Turing, A.M.: Computing machinery and intelligence. Mind **59**(236), 433–460 (1950)
308. Vapnik, V.: Estimation of Dependences Based on Empirical Data (in Russian). Nauka, Moscow (1979). (English translation: Springer (1982))
309. Vapnik, V.: Statistical Learning Theory. Wiley, New York (1998)
310. Vose, M.D.: The Simple Genetic Algorithm: Foundation and Theory. MIT Press, Cambridge, MA (1999)
311. Węglarz, J. (ed.): Project Scheduling: Recent Models. Algorithms and Applications. Kluwer, Boston (1999)
312. Weiss, G. (ed.): Multiagent Systems. A Modern Approach to Distributed Artificial Intelligence. MIT Press, Cambridge (1999)
313. Widrow, B., Hoff, M.E.: Adaptive switching circuits: IRE WESCON convention record. IRE, New York **1960**, 96–104 (1960)
314. Winograd, T.: Thinking machines: can there be? Are we? In: Winograd, T., Flores, C.F. (eds.) Understanding Computers and Cognition: A New Foundation for Design. Addison-Wesley, Reading, MA (1987)
315. Winston, P.H.: Artificial Intelligence. Addison-Wesley, Reading, MA (1993)
316. Wittgenstein, L.: Philosophical Investigations. Wiley-Blackwell, Oxford (2009)
317. Wolf, R.S.: A Tour Through Mathematical Logic. The Mathematical Association of America, Washington, DC (2005)
318. Woods, W.A.: Transition network grammars for natural language analysis. Commun. ACM **13**, 591–606 (1970)
319. Wooldridge, M.: An Introduction to Multiagent Systems. Wiley, New York (2002)
320. Yovits, M.C. (ed.): Advances in Computers 37. Academic Press, San Diego, CA (1993)
321. Zadeh, L.A.: Fuzzy sets. Inf. Control **8**, 338–353 (1965)
322. Zadeh, L.A.: The concept of a linguistic variable and its applications to approximate reasoning—I, II, III. Inf. Sci. **8** 199–249 (1975), **8** 301–357 (1975), **9** 43–80, (1975)
323. Zadeh, L.A. (ed.): Fuzzy Sets. Fuzzy Logic. Fuzzy Systems. World Scientific, Singapore (1996)
324. Żurada, J.: Introduction to Artificial Neural Systems. West Publishing Company, St Paul, MN (1994)

Index

M. Flasiński, *Introduction to Artificial Intelligence*,
DOI 10.1007/978-3-319-40022-8

T

Printed in the United States
By Bookmasters